BIOCHEMISTRY RESEARCH TRENDS

# CHEMICAL AND BIOCHEMICAL REACTIONS: KINETICS AND MECHANISM

# BIOCHEMISTRY RESEARCH TRENDS

Additional books in this series can be found on Nova's website under the Series tab.

Additional E-books in this series can be found on Nova's website under the E-books tab.

BIOCHEMISTRY RESEARCH TRENDS

# CHEMICAL AND BIOCHEMICAL REACTIONS: KINETICS AND MECHANISM

G.E. ZAIKOV
LINSHU LIU
R.Y. DEBERDEEV
AND
O.V. STOYANOV
EDITORS

Nova Science Publishers, Inc.
*New York*

For permission to use material from this book please contact us:
Telephone 631-231-7269; Fax 631-231-8175
Web Site: http://www.novapublishers.com



Additional color graphics may be available in the e-book version of this book.

**LIBRARY OF CONGRESS CATALOGING-IN-PUBLICATION DATA**

Chemical and biochemical reactions : kinetics and mechanism / editors, G.E.
Zaikov ... [et al.].
p. ; cm. -- (Biochemistry research trends)
Includes bibliographical references and index.
ISBN 978-1-61122-860-1 (hardcover : alk. paper)
1. Polymers. I. Zaikov, G. E. (Gennadii Efremovich), 1935- II. Series:
Biochemistry research trends series.
[DNLM: 1. Biochemical Phenomena. QU 34]
QD381.C474 2011
547'.7--dc22
2010048372

*Published by Nova Science Publishers, Inc. † New York*

This volume is dedicated to the memory of Frank Columbus

On December 1st 2010, Frank H. Columbus Jr. (President and Editor-in-Chief of Nova Science Publishers, New York) passed away suddenly at his home in New York.

We lost our colleague, our good friend, a nearly perfect person who helped scientists from all over the world. Particularly Frank did much for the popularization of Russian and Georgian scientific research, publishing a few thousand books based on the research of Soviet (Russian, Georgian, Ukranian etc.) scientists.

Frank was born on February 26th 1941 in Pennsylvania. He joined the army upon graduation of high school and went on to complete his education at the University of Maryland and at George Washington University. In 1969, he became the Vice-President of Cambridge Scientific. In 1975, he was invited to work for Plenum Publishing where he was the Vice-President until 1985, when he founded Nova Science Publishers, Inc.

Frank Columbus did a lot for the prosperity of many Soviet (Russian, Georgian, Ukranian, Armenian, Kazakh, Kyrgiz, etc.) scientists publishing books with achievements of their research. He did the same for scientists from East Europe – Poland, Hungary, Czeckoslovakia (today it is Czeck republic and Slovakia), Romania and Bulgaria.

He was a unique person who enjoyed studying throughout the course of his life, who felt at home in his country which he loved and was proud of, as well as in Russia and Georgia.

There is a famous Russian proverb: "The man is alive if people remember him." In this case, Frank is alive and will always be in our memories while we are living. He will be remembered for his talent, professionalism, brilliant ideas and above all – for his heart.

Professor Gennady Efremovich Zaikov
Honoured Member of Russian Science
Head of Polymer Division, IBCP

O.V. Stoyanov
Kazan State Technological University
Kazan 420015, Russia

Rustam Yabukovich Deberdeev
Kazan State Technological University
Kazan 420015, Russia

LinShu Liu
United States Department of Agriculture
Agricultural Research Service
Wyndmoor, PA, USA

# CONTENTS

# PREFACE

This book presents and discusses research in the study of chemical and biochemical kinetics – the science which deals with the changes over time. Topics discussed include the mechanism of catalysis with nickel (II) complexes in toluene and cumene oxidations by dioxygen; the effect of two oxazoles on membrane lipid structure of liver microsomes calculations of bond energy and the structure and properties of polymer adsorption layers on inorganic pigments surface in aqueous dispersion formed under ultrasonic treatment.

Chapter 1 - By the examples of polycaproamide, polyvinylpyrrolidone, poly-*m*-phenylene isophthalamide and polypyromellitimide, the mechanism of interaction of nitrogen dioxide with polymers containing amide and imide groups is considered. The initiators of conversions of polymers of the given classes are not monoradicals of nitrogen dioxide rather than its equilibrium dimers in the form of nitrosyl nitrate. In primary oxidizing reaction of an electron transfer from donor groups of macromolecules to nitrosyl nitrate, radical cations, nitric oxide and nitrate anion are formed, which in the subsequent reactions give products of nitration and nitrosation. These products are precursors of stable acylalkylaminoxyl, acylarylaminoxyl, iminoxyl macroradicals. The specific interactions of nitrogen dioxide dimers with functional groups of macromolecules determine features of the mechanism of ion-radical conversions of the polymers and the composition of radical and molecular products. The direct detection of radical cations has been realized by ESR method for confirmation of the ion-radical initiation concept using model reaction of nitrogen dioxide with triphenylamine.

Chapter 2 - The analysis of the mechanism of oxidation of alkylarenes (toluene and cumene) with dioxygen, catalyzed by catalytic system $\{Ni(II)L^1_2 + L^2\}$ ($L^1$ = acac$^-$, $L^2$ = ligand-modifier N-methylpirrolidon-2) is resulted for the first time. Comparison with the mechanism of nickel catalyzed ethylbenzene oxidation with dioxygen is spent. It has been established that the mechanism of formation of products of catalytic oxidation of alkylarenes is connected with various activities of complexes of the nickel, formed during oxidation, in micro steps of radical-chain processes of oxidation.

Chapter 3 - The effect of 1,2-diphenyl-1,3-oxazole (DPhO) and its derivative – iod-methylate (IM-DPhO) in the concentration range of $10^{-2}$ -$10^{-13}$M were studied *in vitro* on the microviscosity ($\tau$) and order parameter (S) of membrane lipids

of microsomes isolated from the liver cells of Balb mice. By EPR-technique using of spin probes 5-and 16-doxyl-stearic acids (5-and 16-DSA) localized in the surface and depth of membrane lipid regions respectively it was found that both oxazoles caused an increase in parameters $\tau$ and S, but the effect of DPhO was higher than its derivative. It was shown that the value of the effect depends on the concentrations of oxazoles by nonlinear and different manner for DPhO and IM-DPh-O. It was obtained a bimodal curve of change of parameter S depending on concentration of IM-DPhO with the maxima at $10^{-4}$ and $10^{-9}$ M The results obtained were discussed concerning radio-protecting properties of these oxazoles observed previously.

Chapter 4 - The present paper focuses on the study of novel blends based on poly(3-hydroxybutyrate) (PHB) and polymers with different hydrophilicity, namely low density polyethylene and polyamide (LDPE and PA), respectively. Polymer blends were produced from five ratios of PHB/LDPE in an effort to regulate the resistance to hydrolysis or (bio)degradation through the control of water permeability. The relation between the water transport and morphology (TEM data) shows the impact of polymer component ratio on the water flux regulation in the hydrophobic matrix. To elucidate the role of hydrophilicity of the second component present in the PHB-based blends, we studied the PHB/PA blends where PA is the polyamide resin composed of statistical copolymer of hexamethyleneadipinate and ε-caprolactam in the ratio 1:1. The use of techniques such as DSC and FTIR-imaging (for T-scale) demonstrates the interaction between PHB and PA in the temperature ranges of crystallization and melting. The general approach based on Flory-Huggins equation is presented as the way to choose the pairs of compatible or partly compatible polymers. Blending PHB with PA could be a simple and effective method to design new matrices for drug delivery in medicine, while the same procedure applied for PHB-LDPE system with better resistance against hydrolysis and lower price than the initial PHB, could be used as novel bioerodible packaging materials.

Chapter 5 - Scanning electron microscopy was used to study how the surface architectonics of erythrocytes is affected by new-generation drugs, Ichphans, which have a combined antioxidant and anticholinesterase effect. It was found that intercalation of each of the studied compounds, which have a positive charge on a quaternary nitrogen atom, in the erythrocyte membrane and its distribution in the intramembrane space are accompanied by manifestations of the echinocytogenic, stomatocytogenic, and compensatory effect on the erythrocyte shape. Time-dependent morphological transformations of erythrocytes are likely to be caused by changes in the preferential distribution of the compounds in the outer, inner, or both monolayers of the erythrocyte membrane, respectively. Differences in the transforming effect of the drugs with different hydrophobic properties were revealed.

Chapter 6 - pH-dependent detection of the EPR spectra of atomic hydrogen $H^{\bullet}$ ($g$ = 2.0023, $A$ = 50.6 mT) in frozen aqueous solutions of oxyacids AOH initiated by photoinjected low energy electrons upon energy impact below the excitation thresholds of water or oxyacids (photon energy of not more than 4.5 eV) was crucial for identification of the dissociative electron capture by bound acidic proton. The data on direct electron attachment to an oxyacid that can create an acidic bound [H+] network under certain conditions allows one to consider direct participation of the acidic bound proton in the intramolecular ET as one electron relay being highly effective in mediating electron transport in acid catalysis of redox reactions.

Chapter 7 - Quantum-chemical research of the mechanism of synthesis of 5-acetyloxymethyl-2-chloro-5-ethyl-1,2,3-dioxaphosphorynane for the first time is executed by classical method MNDO. Synthesis of this compound is the result of interaction of acetyl chloride and 4-ethyl-2,6,7-trioxa-1-phosphabicyclo[2.2.2]octane. Interaction proceeds on the mechanism of bimolecular nucleophylic substitution $S_N2$. Reaction is endothermic and has barrier character. The size of a power barrier makes 178 kJ/mol.

Chapter 8 - Quantum-chemical calculation of molecule of 5-acetyloxymethyl-2-chloro-5-ethyl-1,2,3-dioxaphosphorynane (I) and components of its synthesis: acetyl chloride (II) и 4-ethyl-2,6,7-trioxa-1-phosphabicyclo[2.2.2]octane (II) for the first time is executed by method AB INITIO in basis 6-311G**. The optimized geometrical and electronic structures of these compounds are received. Their acid force is theorized. All these compounds concern to a class very weak C-H-acids (pKa > 14, pKa- the universal parameter of acidity). Dependence between acid force of components pKa (II) and pKa (III) of synthesis and acid force pKa (I) a required product is positioned: pKa (I) = pKa (II) - $\sqrt{pKa(III)+14.7}$

Chapter 9 - The course and the result of the surface modification of titanium dioxide and ferrous oxide pigments in aqueous dispersion by ethylhydroxyethylcellulose (EHEC) without and with mechanical treatment of the dispersion by ultrasonic power was studied by the electrokinetic sonic amplitude (ESA) method. The evaluation of the ESA data showed that the ultrasonic treatment causes a significant thickness increase of the EHEC layer on the pigment which is primarily attributed to the ultrasonically induced activation of the pigment surface. The thickness of the polymer adsorption layer derived from ESA measurements was confirmed by TEM investigations. The ultrasonic treatment leads to significant changes of the properties of the adsorption layers and is a promising method for the surface modification of pigments with regard to dispersion stability.

Chapter 10 - The process of surface modification of hydrophobic organic pigments (copper phthalocyanine (CuPc) and carbon black) as well as a hydrophilic inorganic pigment (titanium dioxide) in aqueous dispersions by employing tailor-made thermo-responsive copolymers, and the colloidal stability have been studied. The pigment surface modification is achieved by conventional adsorption and by (thermo)precipitation of amphiphilic methyl vinyl ether (MVE) containing polyvinylether block and PMVE graft copolymers with poly(ethylene oxide) side chains exhibiting a lower critical solution temperature (LCST). The effect of mechanical treatment of the pigment dispersion by ultrasonic power alone or in combination with the LCST property was investigated. The course of the pigment surface coating process was followed by the Electrokinetic Sonic Amplitude (ESA) method. The temperature-controlled sorption of PMVE-g-PEO graft copolymers on both inorganic and organic pigment surface was investigated. It was found that ultrasonic treatment together with LCST thermoprecipitation is a promising method for the surface modification of pigments with regard to dispersion stability.

Chapter 11 - This paper represents the results of the investigations of bifunctional (meth)acrylates copolymerization kinetics by the method of lazer interferometry. The kinetics of the stationary photoinitiated copolymerization has been studied in the system: 1,6-hexanediol diacrylate – triethyleneglycol dimethacrylate (HDDA – TGM-3) till high conversions depending on the system composition, concentration of photoinitiator and UV – irradiation intensity. The obtained experimental data have been interpreted in accordance with the conception of microheterogeneity of the polymerizing system. In the concrete variant of this conception which takes into account a presence of at least two different reaction zones, where the process of copolymerization occurs, the general kinetic model of photoinitiated copolymerization of bifunctional (meth)acrylates till high conversions has been offered.

Chapter 12 - In the present work the total content of antioxidants and their activity relatively oxygen and its radicals are measured in water extract of mint peppery (*Mentha piperita L.*) by two electrochemical methods: ammetric and voltammetric. It is established that these parameters, and also voltammogramms, defining mechanism of oxygen reduction, are essentially changed during the first hours after extraction. The explanation of the received results from the point of view of mint structure is offered.

Chapter 13 - There was offered the quantitative structural model of particulate-filled polymer composites impact toughness, based on the fractal analysis ideas. The indicated model demonstrated good correspondence to the experimental data. It has been shown, that the action of nanofiller as nucleator, resulting to crystallinity degree and amorphous phase structure change, exercises the main influence on impact toughness value.

Chapter 14 - The most typical behaviour features of polymer nanocomposites filled with disperse calcium carbonate were considered. The quantitative analysis was carried out within the frameworks of structural models: the cluster model of polymers amorphous state structure and fractal analysis. It has been shown that all the considered nanocomposites properties changes were defined by polymeric matrix structure variations, which are due to nanofiller introduction.

Chapter 15 - The results of the present paper have shown the existence in polymer particulate-filled nanocomposites nanoadhesion effect, which consists in enhanced level of polymeric matrix-nanofiller interaction. This effect has purely dimensional origin and was realized for nanofiller particles with diameter less than 80 nm. The nanoadhesion effect influences strongly on macroscopic properties of those nanocomposites, in which it was realized. So, essential thermal stability enhancement and considerable thermal expansion coefficient reduction of nanocomposites on the base of phenylone at introduction 1 mass. % aerosil only are due to the nanoadhesion effect. Appreciation of nanoadhesion effect allows qualitatively and quantitatively to describe exactly nanocomposites reinforcement degree.

Chapter 16 - The analysis of charge transfer processes in the electrical conducting solid polymer systems has been carried out. The processes in these systems are divided on two types. First type- the processes of charge transfer between electrodes and particles, second one –the processes of the charge transfer between conductive particles.

The describing of medium is carried out with using of Green temperature functions of polarization operators for the medium molecules. It permits taking into account the effects of frequency and space dispersion.

The analytical expressions for kinetic parameters of the charge transfer processes from electrodes to particles and between particles in condensed matter have been obtained. The comparison of the general theoretical dependence of the current in ECPC on the average distance between conducting particles with analogical dependence for some real ECPC are presented in the paper.

Chapter 17 - The bond energy in some cluster aqueous structures has been calculated using spatial-energy notions and compared with quantum-mechanical methods. It is demonstrated that the variety and specifics of aqueous solutions are determined by the changes in structural energy characteristics of these cluster compounds.

Chapter 18 - The dependence of some thermodynamic characteristics upon initial spatial-energy parameters of free atoms has been analyzed. The corresponding equations have been obtained, the dissociation energies of binary molecules and enthalpy of single-atom gas formation have been calculated based on them.

Chapter 19 - It is ascertained that struggle and opposition between two basic political and economical systems do not have sense since the most progressive system is the combination of the two. One of the possible variants of such formation is pro-posed. It is shown that periods of resonances of solar activity have multipronged importance for our planet.

Chapter 20 - The surface structures of asymmetric cellulose acetate membranes prepared by phase inversion method are investigated by atomic force microscopy (AFM). The differences in the surface morphology have been expressed in terms of four types of roughness parameters such as; roughness average ($S_a$), root mean square roughness ($S_q$), surface skewness ($S_{sk}$) and surface kurtosis ($S_{ku}$). The association of the roughness characteristics with the salt rejection performance has been also investigated. The results show that the increase in the evaporation time results in the change of the membrane morphology and rejection efficiency.

Chapter 21 - The aim of this study is to produce poly(vinylchloride) (PVC) based strong acidic cation exchange membranes and to characterize them. Recent studies show hopeful results regarding ionic liquids, which are one of novel alternatives, since they have low volatility, low melting points, thermal stability and nonflammablity. In this research, membranes were prepared by following a heterogeneous method and characterized in physicochemical, morphological and electrochemical ways. Physicochemical characterization consisted of water contents, dry and wet densities, porosities, ion exchange capacities of membranes and isotherm studies. Scanning Electron Microscopy (SEM) analysis was used to perform morphological characterization. Water contents of the membranes were found between 0.3-0.4 while density values were in a range of 0.7-1.2 g/cm$^3$. Porosities were found between 0.2-0.4 and ion exchange capacities changed between 1.2 and 2.8 meq $Na^+$/g-dry. Electrochemical characterization included transport numbers of membranes determined by membrane potential method. They were found around 0.5 for the membranes prepared with dioctyl phthalate (DOP) whereas around 1.0 for the membranes prepared with IL. Diffusivity values of $Na^+$ ions across the membrane were calculated by using Donnan dialysis method and found in order of magnitude -5.

Chapter 22 - In the work [1], the regularities of chemical crosslinking of HDPE by some maleimide derivatives (MI) have been investigated. It was shown that the crosslinking occurs in the process of polymer oxidation by the air oxygen. In this case, the network junctions are formed directly by the MI molecules.

The goal of this work is the investigation of the structure-mechanical characteristics of HDPE chemically crosslinked by MI.

Chapter 23 - The simplest and commonest method for the description of the correlation between the structure of a network polymer and the kinetics of a chemical reaction is based on the statistical approach. According to this approach- the structure is defined as a single-valued function of the depth of conversion and the initial composition of the system. This method of calculating the structural parameters was described in most detail in monographs [1- 2] and reviews [3- 4]. However as is well known [2-4] when functional groups change their reactivity (substitution effect) during the process (for example during polycondensation with unequal reactivities of primary and secondary amino groups) the polymer structure becomes dependent on the reaction route- that is- the above unambiguity of the structure-conversion relationship is not satisfied. In this case- a rigorous kinetic calculation for all structural elements is required. One should mention that this fact is not yet universally recognized and probabilistic approaches are rather popular (see- for example- [5]).

Chapter 24 - NAROSSA® 2010 conference was held in Maritim Hotel (Magdeburg, Germany) in the period June 7 – 8, 2010. The organizers of conference were the association

Pilot Pflanzenöltechnologie Magdeburg e. V. (PPM) and the sponsor of this conference was OHMI Consulting GmbH.

Chapter 25 - Professor Rustam Yakubovich Deberdeev was born in 1940 in Arsk city, Tatarstan. He graduated from a secondary school in 1958 in the city Naberezhnye Chelny and used to work as a metalworker and a driver. He entered military service and entered the Kazan State Technological University after its completion in 1962. (formerly titled as the Chemical and Technological Institute).

In: Chemical and Biochemical Reactions
Editors: G. E. Zaikov, L. Liu et al.

ISBN 978-1-61122-860-1


*Chapter 1*

# Ion-Radical Conversions of Polyamides and Polyimides in the Nitrogen Dioxide Atmosphere

***E. Ya. Davydov, I. S. Gaponova, T. V. Pokholok, G. B. Pariiskii* and G. E. Zaikov***

Emanuel Institute of Biochemical Physics, Russian Academy of Sciences, 4 Kosygin St., Moscow, 119334, Russia

## Abstract

By the examples of polycaproamide, polyvinylpyrrolidone, poly-*m*-phenylene isophthalamide and polypyromellitimide, the mechanism of interaction of nitrogen dioxide with polymers containing amide and imide groups is considered. The initiators of conversions of polymers of the given classes are not monoradicals of nitrogen dioxide rather than its equilibrium dimers in the form of nitrosyl nitrate. In primary oxidizing reaction of an electron transfer from donor groups of macromolecules to nitrosyl nitrate, radical cations, nitric oxide and nitrate anion are formed, which in the subsequent reactions give products of nitration and nitrosation. These products are precursors of stable acylalkylaminoxyl, acylarylaminoxyl, iminoxyl macroradicals. The specific interactions of nitrogen dioxide dimers with functional groups of macromolecules determine features of the mechanism of ion-radical conversions of the polymers and the composition of radical and molecular products. The direct detection of radical cations has been realized by ESR method for confirmation of the ion-radical initiation concept using model reaction of nitrogen dioxide with triphenylamine.

**Keywords:** nitrogen dioxide, nitrosyl nitrate, polyamides, polyimides, stable radicals, ESR spectra

* E-mail: pgb@sky.chph.ras.ru, Fax: (7-495)1374101

## INTRODUCTION

In comparison with the large number of the publications devoted to researches of interactions of aggressive nitrogen oxides with various low-molecular organic compounds [1], the information on the mechanism of similar reactions with polymeric materials are not so numerou[s] [2,3]. The researches of the polymer conversions in an atmosphere of nitrogen oxides, as well as under their combined action with oxygen and UV light, were carried out for the first time by Jelline[k] [2]. In these researches as the basic characteristics of efficiency of polymeric materials interaction with $NO_2$, the changes of their mechanical properties, IR spectra and characteristic viscosity of solutions have been considered. Polyamides demonstrated sufficiently high sensitivity with respect to nitrogen dioxide.

Under the action of $NO_2$ ($2.5 \cdot 10^{-5}$ – $10^{-3}$ mol·l$^{-1}$), films of nylon-66 are efficiently deteriorated [4]. This process is diffusion-controlled, and its rate essentially depends on a degree of the sample crystallinity. The degradation can be inhibited by the small additives of benzaldehyde or benzoic acid. The interaction with $NO_2$ is accelerated by air and simultaneous irradiation by UV light. It was assumed that benzoic acid blocks amide groups, and only small quantity of the groups unconnected by hydrogen bonds reacts by the following mechanism:

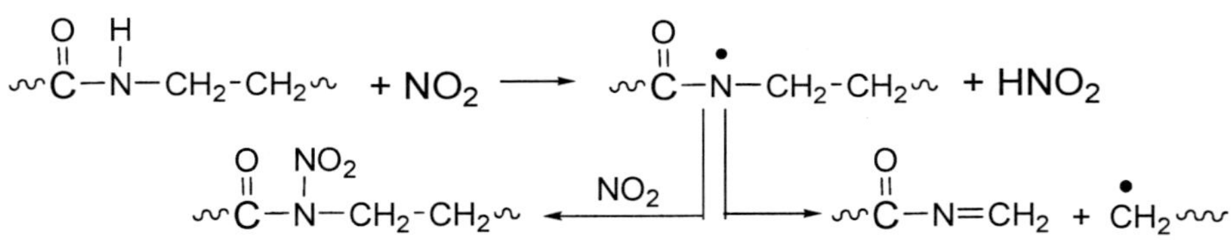


(1)

Thus, in Jellinek's works, the free-radical mechanism of degradation of polyamides has been postulated. According to this mechanism, amidyl macroradicals are formed by abstracting hydrogen atoms from amide groups by $NO_2$. The subsequent decomposition of these radicals causes degradation of macromolecules. However, such mechanism for polyamides is not enough proved. From the data available in the literature it follows that least strong bonds in aliphatic polyamides are C–H adjacent to NH groups, and degradation of macromolecules occurs as a result of decomposition of corresponding macroradicals [5]. In addition, from researches of interaction of $NO_2$ with low-molecular amides in a liquid phase, the conclusion was made that the given process proceeds by the ionic mechanism and represents nitrosation of amides by nitrogen dioxide [6]. The high reactivity of polyamides in respect to nitrogen dioxide cannot be explained by initiating reaction of the hydrogen atom abstraction from macromolecules by radicals $NO_2$. It is well to bear in mind that $NO_2$ is a radical of average reactivity; the dissociation energy for

H–ONO makes up 327 kJ·mol [7]. These radicals are capable of abstraction hydrogen atoms only from the least strong, for example allyl C–H, bonds [8]. Because of this, interaction of polyamides and polyimides with nitrogen dioxide should be considered on the basis of other plausible mechanism of initiation associated with the presence of the nitrosyl nitrate $ONONO_2$ dimeric form of $NO_2$ (NN) [9].

$$O_2NNO_2 \rightleftarrows 2NO_2 \rightleftarrows ONONO_2 \quad (2)$$

The NN has strong oxidizing properties [6] and is capable of generating macroradicals in the electron transfer reaction:

$$RH + ONONO_2 \longrightarrow [\overset{\bullet +}{RH}\ (NOONO_2^{\ -})] \longrightarrow \overset{\bullet}{R} + NO + H^+ + NO_3^- \quad (3)$$

The subsequent recombination of $R^{\bullet}$ and nitric oxide gives macromolecular nitroso compounds which are the precursors of stable radicals [3]. This concept has been confirmed by results of investigation of nitrogen dioxide interaction with poly-2-vinylpyridine (PVPR) [10]. The decomposition of radical cations results in elimination of a proton which is accepted by phenyl rings of PVPR. Pyridinium cations formed were detected by their characteristic bands in the IR spectrum (2200 and 2400-2600 $cm^{-1}$). The efficiency of initiation by the ion - radical mechanism (3) is determined by donor properties of functional groups and also opportunity of conversion of intermediate radical cations into free radicals. These conditions are realized in polyamides in which amide groups transform by the following scheme:

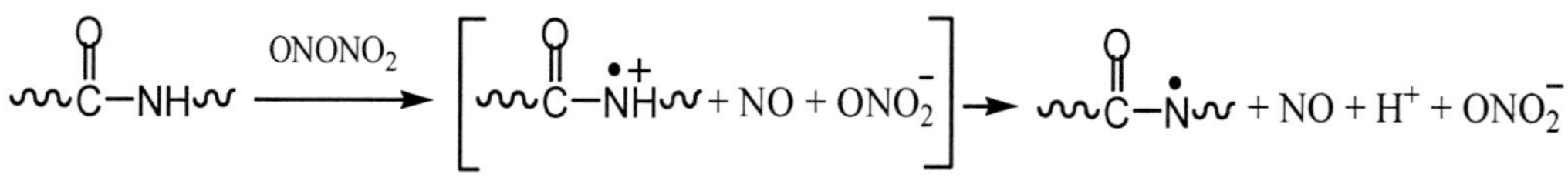

(4)

However chemical structure of aromatic polyimides, for example polypyromellitimide (PPI) does not assume decomposition of radical cations with detachment of a proton by the mechanism similar to (4), and PPI does not contain any chemical bonds reacting with monoradicals $NO_2$ (double carbon - carbon bonds, weak allyl C–H bonds) [8,9]. Thus, the study of interaction of aromatic polyimides with nitrogen dioxide permits to establish other probable ways of conversions by the ion - radical mechanism.

The results of the present work are considered in the context of the mechanism of ion - radical reactions initiated by NN. This mechanism allows explaining structures of radicals and molecular products of interaction of polyamides and polyimides with $NO_2$ as well as kinetic features of this process.

## EXPERIMENTAL

Experiments were carried out on polycaproamide (PCA) with $M_\eta = 2.3\cdot10^4$, polyvinylpyrrolidone (PVP) with $M_\eta = 3.0\cdot10^5$, poly-*m*-phenylene isophthalamide (PPIA), PPI synthesized by polycondensation of pyromellitic dianhydride and 4,4′–diaminodiphenyl ether with subsequent imidization of polyamide acid at 250 °C. The imidization has been followed using IR spectroscopy by disappearance of the N–H (3250 $cm^{-1}$) and appearance of imide bands (1780 $cm^{-1}$, 720 $cm^{-1}$) [11]. As is evident from IR spectra, the imidization is completely carried out in these conditions. Nitrogen dioxide was obtained by the thermal decomposition of lead nitrate. Samples of powder-like polymers (0.06 g.) were placed to quartz tubes for EPR measurements supplied with a stopcock. The tubes were connected to a glass flask of volume 0.5 l attached to vacuum line. After pumping a pressure of ~ $10^{-3}$ mm Hg, the stopcock was closed and the flask was filled with $NO_2$ up to concentration of ~ $10^{-3}$ mol·$l^{-1}$. ESR spectra were directly recorded during exposure of samples to $NO_2$ and after pumping out the gas from tubes. The $NO_2$ concentration in a gas phase was determined by optical density at 410 $n^m$ [12] using Shimadzu UVmini-1240 spectrophotometer. EPR spectra were recorded on X-band Mini-EPR 10 SPIn Co., Ltd. with 100 kHz modulation. After exposure to $NO_2$, samples were pumped out up to a pressure of ~ $10^{-3}$ mm Hg. Then these samples were heated at 373 K, and the EPR records were carried out at room temperature. The values of *g*-factors of radicals and HFI constants were measured by comparison with the standard spectrum of $Mn^{2+}$ ions. For research of molecular products of interaction of the polymers, UV and IR methods were used. In PCA, UV spectra were recorded in films of 1–2 μ thickness put on quartz plates and obtained from a solution of the polymer in concentrated formic acid. IR spectra of PCA films put on KRS-5 plates were recorded using spectrometer Specord IR-75. The products of reactions of PVP with $NO_2$ were identified by UV spectra of the polymer solution in chloroform. Films of original and exposed PPIA were prepared from a solution in dimethylacetamide and analyzed by IR technique on a surface of a KRS crystal. For the confirmation of proposed reaction mechanism, the action of $NO_2$ on low-molecular model compound, namely, triphenylamine (TPA) was used by ESR in the present work. The experiments were carried out on commercial TPA (Eastman Kodak Co). The reagent was further purified by vacuum sublimation.

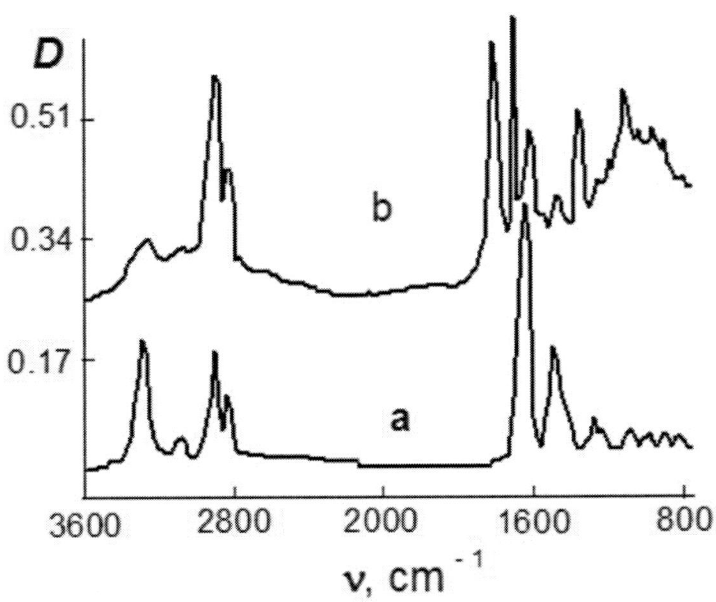


Figure 1. IR spectra of original PCA films (a) and exposed to $NO_2$ (b).

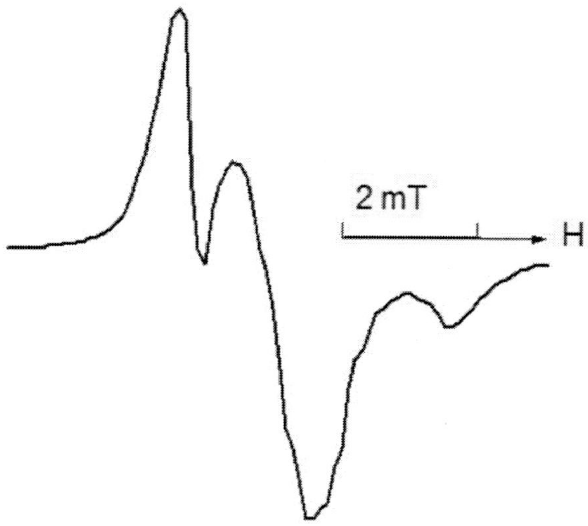


Figure 2. ESR spectrum of acylalkylaminoxyl radicals generated by $NO_2$ in PCA.

# RESULTS AND DISCUSSION

## Interaction of Nitrogen Dioxide with Polycaproamide

On exposure of PCA films at room temperature to $NO_2$ in concentrations of $4{\cdot}10^{-5} - 10^{-3}$ $mol{\cdot}l^{-1}$, the optical absorption bands in the range of 390 – 435 nm are appeared corresponding to *N*–nitrosoamide group[s] [3 ,13-15]. In parallel with them, IR spectra (Figure 1) show decreasing the intensity of band (3293 $cm^{-1}$) associated with the stretching vibrations of H–bonded N–H groups and occurrence of the band at 1730 $cm^{-1}$ of carbonyl groups of nitrosoamides. The bands at 1504 and 1387 $cm^{-1}$ which correspond to the asymmetric and symmetric stretching vibrations of N=O groups of nitrosoamide[s] [6] appear in the IR spectrum. The ESR spectrum observed in these conditions (Figure 2) represents an anisotropic triplet with $A$= 1.94 mT and $g$ = 2.003 characteristic of stable acylalkylaminoxyl radicals in solid phase [3].

The experimental data obtained by ESR, UV and IR methods cannot be explain in the context of initiation of free radical processes in PCA by hydrogen atom abstracting via $NO_2$ [4]. The occurrence of nitrosoamide groups in PCA [14] is not consistent with the mechanism of interaction of polyamides with $NO_2$ (1) in which the main molecular product of reaction should be *N*–nitroamide. In addition, the square-law dependence of initial rates of accumulation of N–nitroso groups in PC [14] and degradation of macromolecules of nylon-66 [4] on concentration $NO_2$ in a gas phase were observed.

As was shown by UV spectroscopy, NN is capable of forming charge-transfer complexes with a number of compounds: methylbenzenes [16, 17], hydroquinone [18], benzoqiononedioxine [19]. The electron transfer in such complexes should result in the formation of three particles located at a close distance: a radical cation of donor molecules, NO radical and nitrate anion. In polyamides, amide groups can act as electron donors. As a result of decomposition of radical cations with the proton elimination, nitric acid, amidyl macroradical and nitric oxide are formed. The subsequent recombination of such radical pair in "cage" by reaction (5) gives N–nitrosoamide groups as the main molecular product of interaction of nitrogen dioxide with PCA. The features of interaction of $NO_2$ with aliphatic polyamides are reflected by the following scheme:

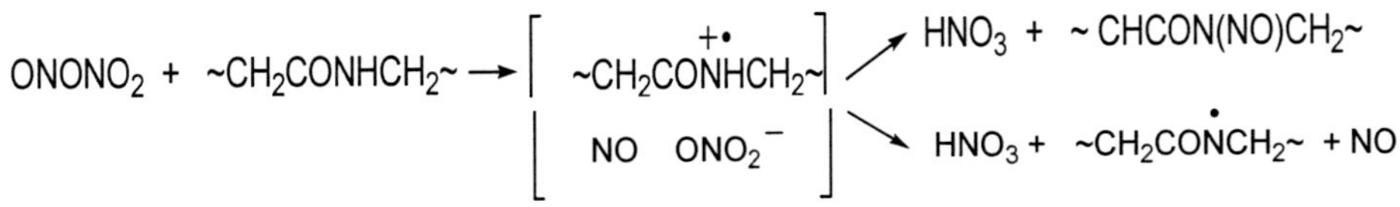

(5)

(6)

(7)

(8)

Reactions (5)–(8) on long exposure lead to practically complete disappearance of NH groups in macromolecules. Amidyl radicals left "cages" by reaction (6) can decompose with breaking of the main chain and also can transform into *N*–nitrites in recombination with $NO_2$. These compounds by reaction (8) give acylalkylaminoxyl macroradicals of PCA (Figure 2).

The kinetic analysis of the reaction scheme (5)–(8) shows that the initial rate of degradation should be proportional to $[NO_2]^2$ at $k_d >> k_1$ $[NO_2]$, and the initial rate of the acylalkylaminoxyl radical accumulation is proportional to $[NO_2]^3$ as it follows from experiments [4, 14].

## Interaction of Nitrogen Dioxide with Polyvinylpyrrolidone

In PVP, as well as in PCA, on exposure to $NO_2$ [13, 14], the UV spectrum demonstrates absorption bands of nitrosoamide (Figure 3). They have three maxima at 386, 410 and 430 nm [15]. In addition to these compounds, alkylnitrites having characteristic system of absorption bands spaced at ~10 nm are formed in the range from 320 to 380 nm [20]. The more complex composition of stable nitrogen-containing radicals is observed in PVP in comparison with PCA. The ESR spectrum of PVP exposed to $NO_2$ represents a superposition of signals of acylalkylaminoxyl radicals (I) with the same parameters as in PCA, and a triplet signal of iminoxyl radicals (II) with $A_{II}^{N} = 4.33$ mT, $A_{\perp}^{N} = 2.44$ mT; $g_{II} = 2.0029$, $g_{\perp} = 2.0053$ [21, 22] (Figure 4). The scheme of PVP conversions in the presence of nitrogen dioxide based on the same conception of the NN initiating action is represented by the following reactions:

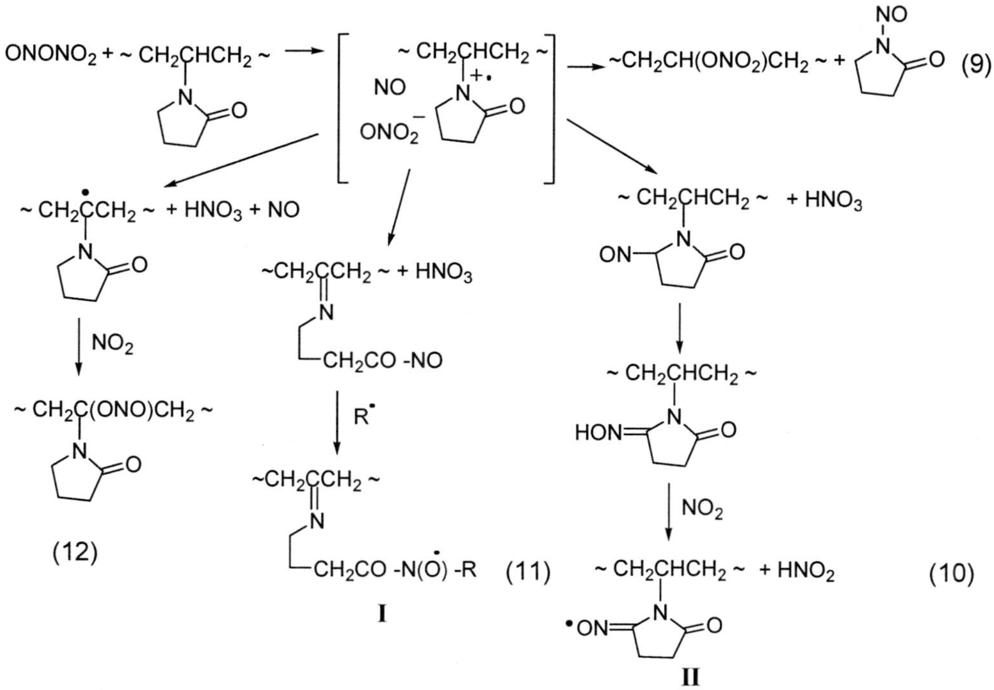

The interaction of NN with tertiary amide groups of PVP generates radical cations and nitric oxide. The probable pathways of conversions of radical cations determining structure of products of interaction with nitrogen dioxide are similar to the processes for radical cations in low-molecular amides (dimethylformamide, dimethylacetamide, tetramethylures) [23]. These processes include decomposition of C–N bonds with the formation of macrocations and free radicals, which recombine in "cage" with $ONO_2^-$ and NO giving are formed nitrates and nitroso compounds. In such a manner, N–nitrosoamides being one of the main products of the PVP conversions are formed. Other reactions of radical cations are associated with detachments of protons from CH or $CH_2$ groups adjacent to α–position of nitrogen atoms. The radicals I and II are formed by reactions (10) and (11). The occurrence of iminoxyl radicals II is conditioned by the isomerization of nitroso compounds into oximes [1] with the subsequent abstraction of mobile hydrogen atoms by $NO_2$ radicals. A condition of the radical I formation is the encounter of corresponding nitroso compounds with alkyl radicals $R^{\bullet}$ formed in reactions (10) or (11).

## Interaction of Nitrogen Dioxide with Aromatic Polyamides

The materials based on aromatic polyamides and polyimides have voluable operational properties [24] and find wide application including for manufacture of products of space engineering. The effect of $NO_2$ on mechanical properties of polyamidoimides (PAI)

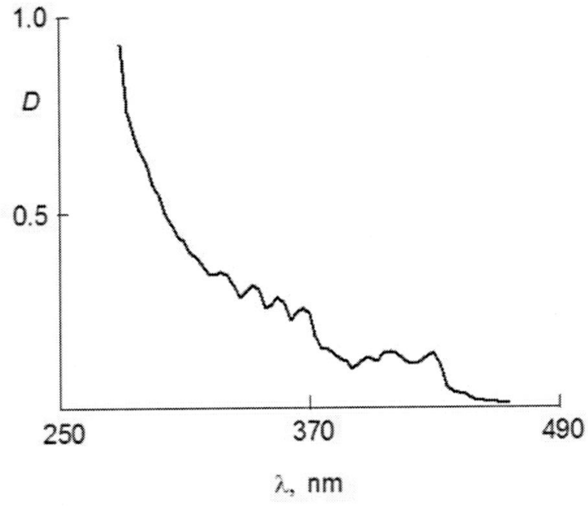


Figure 3. Absorption spectrum of chloroform solution of PVP exposed to $NO_2$.

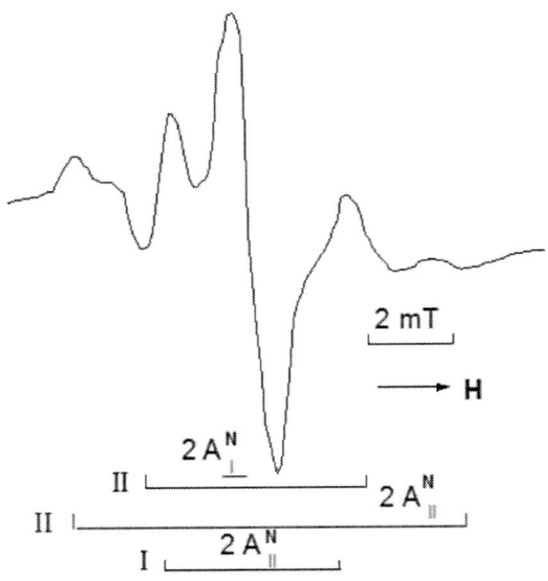


Figure 4. ESR spectrum of radicals generated by $NO_2$ in PVP.

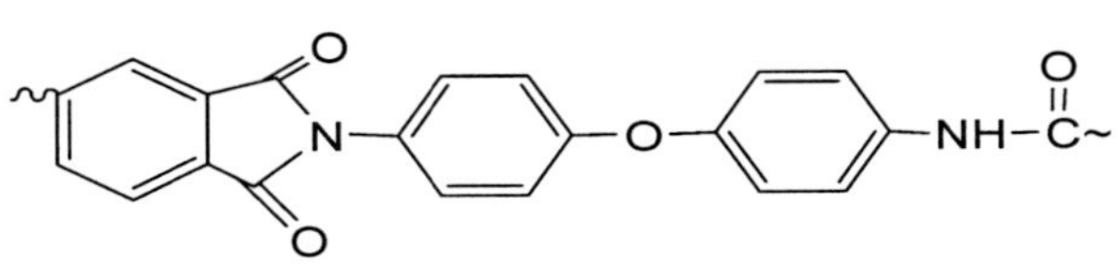

was investigat[ed] [25] at 323 K and $[NO_2] = 5 \cdot 10^{-3}$ mol·l$^{-1}$. Temperature dependencies of the storage modulus *E′* and the loss modulus *E″* were determined for various times of an exposure in an atmosphere of $NO_2$. The sharp drop of *E′* was observed at 473 K. At glass-transition temperature (563 K), the value of *E′* increases for PAI films exposed to $NO_2$ over a period of eight days. When studying the temperature dependencies of the loss modulus *E″*, an absorption peak was found to appear at 473 K, which indicates degradation of the polymer. A decrease in the band intensities of the stretching vibrations of the N–H bonds and the appearance of the peaks associated with the addition of $NO_2$ groups to macromolecules was observed in the IR spectra of the films.

Under the action of $NO_2$, iminoxyl radicals with the ESR spectral parameters $A_{II}^{N} = 4.1$ mT, $g_{II} = 2.0028$, $A_{\perp}^{N} = 2.1$ mT, and $g_{\perp} = 2.0058$ are mainly produced in PAI [26]. The proportion of the acylarylaminoxyl radicals, which are likely produced by the same mechanism as (6), (8) in PCA, is less than 30%.

The features of the mechanism of the iminoxyl radical formation in PAI are determined by nitrosation of phenyl rings. This mechanism has been investigated in detail by the example of PPIA [27]. The ESR spectrum recorded in initial stages of exposure of PPIA samples in an atmosphere of $NO_2$ ($10^{-3}$ mol·l$^{-1}$) at 295 K is shown in Fig 5 a. It corresponds in its shape and parameters ($A_{II}^{N} = 4.8$ mT, $g_{II} = 2.0024$, $A_{\perp}^{N} = 3.2$ mT, $g_{\perp} = 2.0047$) to the spectrum of iminoxyl radicals (III) with the hindered rotary mobility [14, 22, 23]. At later stages of the exposure (Figure 5 b), the radicals of iminoxyl type (IV) with other ESR parameters are accumulated: $A_{II}^{N} = 4.1$ mT, $g_{II} = 2.0024$, $A_{\perp}^{N} = 2.6$ mT, $g_{\perp} = 2.0050$. However, as a result of subsequent pumping out of samples within 3 hours, components of the ESR spectrum of radicals III are again appeared (Figure 5 c). At that, the total concentration of iminoxyl radicals is increased approximately fourfold. Such effect is caused by ability of radicals III to form non-paramagnetic complexes with nitrogen dioxide, which easily dissociate into initial components by pumping out of samples. The appreciable distinctions in anisotropic ESR constants and also in the kinetic of accumulation of these radicals specify different chemical structures of radicals III and IV.

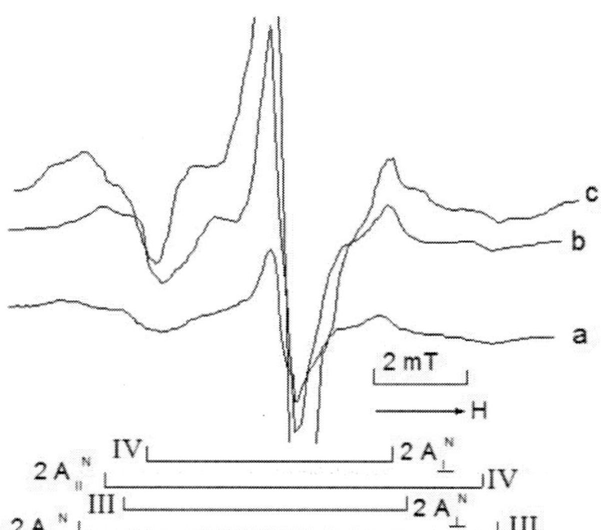


Figure 5. ESR spectra of PPIA after exposure to $NO_2$ at 295 K: 0.5 h (a), 95 h (b) and after consequent pumping for 3 h (c).

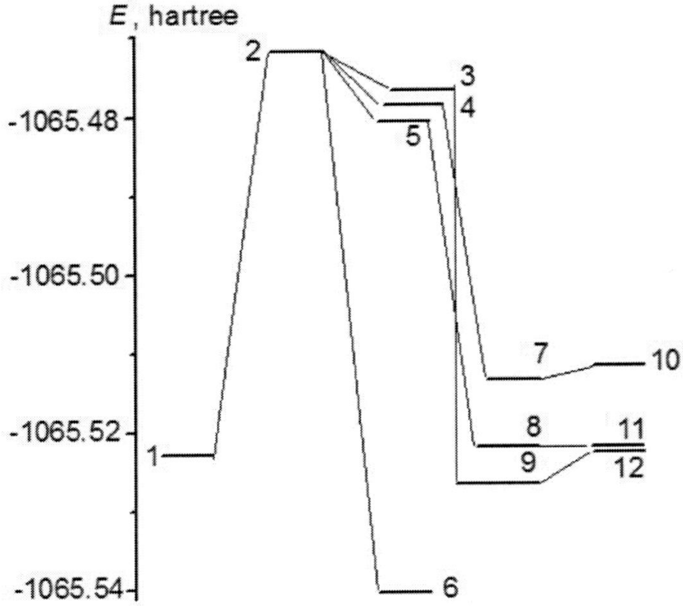


Figure 6. Variations of calculated energies in the iminoxyl radical generation: (1) X + NN +$NO_2$;(2) V (Va, Vб, Vв) + NO + $HNO_3$ + $NO_2$; (3) VIIa +$HNO_3$ + $NO_2$; (4) Vc $HNO_3$ + $NO_2$; (5) Vb+ $HNO_3$ + $NO_2$; (6) VI + $HNO_3$ + $NO_2$; (7) VIIIc+ $HNO_3$ + $NO_2$; (8) VIIIb+ $HNO_3$ + $NO_2$; (9) VIIIa+ $HNO_3$ + $NO_2$; (10) IXc+ $HNO_3$ + $HNO_2$; (11) IXb+ $HNO_3$ + $HNO_2$; (12) IXa + $HNO_3$ + $HNO_2$.

Based on formal mechanism (4), one can assume that the primary act is the oxidizing interaction of amide groups with NN:

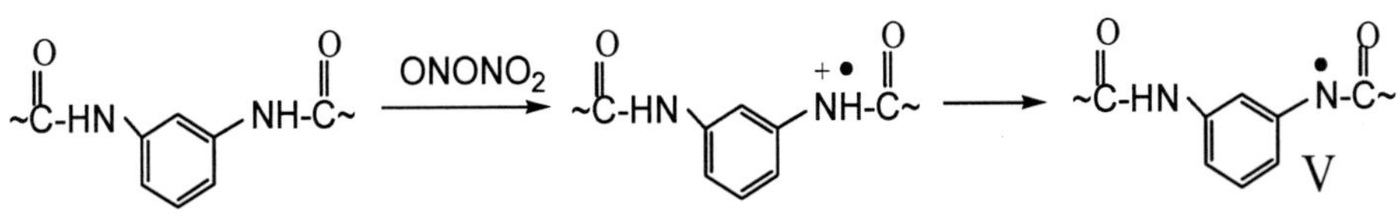


(13)

The recombination of amidyl radical V with nitric oxide gives *N*–nitroso compound VI:

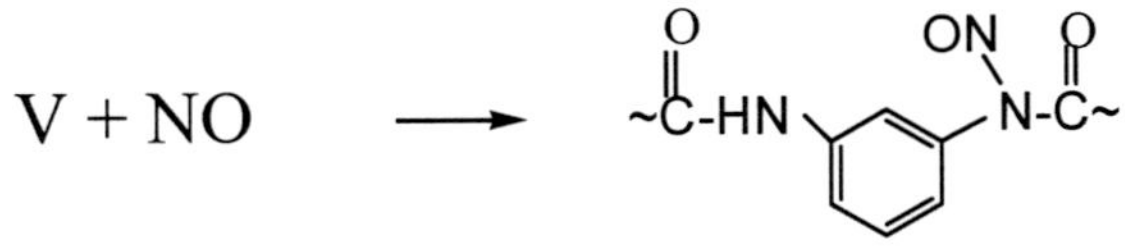


VI

(14)

Isomeric forms of radicals V are radicals of cyclohexadienyl type:

V 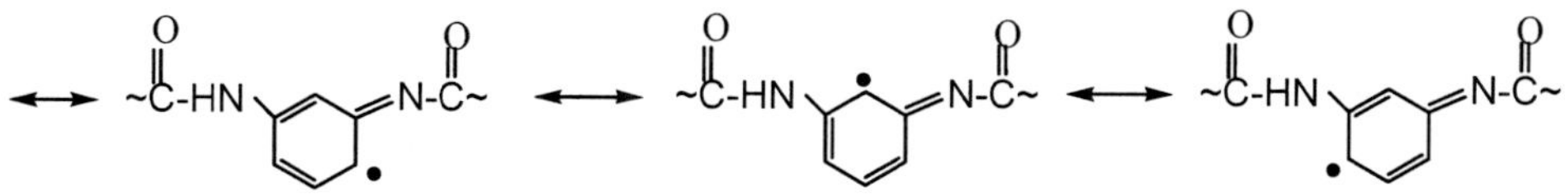

Va Vb Vc

Therefore, along with formation of compounds VI in recombination reaction (14), other nitroso compounds should be expected:

Va +NO 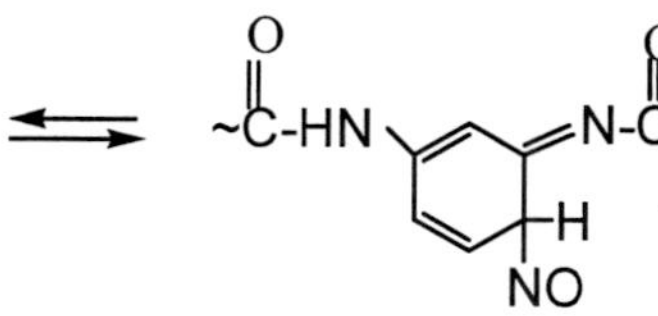 VIIa (15)

Vb +NO 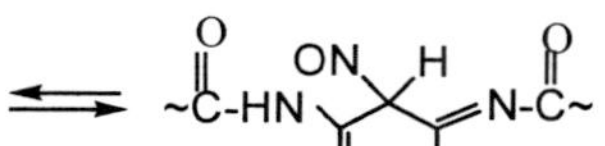 VIIb (16)

Vc +NO 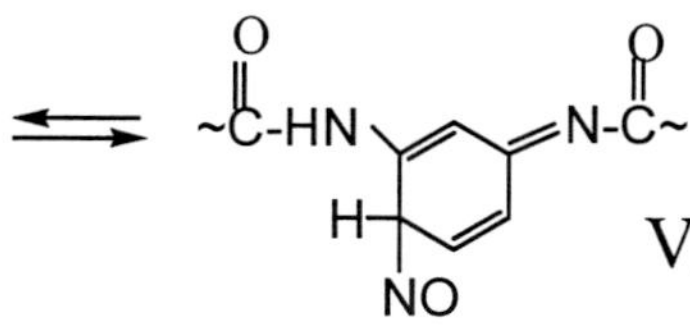 VIIc (17)

Nitroso compounds VIIa, VIIb and VIIc containing α–hydrogen are easily isomerized to the respective oximes [1]:

VIIa 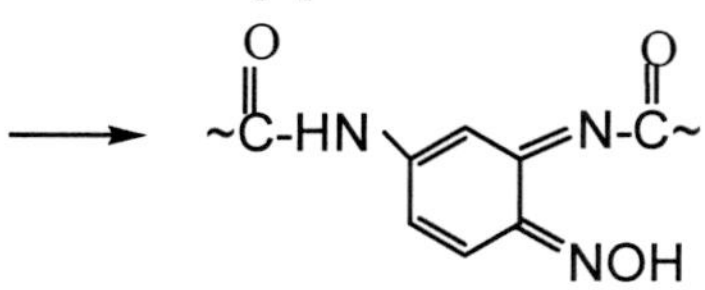

VIIIa (18)

VIIb 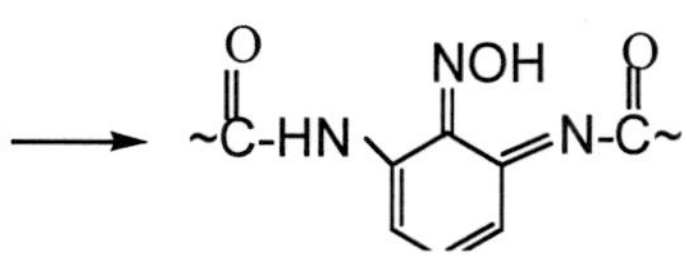 (19)

VIIIb

VIIc 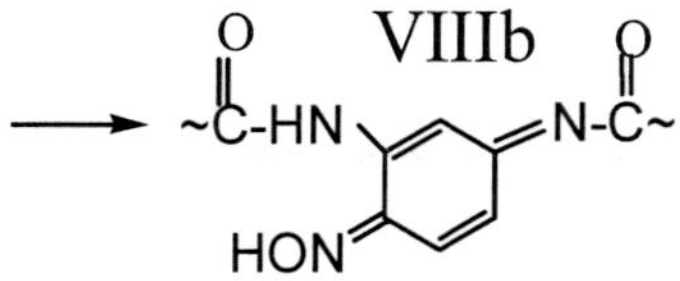

VIIIc (20)

In the following stage, the hydrogen atom abstraction from oximes leads to stable iminoxyl radical formation:

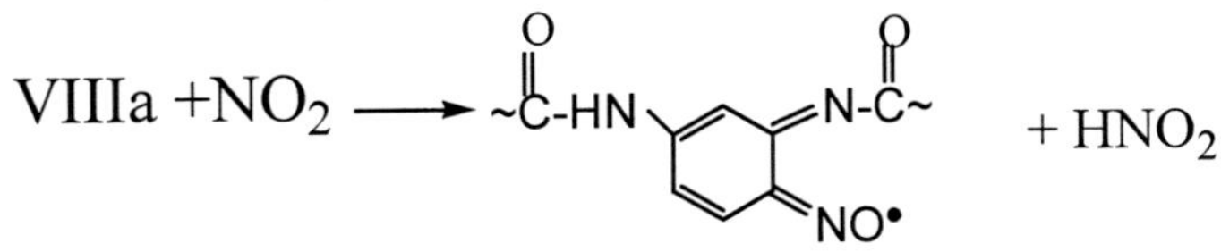


(21)

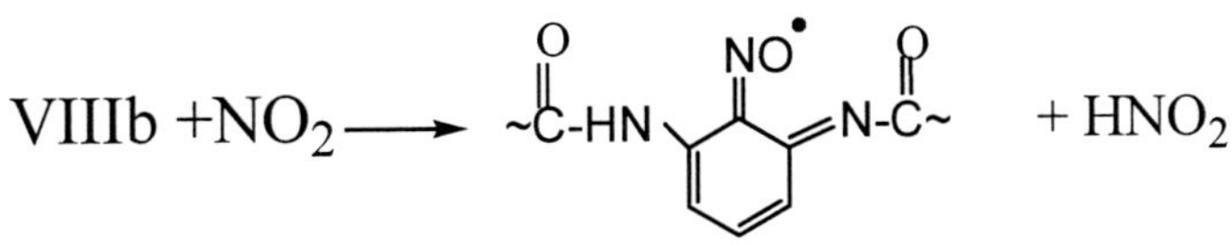


(22)

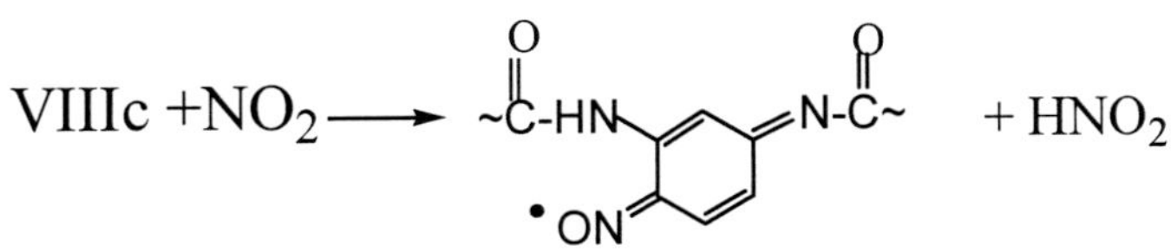


(23)

Thus, the interaction of PPIA with $NO_2$ by the ion-radical mechanisn (4) is capable of initiating conversions of chemically inert phenyl groups into much more reactive cyclohexadiene groups.

The submitted formal scheme of reactions (13) – (23) explains the appearance of stable radicals III and IV (Figure 5) in the given system. However, in order to attribute concrete structures to these radicals, ab initio calculations of isotropic hyperfine coupling constants were performed for following model iminoxyl radicals:

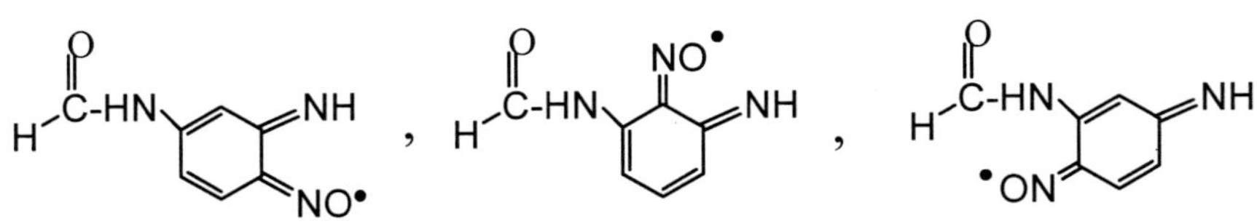

IXa IXb IXc

Calculations were accomplished by the Gaussian 98W progra$^{m}$ [28, 29] using density functional theory in B3LYP approximation. From the results obtained follows that the modelling structures IXb and IXc correspond to radicals III, whereas the radicals IV have structure IXa.

An application of B3LYP method also allows peculiarities of accumulations of iminoxyl radicals in PPIA under the action of $NO_2$, namely, a later appearance of the radicals IV signal in ESR spectrum as compared with radicals III [27]. For that the energy variations in a model reacting system in the course of consecutive conversions to iminoxyl radicals IXa, IXb and IXc (structures analogous of radicals III and IV) by reactions similar to (13) – (23) were calculated. As a primitive model of aromatic amide, compound (X) has been chosen:

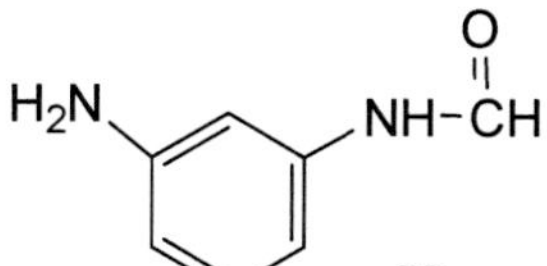


**X**

The results obtained are represented in Figure 6. As is evident from the figure, the first stage similar to reaction (13) requires significant energy consumption. In the subsequent reaction (14), large energy is released (level 6 in Figure 6), and in whole this transformation is an exothermic process. However, the recombination of NO and cyclohexadienyl radical (reactions (15)-(17)) decreases insignificantly the energy (levels 3 – 5) ($-\Delta E$ = 12 – 23 kJ·mol$^{-1}$). Because of this returning back to radicals Va, Vb and Vc is possible until nitroso compounds VIIa, VIIb and VIIc are isomerized to their corresponding oximes (levels 7 – 9) with the large energy release ($-\Delta E$ = 128 kJ·mol$^{-1}$). But for all that the greatest probability of back transition to cyclohexadienyl radicals with an exit of NO from reacting "cage" would be expected for compounds VIIa (level 3), which according to the scheme of the process are the precursors of stable radicals IV recorded at later stages of PPIA exposure to $NO_2$ (Figure 5 b). For compounds VIIb and VIIc (levels 4 and 5), which transform into radicals III, the back transition is least probable, and consequently these radicals are effectively formed at the beginning of exposure (Figure 5 a). As a whole the stable radical generation in PPIA in the presence of $NO_2$ by reactions (13)–(25) is a thermoneutral or weakly endothermic process ($-\Delta E \approx 28$ kJ·mol$^{-1}$).

The formation of unstable compounds of iminoxyl radicals III with $NO_2$ allows us to choose between the structures IXb and IXc having approximately identical ESR spectra. By ESR research on an interaction of *p*–benzoquinone with nitrogen dioxide [30], the capability of $NO_2$ for addition to conjugated double bonds with formation of appropriate oxyaminoxyl radicals has been established:

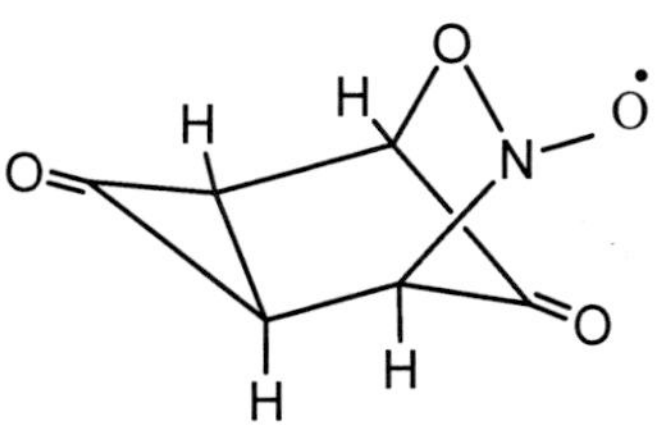


**XI**

Based on this fact, the conversion of radicals III having a system of conjugated double bonds similar to *p*–benzoquinone can be represented as recombining with $NO_2$:

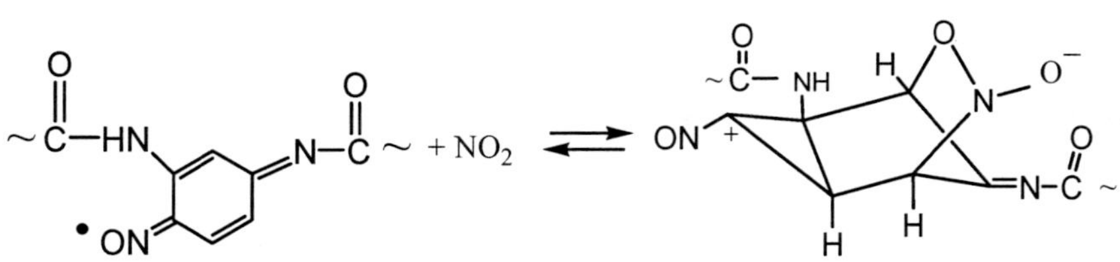


(24)

The strong structural strains cause instability of the compound formed in reaction (24), and it quickly turns into initial iminoxyl radicals during pumping-out of $NO_2$ (Figure 5 c). For radicals IV with conjugated double bonds similar to those of *o*–benzoquinone, the reaction (24), apparently, is less probable, and radicals IV are substantially accumulated on long exposure of samples. On this basis one can conclude that the mechanism of generation of iminoxyl radicals with structure IXc is realized in practice. It is probable that significant steric hindrances created by substituents of a phenylene ring make the mechanism including reactions (16), (19) and (22) energetically less preferable in PPIA.

## Interaction of Polypyromellitimide with Nitrogen Dioxide

Many investigations of thermal and thermo-oxidative stability of polyimides are reporte[d] [31]. However, the mechanism of action of such aggressive gases as nitrogen oxides on these materials remains unexpected. In contrast to PPIA, the exposure of PPI to $NO_2$ at room temperature within 7 days does not result in occurrence of appreciable ESR signals. After pumping out of $NO_2$ from exposed samples, the anisotropic ESR spectrum shown in Figure 7a was observed. The parameters of this spectrum $A_{II}^{N} = 2.0$ mT, $g_{II} = 2.0018$ entirely correspond to acylarylaminoxyl radicals (XII) detected in solid polymers [3, 26]. If the same samples are heated at 373 K in vacuum, additional signals appear in the spectrum (Figure 7 b and c). In this case, the total concentration of radicals increases considerably. The new signals can be divided into two types: four anisotropic side components and a central singlet line. The content of these novel components increases if the exposure of PPI to nitrogen dioxide is longer. Such conclusion is reached comparing ESR spectra obtained after the same time of warming PPI in vacuum (8 h), but durations of exposure of the samples to $NO_2$ are 5 h and 7 days, respectively (Figure 7 c and d).

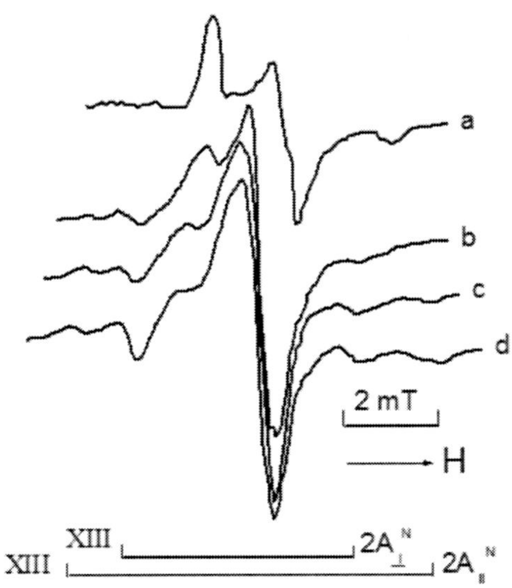


Figure 7. ESR spectra of PPI exposed to $NO_2$ at 295 K for 5 h (a – c) and 7 days (d) after pumping at 295 K (a) and subsequent heating at 373 K in vacuum 1 h (b) and 8 h (c, d).

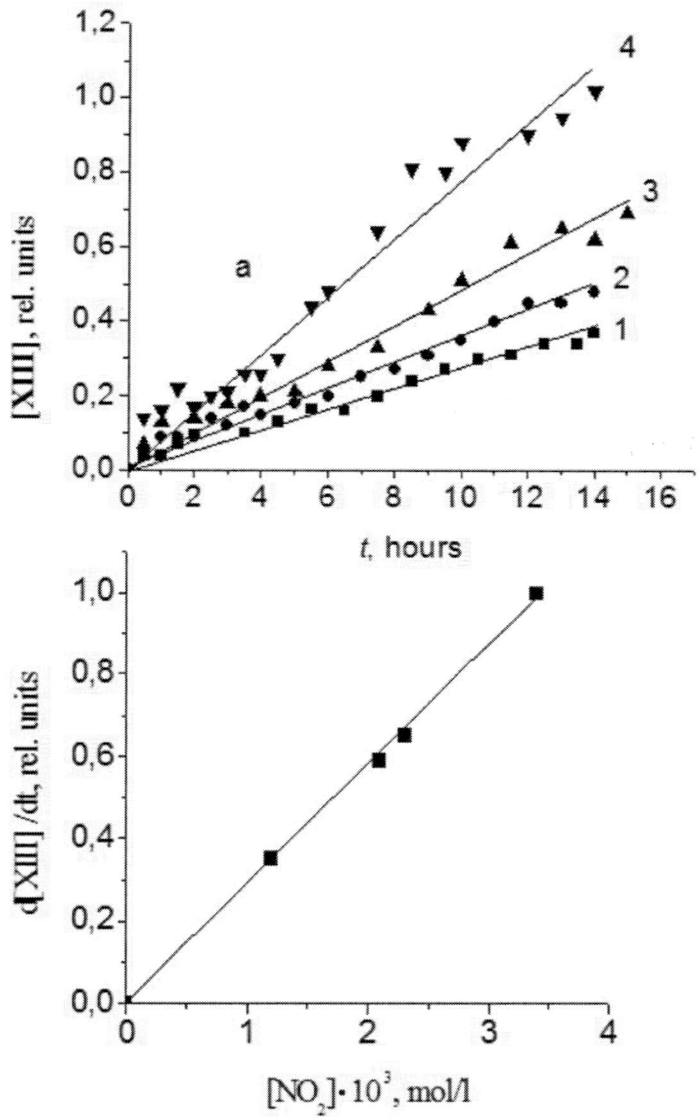


Figure 8. Kinetics of the radical XIII accumulation obtained by hating (373 K) PPI preexposed to $NO_2$ at 295 K for 7 days (a): $[NO_2] = 1.2 \cdot 10^{-3}$ (1); $2.1 \cdot 10^{-3}$ (2); $2.3 \cdot 10^{-3}$ (3); $3.4 \cdot 10^{-3}$ mol $l^{-1}$ (4), and the dependence of rates of the radical XIII accumulation on $NO_2$ concentrations (b).

On the basis of spectral splitting and $g$–factor values, the side components can be unequivocally attributed to a signal of iminoxyl radicals (XIII) with the parameters: $A_{II}^{N}$ = 4.4 mT, $g_{II}$ = 2.0022, $A_{\perp}^{N}$ = 2.4 mT, $g_{\perp}$ = 2.0056 [3, 20, 21]. As to the central singlet component with $g = 2.0045$ and $\Delta H_{1/2} = 0.7$ mT, its assignement to radicals (XIV) is feasible from the analysis of the general mechanism of radical transformations of PPI and results of computer modeling.

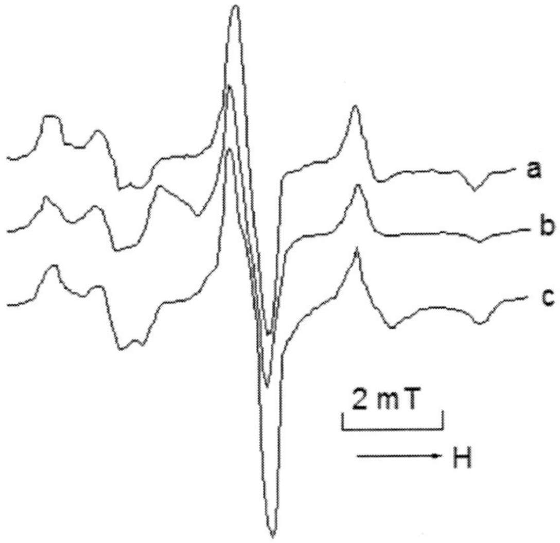


Figure 9. ESR spectra of PPI exposed to $NO_2$ for 5 h at 373 K (a), after subsequent pumping the samples at 295 K (b), and after heating (b) at 373 K in vacuum (c).

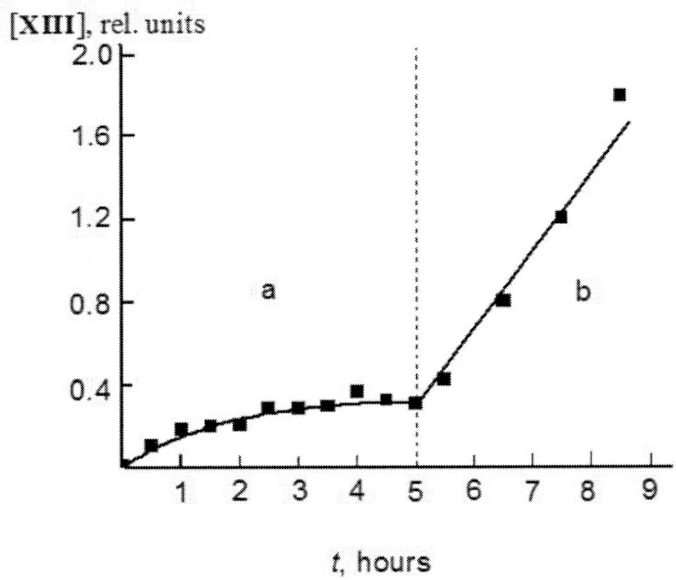


Figure 10. Kinetics of the radical XIII accumulation on exposure of PPI to $NO_2$ ($3.7{\cdot}10^{-3}$ mol·l$^{-1}$) at 373 K (a) and after subsequent heating at 373 K in vacuum (b).

The kinetic dependencies of the radical XIII accumulation in vacuum at 373 K for PPI samples exposed to $NO_2$ for 7 days in conditions of various $NO_2$ concentrations are shown in Figure 8. As indicated in the figure, radicals XIII are formed within 14 h approximately at constant rate.

In contrast to the exposure at room temperature, practically only iminoxyl radicals XIII are accumulated in $NO_2$ atmosphere at 373 K (Figure 9 a). However, the signal intensity of radicals XIII in these conditions becomes steady in 2-3 h (Figure 10 a). After subsequent pumping out of the samples, the signal of acylarylaminoxyl radicals XII is appeared (Figure 9 b). A significant increase of radical XIII concentrations is observed on subsequent heating these samples at 373 K in vacuum as shown in Figure 10 b. Their signal in the ESR spectrum remains prevailing (Figure 9 c).

The mechanism of radical formation in the given system can be represented as follows.

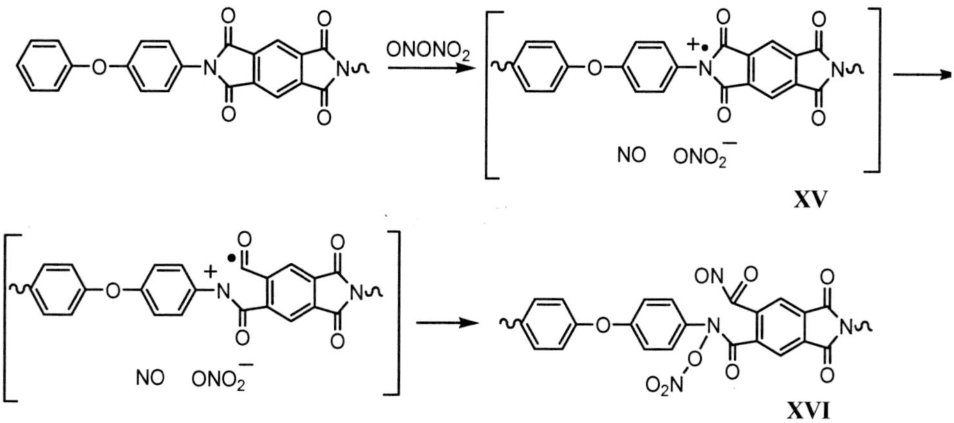


XV

XVI

(25)

As a result of the primary stage, radical cations XV are formed. The dissociating decomposition of radical cations and following cage recombination of the primary stage products give compounds XVI converting into stable radicals XII due to reversible dissociation with detachment of $NO_2$:

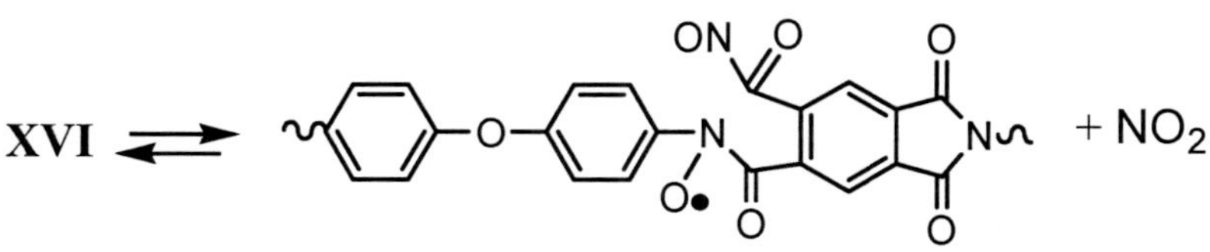


XII

(26)

In $NO_2$ atmosphere at room temperature, the equilibrium (26) is shifted to the left, and any paramagnetic products are not detected in PPI in these conditions. This equilibrium can be shifted to radicals XII by pumping out nitrogen dioxide from samples (Figure 7 a).

The decomposition of radical cations XV arises probably as a result of dissociation of other bonds of imide cycles:

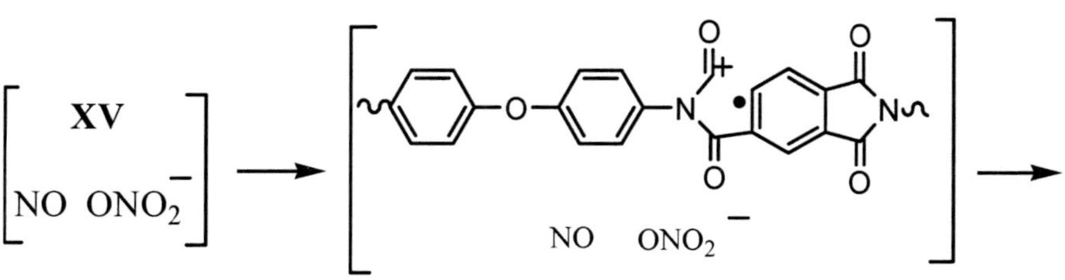

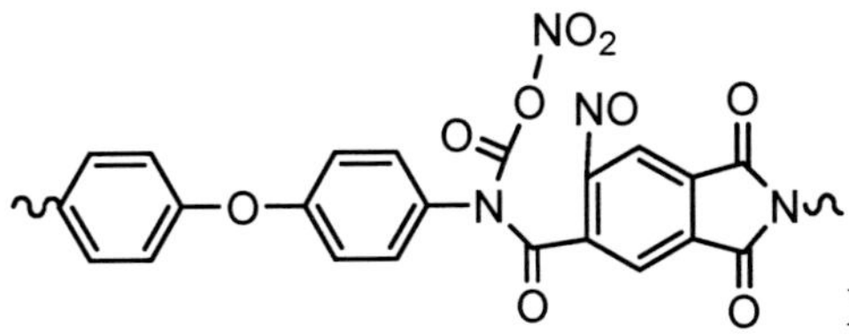


XVII (27)

The compounds XVII can be considered itself as precursors of iminoxyl radicals XIII and radicals XIV characterized by singlet ESR spectrum. In the course of heating in vacuum PPI previously exposed to $NO_2$, amidyl radicals XVIII are formed by thermal decomposition of XVII:

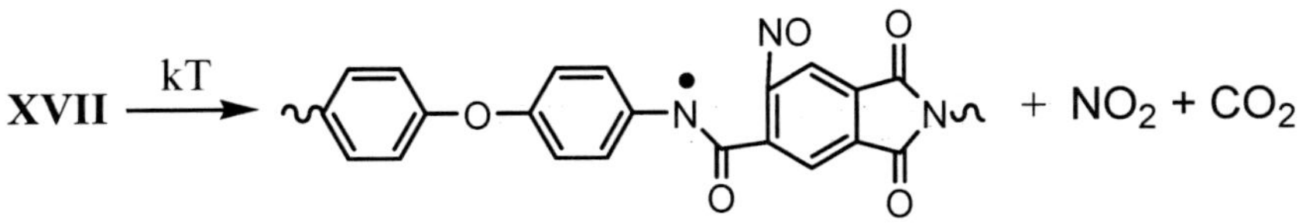


XVIII (28)

The isomerization of radicals XVIII in accordance with the following scheme can give phenoxyl radicals which have rather high thermal stability [32]:

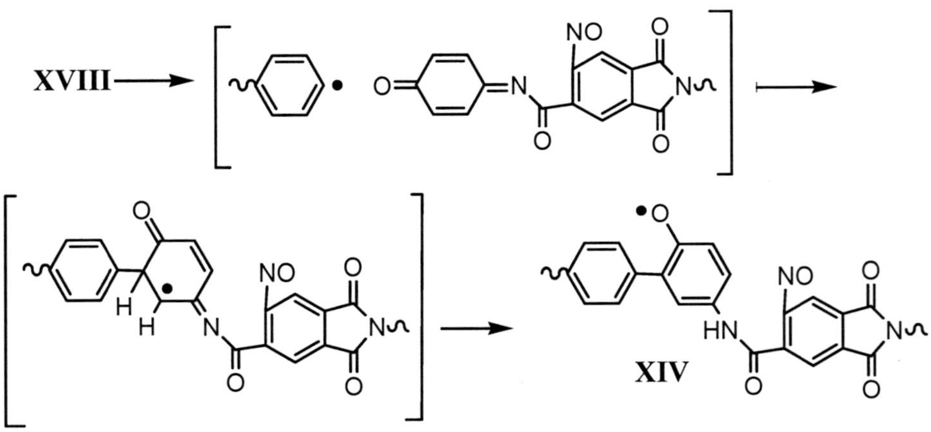


(29)

Similarly, isomeric phenoxyl radical XIVa or XIVb can be formed:

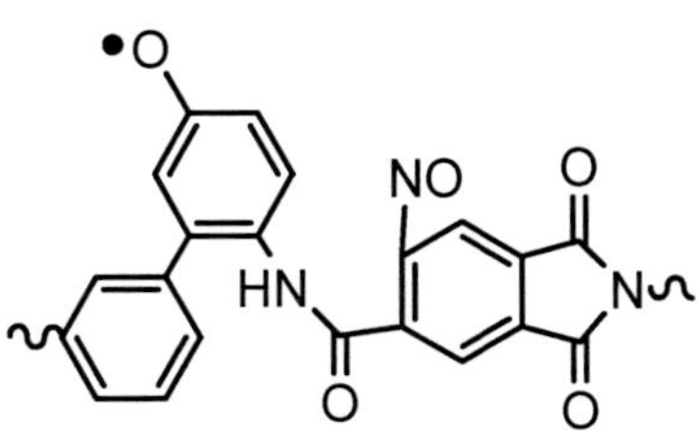


XIVa

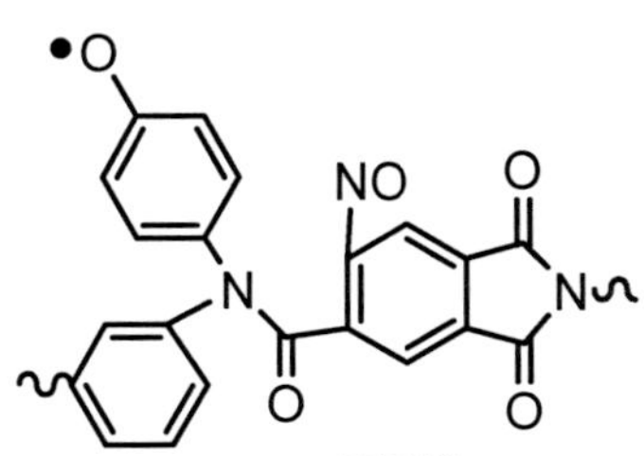


XIVb

**Table 1. Calculated minimum energies of model radicals**

| Radical | *E*, hartrees |
|---|---|
| $XVIII_m$ | 702.65904 |
| $XIV_m$ | 702.68949 |
| $XIV_m(a)$ | 702.69134 |
| $XIV_m(b)$ | 702.68100 |

Thus the intense singlet central component of ESR spectra represented in Figure 7 b-d corresponds to phenoxyl radicals. This conclusion is indirectly confirmed by ESR spectra obtained during exposure of PPI to $NO_2$ at 373 K and subsequent heating of the samples at the same temperature in vacuum. It is known that phenoxyl radicals effectively react with nitrogen dioxide [34]. Therefore thermal reactions at 373 K in $NO_2$ atmosphere are also accompanied by decay of the formed phenoxyl radicals and an ESR signal of the radicals is not observed in these conditions.

Ab initio calculations show that the thermal conversions of XVIII into phenoxyl radicals are energetically efficient. The calculations were carried out by Gaussian 98W program within the framework of density functional theory using UB3LYP method. As models of XVIII, XIV, XIVa and XIVb the following structures were considered:

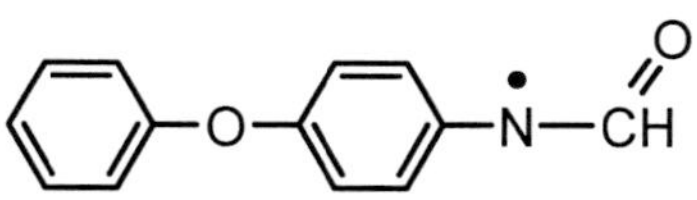


**$XVIII_m$**

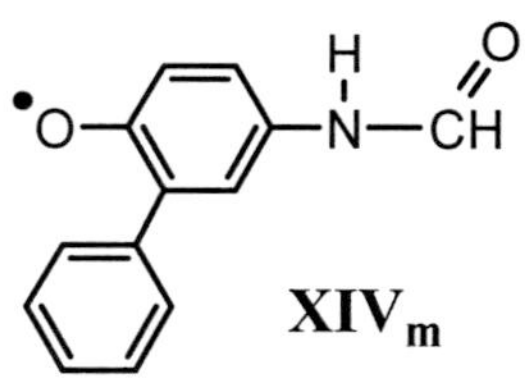


**$XIV_m$**

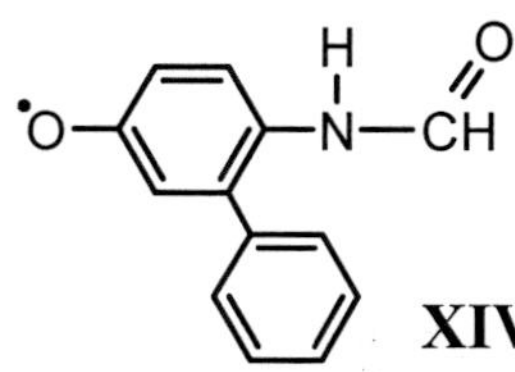


**$XIV_m(a)$**

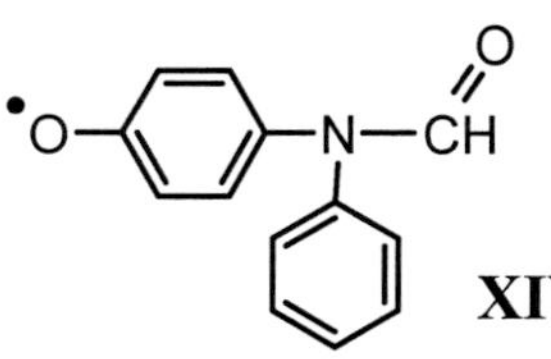


**$XIV_m(b)$**

The calculated minimum energies are given in Table 1. These calculations clearly show that the energy of amidyl radicals $XVIII_m$ far exceeds that of phenoxyl radicals ($\Delta E = 58 - 85$ $kJ \cdot mol^{-1}$).

The generation of iminoxyl radicals XIII after exposure of PPI to $NO_2$ at room temperature followed by warming samples in vacuum at 373 K (Figure 7 b-d) or in the course of exposure to $NO_2$ at the same temperature (Figure 9 a) is conditioned by specific products. In the primary oxidative reaction, the formation of radical cations of PPI can take place also owing to an electron transfer from phenyl rings to nitrosyl nitrate. Such a possibility follows from the available data of UV spectroscopy according to which nitrosyl nitrate is capable of forming charge transfer complexes with aromatic compounds [16, 17] As a result of the oxidative reaction, aryl radical cations appear. These species give intermediate aryl nitroso compounds after detachment of protons. If the nitrosation process takes place in monomer units modified by nitrogen-containing groups as in XVII, these consecutive reactions can be represented as follows:

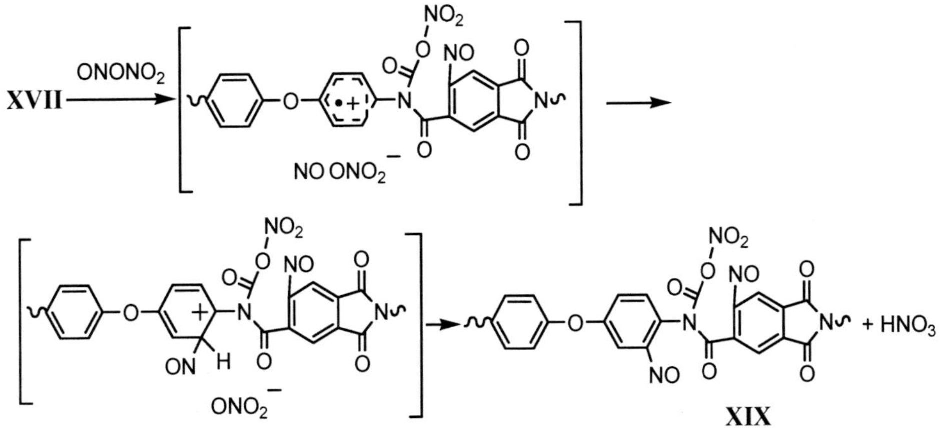


(30)

Due to thermal decomposition of compounds XIX during heating, amidyl radicals XX are formed:

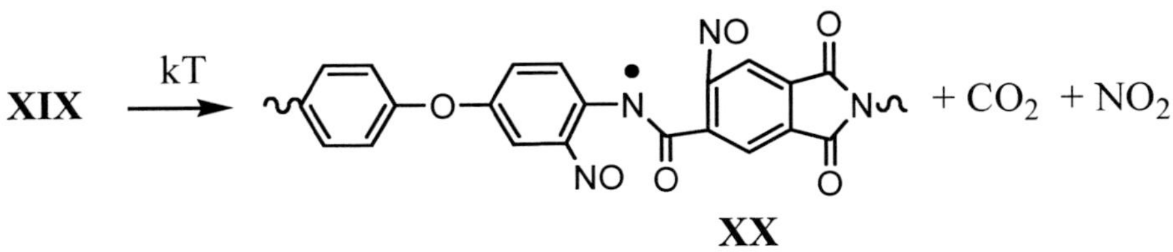


(31)

The radicals XX may also present in other isomeric forms, for instance, as nitrosoalkyl radicals which in turn isomerized into iminoxyl radicals XIII [34]:

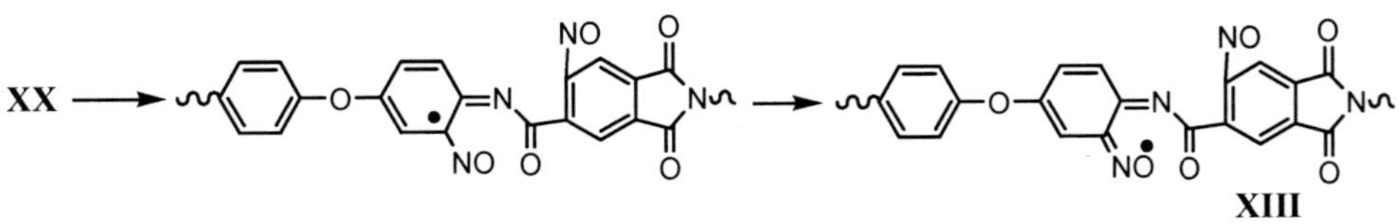


(32)

The steady-state level of the radical XIII concentration on exposure of PPI to nitrogen dioxide at 373 K (Figure 10 a) is apparently caused by the recombination of formed iminoxyl radicals with $NO_2$:

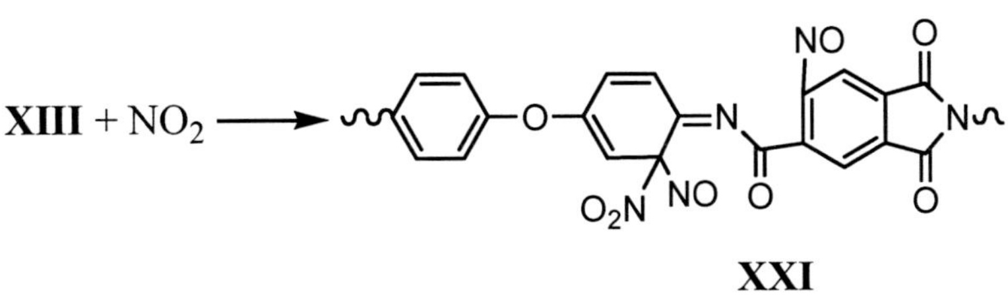


(33)

The products XXI may be thermally less stable comparatively with XIX and decompose into initial components. In this case, replacing XIX by XXI takes place during exposure of PPI to $NO_2$ at 373 K. Therefore, the rate of accumulation of XIII at subsequent heating PPI in vacuum (Figure 10 b) is four times greater than that for samples exposed at room temperature (Fig 8 (4)) when products XX are not generated and XIX is the sole precursor of iminoxyl radicals.

As is seen from Figure 8 b, the rate of accumulation of radicals XIII during thermolysis of PPI preexposed to nitrogen dioxide linearly depends on concentrations of $NO_2$ in gas phase. At first glance, this result contradicts the mechanism with participation of nitrosyl nitrate as an initiator, according to which the square-low dependence should be observed for the rate. However, if we take into account that aromatic nitroso compounds are oxidized by monoradicals of nitrogen dioxide to nitro compounds [35], one can explain the experimentally observable linear dependence. Indeed, if nitroso compounds XIX are formed by consecutive reactions (27), (30), and concentrations of XVII and XIX become stationary during long exposure of PPI to $NO_2$ at room temperature, then $k_1\,[NO_2]^2 = k_2\,[XIX]\,[NO_2]$ and $d[XIII]/dt = k_1 k_3 k_2^{-1}[NO_2]$ where $k_1$ and $k_2$ are effective rate constant of formation and decay of XVIII in $NO_2$ atmosphere, $k_3$ is the rate constant of thermal conversion of XIX into XIII. Let's assume other parallel process when aromatic nitroso compounds are initially formed and transformed later into XIX in the reaction with nitrosyl nitrate:

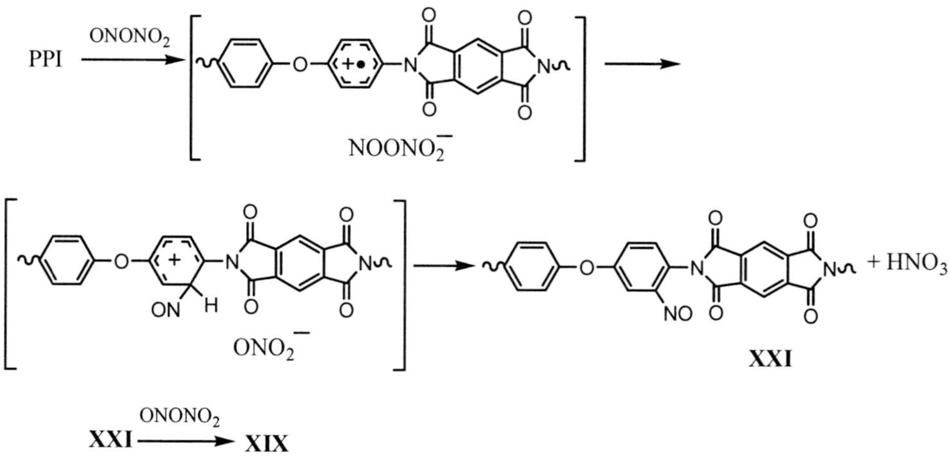


(34)

The formal kinetic scheme for the sequence of reactions (34) taking into account $NO_2$ reactions can be represented as follows:

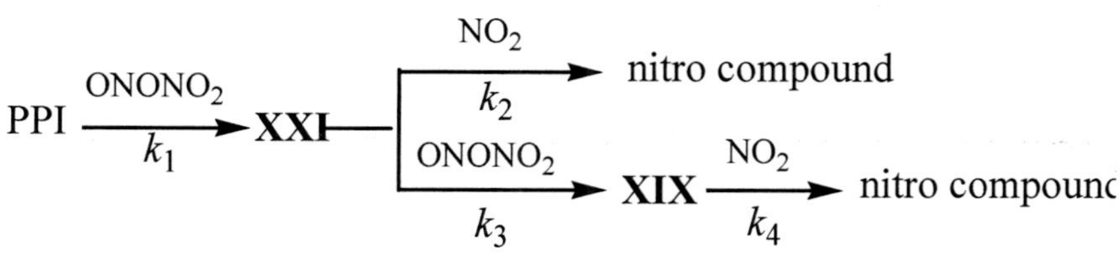


(35)

Then the equation for stationary concentrations of XIX is

$$[\mathrm{XIX}] = \frac{k_1 k_3 [\mathrm{NO}_2]^3}{k_4 (k_2 [\mathrm{NO}_2] + k_3 [\mathrm{NO}_2]^2)} \qquad (36)$$

If the conversion of nitroso compound XXI takes place enough efficiently, that is $k_3[\mathrm{NO}_2]^2 >> k_2[\mathrm{NO}_2]$, then $[\mathrm{XIX}] = \dfrac{k_1[\mathrm{NO}_2]}{k_4}$. In this case, too, the linear dependence on $NO_2$ concentrations should be observed for the kinetics of the iminoxyl radical XIII formation by reaction (32).

Direct detection of radical cations in reactions initiated by nitrogen dioxide

The ion-radical mechanism formally explains the generation of various stable nitrogen-containing radicals in polymers free of weak C – H or double carbon – carbon bonds capable of reacting with $NO_2$ monoradicals. Nevertheless, the direct detection of radical cations by ESR method is necessary for confirmation of the ion-radical initiation concept of conversion of compounds under the action of nitrogen dioxide. For this purpose, low-molecular amine TPA was used in the present work. The formation of radical cations in this compound has been revealed in reactions with some Lewis acids as salts [36]. On exposure of TPA crystals at room temperature during 1 min, the very narrow single line with $\Delta H_{1/2}$ = 0.25 mT and $g$ = 2.0024±0.0005 is observed in ESR spectrum (Figure 11 a). Taking into consideration that radical cations of TPA formed by the reaction

$$(\mathrm{Ph})_3\mathrm{N} + \mathrm{NOONO}_2 \rightarrow \underset{\mathbf{XXII}}{(\mathrm{Ph})_3\mathrm{N}^{+\bullet}} + \mathrm{NO} + \mathrm{ONO}_2^- \qquad (37)$$

should have in the isolated state a triplet ESR spectrum due to HFI of unpaired electrons with nitrogen nuclei, the narrow line observed can be caused by an electron exchange between radical cations XXII and neutral molecules of TPA owing to the overlap of molecular orbital [37]. In this case, electron spins are exchanged so rapidly that the time-averaged value of a magnetic field caused by HFI is close to zero. At room temperature in nitrogen dioxide atmosphere, this single line rather quickly disappears. Its intensity decreases for 6 – 7 minutes in half. On the further exposure during 3 days, the ESR spectrum is transformed to an anisotropic triplet signal (Figure 11 b) with the distance between components of 1.86 mT and the same $g$-factor as for the singlet line. If to pump out the sample after exposure to $NO_2$ during 1 min, the disappearance of the signal also takes place. However, the spectrum shown in Figure 11 b does not occur in this case. The kinetics of decay of paramagnetic species in vacuum can be seen in Figure 12. In this figure, the linearization of experimental kinetic dependence in coordinates of the second order equation is given. From the results obtained one can conclude that the main path of decay of radical cations XXII is their recombination:

**XXII** + **XXII** →

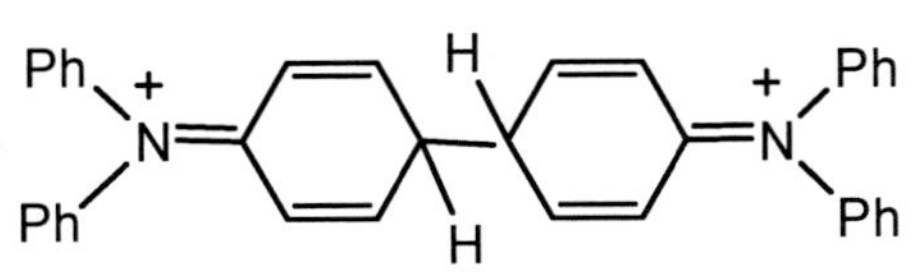


**XXIII**

(38)

The analogous process was spectrophotometrically observed for dimethylaminophenyl radical cations generated by oxidative reactions with *tris* – (*p* – bromophenyl)ammoniumyl perchlorate [38] As a result of reaction (38), the formation of tetraphenylbenzidine along with nitric acid occurs through the deprotonation of dications XXIII:

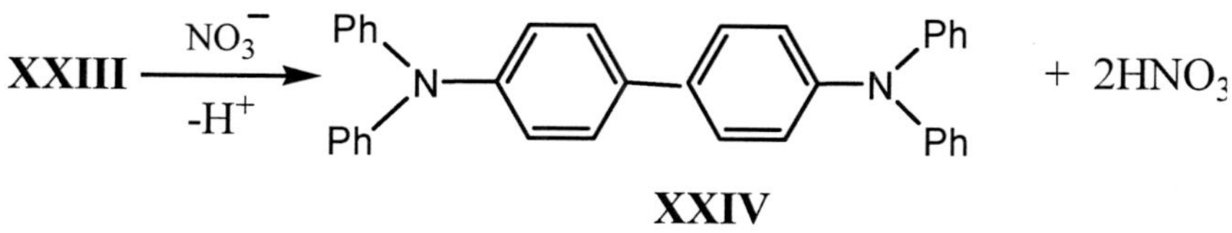


(39)

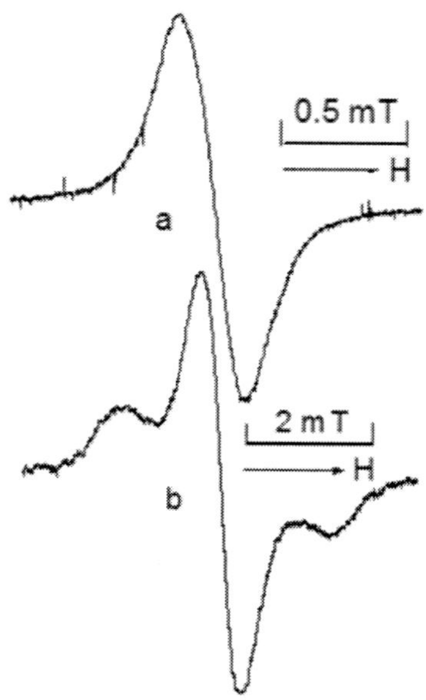


Figure 11. ESR spectra of TPA exposed to $NO_2$ during 1 min (a) and 3 days (b) at 295 K.

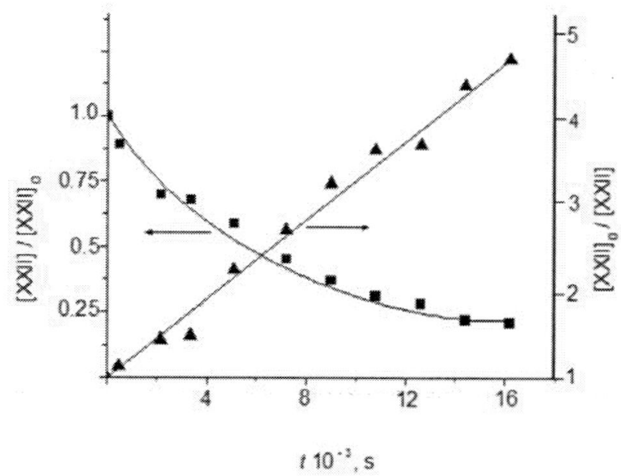


Figure 12. Kinetics of the radical cation XXII decay in vacuum at 295 K (1) and its linearization (2) in the second order equation coordinates.

As to the structure of these particles (Figure 11 b), one can say with reasonable confidence that they are a product of conversion of primary radical cations XXII under the action of nitrogen dioxide. As a result of these reactions, *p*-nitroderivatives of radical cations are sequentially formed:

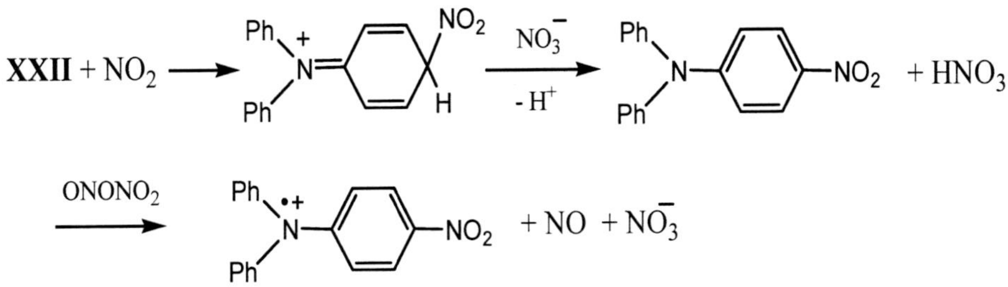


(40)

The derivatives nitrated into all $p$ – positions of phenyl rings are similarly obtained:

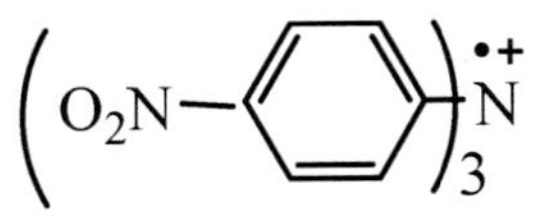


**XXII'**

It is obvious that radical cations XXII′ are not capable of recombining by reaction (38) and are stabilized in the given conditions. At the same time, these nitrosubstituted radical cations cannot fast exchange electrons with TPA molecules because of structural distinctions. Therefore, the triplet ESR spectrum recorded instead of a narrow line on later stages of TPA exposure belongs most likely to radical cations XXII′. Notice that a triplet spectrum outwardly similar to that of Figure 11 b was observed in the course of the TPA interaction with iodine [39].

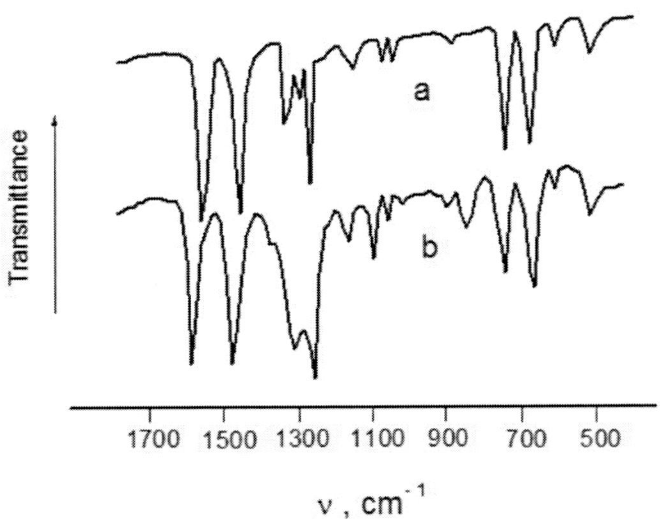


Figure 13. IR spectra of original TPA (a) and after exposure to $NO_2$ for 10 min (b).

The scheme of reactions (40) is confirmed by IR measurements for products of TPA nitration. In Figure 13, an original IR spectrum of TPA and that after exposure for 10 minutes to $NO_2$ are shown. Under the action of nitrogen dioxide, the absorption band corresponding to asymmetric stretch vibrations of N⋯O groups in aromatic nitro compounds appears at 1360 $cm^{-1}$. Two intense bands are also observed at 1320 $cm^{-1}$ and 840 $cm^{-1}$ owing to nitrate anions formed by oxidative reactions (37) and (40).

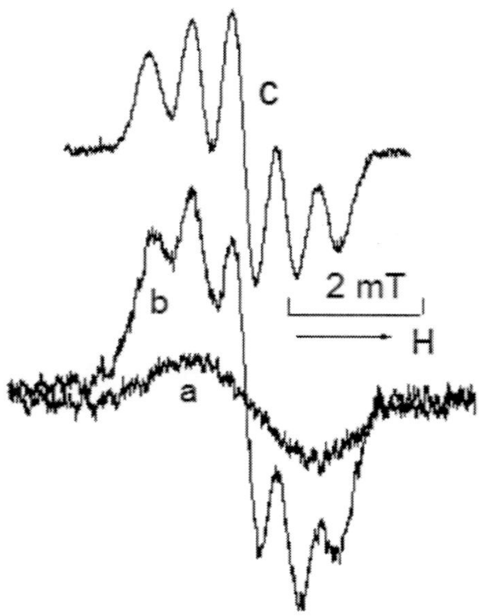


Figure 14. ESR spectra of TPA exposed to $NO_2$ (1 min): immediately after dissolving in $CH_2Cl_2$ (a), after 70 min (b), after dissolving in $CH_2Cl_2$ without oxygen (c).

The singlet ESR spectrum (Figure 11 a) cannot recognize directly chemical structure of radical cations XXII. This information has been obtained via dissolving TPA preliminary exposed to $NO_2$. After one minute of keeping in nitrogen dioxide, a tube with TPA sample was evacuated, and methylene chloride was added to the sample. The ESR spectrum recorded at once after TPA dissolution ([TPA] ~ 0.1 mol·l $^{-1}$ ) is given in Figure 14 a. In this case instead of narrow single line (Figure 11 a), the wide line with $\Delta H_{1/2} = 1.7$ mT is observed. At that, the concentration of paramagnetic species is lowered by a factor 30. The observable effect shows that as a result of TPA dissolving the rate of electron exchange between radical cations XXII and neutral molecules sharply decreases due to a dilution, and the singlet spectrum is broadened. An overwhelming fraction of radical cations XXII recombine yielding dications XXIII. Such conclusion is substantiated by the fact that the ESR spectrum of the solution gradually varies, and after 70 minutes it turns into quintet with a distance between components of 0.57 mT and g = 2.0033 (Figure 14 b). The amount of paramagnetic species for this time becomes four fold as much. The quintet spectrum specifies the presence of two equivalent nitrogen atoms in the given particles which can be attributed to radical cations of tetraphenylbenzidine:

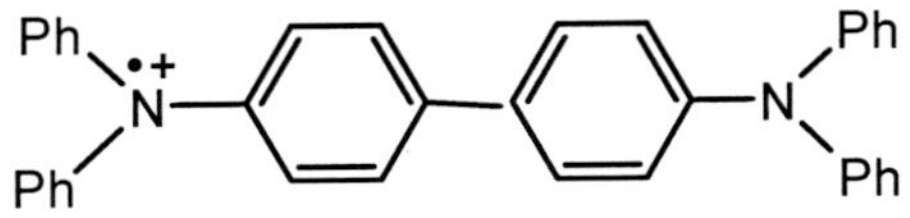


**XXV**

The formation of radical cations XXV in a solution of TPA is conditioned by the oxidative action of dications XXIII:

**XXIII** + **XXIV** ⟶ 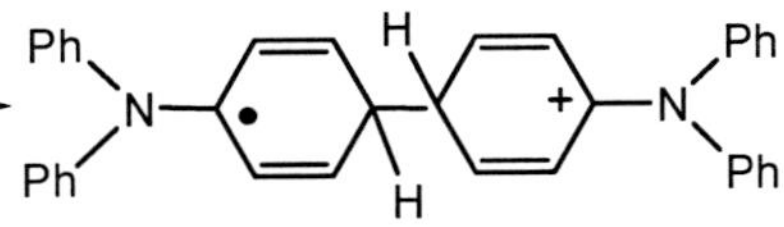

+ **XXV** (41)

**XXVI**

**XXVI** + TPA $\xrightarrow{-H^+}$ 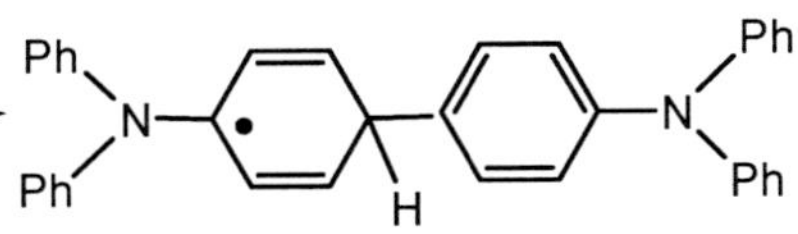

+ $(Ph)_3NH^+$ (42)

**XXVII**

**XXVII**+ **XXVII** ⟶ 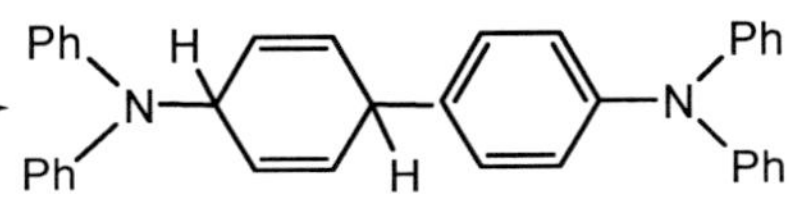

+ **XXIV** (43)

or

**XXVII** ⟶ TPA + 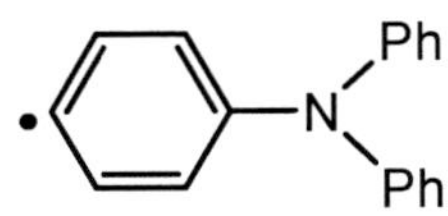


**XXVIII**

(44)

**XXVIII** + **XXVIII** ⟶ **XXIV** (45)

On evidence derived from the literature, organic cations and dications are capable of oxidizing bases with the resulting the formation of radical cations [40]. In the considered system, tetraphenylbenzidine XXIV reacts as the organic base. In the absence of oxygen in similar systems, the oxidation takes place fast and with a high yield. The presence of oxygen influences essentially on the mechanism of the further transformations of radical cations. If to freeze at 77 K the TPA sample exposed to $NO_2$ during one minute immediately after addition of methylene chloride and then to degas the solution by consecutive freezing – pumping out – thawing, the concentration of radical cations XXV in these conditions is approximately 8 times greater than that in the presence of oxygen (Figure 14 b) ($2.8 \cdot 10^{-2}$ mol·l$^{-1}$ and $3.6 \cdot 10^{-3}$ mol·l$^{-1}$ respectively). As this takes place, radical cations XXV appear instantly at room temperature, and they have more resolved ESR spectrum (Figure 14 c). The increase of the spectrum resolution is obviously connected with the absence of broadening effect of paramagnetic oxygen molecules. At that, oxygen evidently effects strongly on the conversions of radical cations in stages (42) – (45), in which free radicals formed can be oxidized. The efficiency of the formation of tetraphenylbenzidine XXIV, which is the precursor of radical cations XXV, should thereby greatly reduce.

Thus, the occurrence of a quintet ESR spectrum in TPA solution testifies redox reactions with participation of non-paramagnetic products of nitrogen dioxide interaction with TPA. Because of high reactivity of the primary radical cations XXII, their ESR spectrum was not observed in solutions. Nevertheless, an occurrence of the five component spectrum unequivocally points to the formation of radical cations XXII in primary reaction of nitrogen dioxide with TPA. The results obtained by a direct ESR technique for modelling reactions of TPA confirm the mechanism of an oxidative ion-radical initiation of conversions of high- and low-molecular organic compounds under the action of nitrogen dioxide.

## Conclusions

Polyamides and polyimides actively interact with nitrogen dioxide. The increased reactivity of these polymers can be explained on the basis of the concept of primary oxidizing reaction with participation of nitrogen dioxide dimers (nitrosyl nitrate) resulting in the formation of radical macrocations, nitric oxide and nitrate anions. The main molecular products of the subsequent reactions of active reactants formed are nitroso and others nitrogen-containing compounds which are then converted into stable macroradicals. The conditions the stable radical generation in polyamides and polyimides on exposure to nitrogen dioxide atmosphere are essentially differ. In polyamides, these radicals are formed at room temperature directly during exposure. In the same conditions, polyimides give products of nitrosation and nitration which give stable radicals only at higher temperatures, and nitrogen dioxide inhibits their formation in these conditions. The observable distinctions are conditioned by features of structure and mechanism of reactions of primary radical cations located on amide or imide groups of macromolecules. The specific ion - radical mechanism allows explaining high reactivity with respect to nitrogen dioxide even such thermally and chemically stable materials as aromatic polyamides and polyimides.

The important question is the estimation of the contribution of the considered reactions to process of aging of polymers of the given classes in real conditions of an atmosphere polluted by nitrogen oxides. Maximum possible concentrations of $NO_2$ allowable in an atmosphere change in the range of $10^{-9} - 10^{-7}$ mol·l$^{-1}$. These concentrations are in several orders less than those used usually in experimental researches. Therefore one can expect that in natural conditions other factors of aging (light, oxygen, heat, ozon, etc.) play more appreciable influence on polymers in comparison with action of atmospheric impurity of radicals $NO_2$ and their dimers in such insignificant concentrations. However in the course of $NO_2$ diffusion to accessible areas of a structurally heterogeneous solid polymeric matrix with the strongly hindered molecular mobility, local concentrations of nitrogen dioxide can considerably exceed those in an atmosphere. In these conditions, the dimerization of $NO_2$ is possible with formation of nitrosyl nitrate initiating consecutive ion - radical transformations of macromolecules.

## REFERENCES


[1] Feuer, H. (ed.) (1969) *The Chemistry of the Nitro and Nitroso Groups;* Wiley: New York.

[2] Jellinek, H. H. G. *Aspects of Degradation and Stabilization of Polymers*; Elsevier: Amsterdam, 1978.

[3] Pariiskii, G. B.; Gaponova, I. S.; Davydov, E. Ya. *Russian Chem Rev* 2000, 69, 985.

[4] Jellinek, H. H. G.; Chaudhuri, A. *J Polym Sci* A-1 1972, 10, 1773.

[5] Makarov, G. G.; Mikheev, Yu. A.; Pariiskii, G. B.; Postnikov, L. M.; Toptygin, D. Ya. *Vysokomolek Soed A* 1982, 24, 2601.

[6] White, E. H. *J Am Chem Soc* 1955, 77, 6008.

[7] Rånby, B.; Rabek, J. F. *Photodegradation, Photo-oxidation and Photostabilization of Polymers;* Wiley: London, 1975.

[8] Giamalva, D. H.; Kenion, G. B.; Church, D. F.; Pryor, W. A. *J Am Chem Soc* 1987, 109, 7059.

[9] Colding, P.; Powell, J. L.; Ridd, J. H. *J Chem Soc Perkin Trans* 2 1996, 813.

[10] Davydov, E. Ya.; Gaponova, I. S.; Pariiskii, G. B.; Pokholok, T. V. *Polymer Sci A* 2006, 48, 375.

[11] Sroog, C. E.; Endrey, A. L.; Abramo, S. V.; Edwards, W. N.; Olivier, K. L. *J Polym Sci A* 1965, 3, 1373.

[12] Hall, T.G.; Blacet, P. J. *J Phys Chem* 1952, 20, 1745.

[13] Gaponova, I. S.; Davydov, E. Ya.; Makarov, G. G.; Pariiskii, G. B.; Pustoshnyi, V. P. *Polymer Sci A* 1998, 40, 309.

[14] Pariiskii, G. B.; Gaponova, I. S.; Davydov, E. Ya.; Pokholok, T. V. *Aging of Polymers, Polymer Blends and Polymer Composites.* V 2, Eds Zaikov, G. E.; Buchachenko A. L.; Ivanov V. B; Nova Science Publischers: New York, 2002, 31.

[15] Djerassi, C.; Lund, L.; Bunnenberg, E.; Sjoberg, B. J Am Chem Soc. 1961, 83, 2307.

[16] Bosch, E.; Kochi, J. K. *J Org Chem* 1994, 59, 3314.

[17] Bockman, T. M.; Karpinski, Z. J.; Sankararaman, S.; Kochi, J. K. *J Am Chem Soc* 1992, 114, 1970.

[18] Bosch, E.; Rathore, R.; Kochi, J. K. *J Org Chem* 1994, 59, 2529.

[19] Rathore, R.; Kim, J. S.; Kochi, J. K. *J Chem Soc Perkin Trans* 1 1994, 2675.

[20] Ungrande, H. E.; Smiley, R. A. *J Org Chem* 1956, 21, 993.

[21] Lakkaraju, P. S.; Zhangh, Z.; Roth, H. D. *J Phys Chem* 1994, 98, 2722.

[22] Fox, W. M.; Symons, M. C. R. *J Chem Soc* A 1966, 1503.

[23] Belevsky, V. N.; Tyurin, D. A.; Chuvylkin, N. D. *Khimiya Vys Energii* 1998, 32, 342.

[24] Bühler, K. U. *Spezialplaste;* Academie Verlag: Berlin, 1978.

[25] Kambe, H.; Yokata, R. *The 2nd International Symposium on Degradation and Stabilization of Polymers,* Dubrovnik, 1978. DPG: Zagreb, 1978, 39.

[26] Gaponova, I. S.; Davydov, E. Ya.; Pariiskii, G. B.; Pustoshnyi, V. P. *Vysokomolek Soed A* 2001, 43, 98.

[27] Pokholok, T. V.; Gaponova, I. S.; Davydov, E. Ya.; Pariiskii, G. B. *Polym Degrad Stab* 2006, 91, 2423.

[28] Frisch, M. J.; Trucks, G. W.; Schlegel, H.B.; Scuseria, G. E.; Robb, M. A.; Cheeseman, J. R.; et al. *Gaussian 98;* Gaussian Inc.: Pitsburgh PA, 1998.

[29] Jaszewski, A. R.; Jeziewska, J.; Jezierski, A. *Chem Phys Lett* 2000, 319, 611.
[30] Davydov, E.Ya.; Gaponova, I. S.; Pariiskii, G. B. *J Chem Soc Perkin Trans* 2 2002, 1357.
[31] Xie, W.; Helstley, R.; Cai, X.; Deng, F.; Lieu, J.; Lee, C.; Pan, W. P. *J Appl Polym Sci* 2002, 83, 1219.
[32] Nonhebel, D. C.; Tedder, J. M.; Walton, J. C. *Radicals; Cambridge University Press:* London, 1979.
[33] Brunton, G.; Cruse, H. W.; Riches, K. M.; Whittle, A. *Tetrahedron Lett* 1979, 1093.
[34] Bordwell, F. G.; Zhang, S. *J Amer Chem Soc* 1995, 117, 4858.
[35] Gowenlock, B. G.; Pfab, J; Young, V, M.; *J Chem Soc Perkin Trans* 2 1997, 1793.
[36] Koshechko, V. G.; Titov, V. E.; Pohodenko,V. D. *Theor Exper Chem* 1983, 19, 161.
[37] Wertz, J. E.; Bolton, J. R. *Electron Spin Resonance;* McGrow – Hill: New York, 1972.
[38] Bell, F. A.; Barestford, P.; Kricka, L. J.; Ledwith, A. *J Chem Soc Perkin Trans* 1 1974, 1788.
[39] Stamires, D. N.; Turckevich, J. *J Am Chem Soc* 1963, 85, 2557.
[40] Ledwith, A. *Acc Chem Res* 1072, 5, 133.

In: Chemical and Biochemical Reactions
Editors: G. E. Zaikov, L. Liu et al.

ISBN 978-1-61122-860-1


*Chapter 2*

# MECHANISM OF CATALYSIS WITH NICKEL (II) COMPLEXES IN TOLUENE AND CUMENE OXIDATIONS BY DIOXYGEN

***L. I. Matienko** and L. A. Mosolova***
Emanuel Institute of Biochemical Physics, Russian Academy of Sciences,
4 Kosygin str., Moscow, 119334 Russia

## ABSTRACT

The analysis of the mechanism of oxidation of alkylarenes (toluene and cumene) with dioxygen, catalyzed by catalytic system $\{Ni(II)L^1{}_2 + L^2\}$ ($L^1$ = acac$^-$, $L^2$ = ligand-modifier N-methylpirrolidon-2) is resulted for the first time. Comparison with the mechanism of nickel catalyzed ethylbenzene oxidation with dioxygen is spent. It has been established that the mechanism of formation of products of catalytic oxidation of alkylarenes is connected with various activities of complexes of the nickel, formed during oxidation, in micro steps of radical-chain processes of oxidation.

**Keywords:** oxidation, dioxygen, alkylarens (toluene, cumene, ethylbenzene), homogeneous catalysis, Ni(II) bis(acetylacetonate), ligand-modifier N-methylpirrolidon-2

## 1. AIM AND BACKGROUND

The effect of modifying ligands, which usually occupy the axial position in a complex, was as a rule studied using model compounds that mimic active sites of enzymes, namely, mono- and dioxygenases. At present, examples of catalytic systems are known that substantially increase the rates of catalyzed reactions and, sometimes, the yields of products upon the addition of modifiers. As a rule, the mechanisms of action of additives were not

* E-mail:matienko@sky.chph.ras.ru

established, although possible schemes were put forward to explain the action of modifying ligands [1].

Oxidation of alkylarenes with dioxygen catalyzed by model systems mimicking biological systems (mono- and dioxygenases) that favors the selective insertion of oxygen atoms into C— H bonds of organic molecules affords alcohols and carbonyl compounds [2-6]. For example, toluene 2- or 4-monooxygenase performs aromatic hydroxylation of toluene to corresponding cresols [7]. Unfortunately, dioxygenases capable of effecting alkane dioxygenation to hydroperoxides are so far unknown [3].

The method of transition metal catalysts modification by additives of electron-donor monodentate ligands for increase in selectivity of liquid-phase alkylarens oxidations into corresponding hydroperoxides was proposed by us [8] for the first time.

Based on the established (for Ni complexes) and hypothetical (for Fe complexes) mechanisms of formation of catalytically active species and their operation, efficient catalytic systems $\{ML^1_n + L^2\}$ ($L^2$ are crown ethers or quaternary ammonium salts) of ethylbenzene oxidation to α-phenyl ethyl hydroperoxide were modeled. Selectivity $\mathbf{S_{PEH, max}}$, conversion, and yield of PEH in ethylbenzene oxidation catalyzed by these systems were substantially higher than those observed with conventional catalysts of ethylbenzene oxidation to PEH [8]. The high activity of systems $\{ML^1_n + L^2\}$ ($L^2$ are crown ethers or quaternary ammonium salts) is associated with the fact that during the ethylbenzene oxidation, the active primary $(M^{II}L^1_2)_x(L^2)_y$ complexes (I macro step of oxidation) and heteroligand $M^{II}_xL^1_y(L^1_{ox})_z(L^2)_n$ complexes (II macro step of oxidation) are formed to be involved in the oxidation process [8].

The involvement of transition metal complexes in the chain propagation step of ethylbenzene oxidation was considered exclusively in our studies [8-14] and in theoretical publications [15, 16]. The attention of researchers is concentrated mainly on the role of transition metal compounds in the steps of generation of free radicals (in chain initiation and hydroperoxide decomposition), while the reaction of a catalyst with peroxide radicals (chain propagation) in the catalytic hydrocarbons oxidation with dioxygen still remains insufficiently explored.

In our previous work [17] it has been established, that electron-donating exo ligand-modifiers $L^2$ can be used for modification of $Ni(II)(acac)_2$ activity as catalyst of oxidation with dioxygen not only of ethylbenzene, but and other alkylarens with primary and tertiary α-CH-bonds, namely, toluene and cumene, aimed the increase in yield of ROOH. However ligand-modifier influence on the mechanism of toluene and cumene oxidation, catalyzed with $Ni(II)(acac)_2$, has not been established. Here the analysis of the mechanism of oxidation of toluene and cumene with dioxygen, catalyzed by catalytic system $\{ML^1_n + L^2\}$ ($L^2$ = N-methylpirrolidon-2) is resulted.

## INTRODUCTION

Earlier by us it has been established that oxidation reactions of considered hydrocarbons (ethylbenzene, toluene and cumene) by molecular oxygen, catalyzed with system $\{Ni(II)(L^1)_2+L^2\}$ ($L^2$ = N-methylpirrolidon-2 (MP)) are characterized with general kinetic regularities. As well as at oxidation of ethylbenzene, synergic effects of growth of initial rates of oxidation of toluene and cumene, and also $S_{ROOH}$ and degree of conversion of oxidation in

ROOH in presence of system $\{Ni(II)(L^1)_2+L^2\}$ are observed, and maximum possible values of these parameters are received in the optimal conditions of researched oxidations. However there are differences which are connected with specificity of each hydrocarbon. [17].

In this work features of oxidation of toluene and cumene together with features of oxidation of ethylbenzene are considered in more details with addition of new our and literature data. For the first time the effect of coordination of ligand-modifier to metal (Ni(II)) complex catalyst on mechanism of formation of all analyzed products (toluene, ethylbenzene and cumene) is established and the schemes of oxidation resulted in initial research [17] are added.

It is known that aliphatic C-H bond energy is 87 kcal/mol (ethylbenzene, cumene), and 90 kcal/mol (toluene) [18].

It is well-known, what toluene, ethylbenzene and cumene, which are monoalkylbenzenes with primary, secondary and tertiary α-CH-bonds, on "Oxidability" with *molecular oxygen,* are situated in a number:

toluene < ethylbenzene < cumene [19] (Table 1).

## 3. Experimental

Alkylarenes (RH) oxidation was studied at 100°C (toluene, cumene), at 120°C (ethylbenzene), in glass bubbling-type reactor at the absence catalyst and in the presence of $Ni(II)(acac)_2$ and $\{Ni(II)(acac)_2+L^2\}$($L^2$= N-methylpirrolidone-2). The selectivity $\mathbf{S_{ROOH}}$ and the RH conversion **C** of ethylbenzene into PEH in the alkylarenes oxidation were determined using the formulas: $\mathbf{S_{ROOH}} = [ROOH] / \Delta[RH]\cdot 100\%$ and $\mathbf{C} = \Delta[RH] / [RH]_0\cdot 100\%$.

### Analysis of Oxidation Products

Hydroperoxides of α-phenylethyl and cumyl, and also the sum of benzyl hydroperoxide (BH) and perbenzoic acid (PBA) at toluene oxidation were determined by iodometry titration [9-14, 17].

**Table 1. The values of parameters of $k_2 / \sqrt{k_5}$ ((l mol$^{-1}$)$^{1/2}$s$^{-1/2}$) and chain propagation ($RO_2^{\bullet}$ +RH→) $k_2$ and chain termination ($RO_2^{\bullet}$ + $RO_2^{\bullet}$→) $k_5$ rate constants (l mol$^{-1}$ s$^{-1}$) for monoalkylbenzenes oxidations at 30$^0$C**

| Hydrocarbon | $k_2 / \sqrt{k_5} .10^3$ | $k_2$ | $k_5.10^{-6}$ |
|---|---|---|---|
| **Toluene** | 0.014 | 0.24 | 150 |
| **Ethylbenzene** | 0.21 | 1.3 | 20 |
| **Cumene** | 1.5 | 0.18 | 0.0075 |

At separate definition of benzyl hydroperoxide (BH) and perbenzoic acid (PBA) have taken advantage of their various activities in reaction with KJ. Thus sum of BH and PBA was determined using ethanol solution of KJ after keeping in the dark within several hours. Aqueous KJ was used for determination for perbenzoic acid, thus $J_2$, formed on reaction PBA with KJ, was titrated right after. Accuracy of definition of PBA was ±5 %. The total maintenance of benzoic and perbenzoic acids was determined alkalimetrcally [17].

Benzaldehyde (BAL), benzyl alcohol (BZA), acetophenone (AP), methylphenylcarbinol (MPC), dimethyl phenylcarbinol (DMPC), and phenol (PhOH), as well as the RH content in cumene and ethylbenzene oxidation processes were examined by GLC [9-14, 17]. The overall rate of the process was determined from the rate of accumulation of all oxidation products (toluene, ethylbenzene), or consumption of starting hydrocarbon (cumene). A correlation between RH consumption and product (P) accumulation was established: $\Delta[RH] = \Sigma [P]$.

The catalytic alkylarens oxidation with dioxygen was carried out in the $O_2$ – solution two phase systems under kinetic control.

The order in which products P formed was determined from the time dependence of product accumulation rate rations at **t → 0**. The variation of these rations with time was evaluated by graphic differentiation $\Delta[P_1]_{ij}/\Delta[P_2]_{ij}$ on $t_j$, where $\Delta[P_1]_{ij}$ and $\Delta[P_2]_{ij}$ - increase in concentration $P_1$ и $P_2$ during $\Delta t = t_j - t_i$ [8, 39].

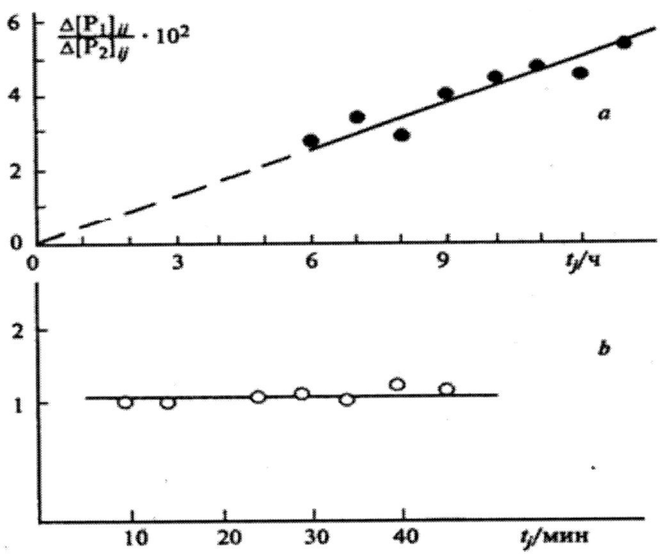


## 4. Results and Discussion

### Toluene Oxidation, Catalyzed with System {Ni(II)(acac)$_2$ + MP}

Due to the low values of kinetic parameter $k_2 / \sqrt{k_5}$ (Table 1) toluene is virtually not oxidized in the absence of catalyst or initiator. The major products of the initiated and most catalytic oxidations of toluene are benzaldehyde and *benzoic acid* [20-23]. For example, high-selective catalysts of oxidation of methylbenzenes (in that number of toluene) are so called *MC*-systems (Co-Mn-Br-catalytic systems) [20].

The high recombination rates of benzylperoxy radicals (Table of VII. 1), resulting in the formation of benzyl alcohol and benzaldehyde (∗) [24, 25], and low stability of benzyl hydroperoxide, BH, do not permit significant selectivity in BH formation already at low extents of oxidation. Benzaldehyde, formed at decomposition of BH and in chain termination reaction (∗), is then oxidized at high rates to benzoic acid.

$$2PhCH_2O_2^{\cdot} \rightarrow PhCHO + PhCH_2OH + O_2 \ (*)$$

*Thus, in the case of use of tert.butylperbenzoate as initiator, sufficient high selectivity suff of BH formation (62,5 % in calculation on swallowed up oxygen)* are reached at degrees of conversion *C* only ~ 0.05 % [26]. At oxidation of toluene by $O_2$ in presence of manganese resinate and peroxide of isopropyl benzene at 110°C the content of BH hydroperoxide was 0.6 % only [23].

The toluene oxidation proceeds at measurable rates ($\geq 10^{-7}$ mol/l $s^{-1}$) at catalyst concentrations $[Ni(II)(acac)_2] \geq 3 \cdot 10^{-3}$ mol/l. In addition, in the toluene oxidation, catalyzed with $Ni(II)(acac)_2$, without donor additives, and in the presence of MP, the next products, namely, benzyl hydroperoxide (BH), benzaldehyde (BAL), benzyl alcohol (BZA), and acids: benzoic (BA) and perbenzoic (PBA) acids, − are formed (Figure 1). Benzaldehyde and benzoic acid are formed in great concentrations. Similar products, BH, BAL and BA, moreover BAL in great quantity, were found in toluene oxidation, catalyzed with resinate of manganese [23].

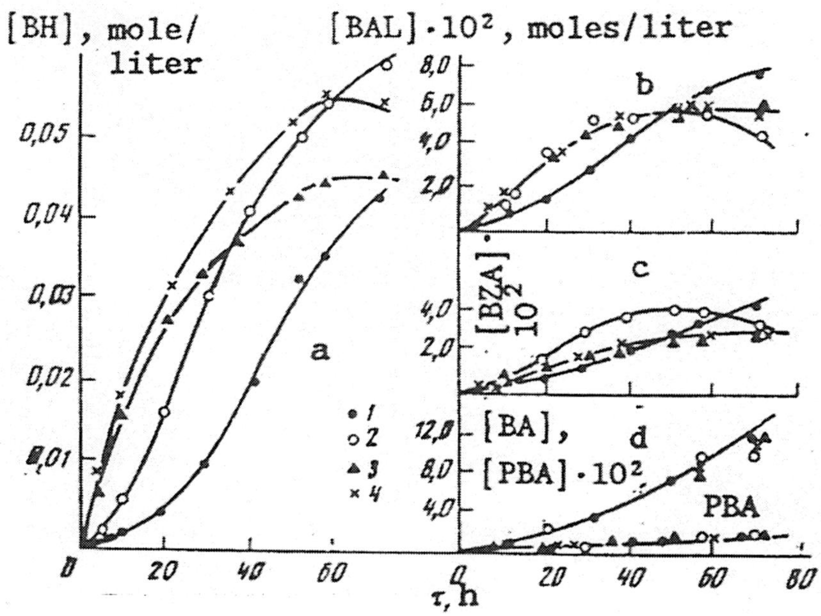


Figure 1. Kinetic curves for the accumulation of the oxidation products: BH (mol/l) (a); and BAL (b), BZA (c), BA and PBA (d) (X10$^2$, mol/l) in the oxidation of toluene at 100°C in the presence of $Ni(II)(acac)_2$ with 0 (1), 0.03 (2), 0.078 (3), and 0.15 mol/l MP.

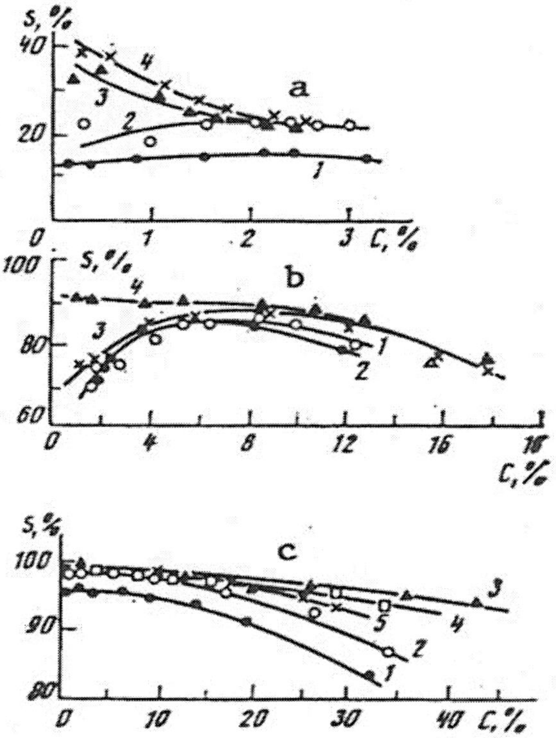


Figure 2. Dependence of the selectivity $\boldsymbol{S}_{ROOH}$ of the oxidation of monoalkylbenzenes on hydrocarbon conversion $\boldsymbol{C}$: a) toluene (designations and conditions in Figure 1), b) ethylbenzene at 120°C in the presence of $Ni(II)(acac)_2$ ($3 \cdot 10^{-3}$ mol/l), and 0.03 (1), 0.05 (2), 0.07 (3), and 0.15 mol/l MP (4), and c) cumene (1), cumene in the presence $Ni(II)(acac)_2$ ($6 \cdot 10^{-6}$ mol/l) without donor additives (2) and with $6 \cdot 10^{-5}$ (3), $1.5 \cdot 10^{-4}$ (4), $3 \cdot 10^{-4}$ (5) mol/l MP (100°C).

In [21] is established that with an anisole-containing polypyridylamine potential ligand °L, a μ-1,2-peroxodicopper(II) complex $[\{°LCu^{II}\}_2(O_2^{2-})]^{2+}$ forms from the reaction of the mononuclear compound $[CuI(°L)(MeCN)]B(C_6F_5)_4$ ($°LCu^I$) with $O_2$ in no coordinating solvents at -80°C. Thermal decay of this peroxo complex in the presence of toluene or ethylbenzene leads to rarely seen C-H activation chemistry; benzaldehyde and acetophenone/ 1-phenylethanol mixture (~1:1), respectively, are formed. Very similar toluene oxygenation chemistry occurs with dicopper(III)-bis-μ-oxo species $[\{^{Bz}LCu^{III}\}_2(\mu\text{-}O^{2-})_2]^{2+}$.

It can how see from Figure 1 (a, в, c), additives MP into oxidation of toluene in presence $Ni(acac)_2$ result in increase in initial rates of accumulation of BH, BZA and BAL. This allows to assume increase in the activity of $Ni(acac)_2$ upon its complexation with $L^2$ in chain initiation (1.1), BH decomposition (1.3-1.3'), and free radical termination (1.5.) (Scheme 1).

*Scheme 1.*

$PhCH_3 \xrightarrow{O2Ni(acac)2L2} PhCH_2O_2^{\cdot}$ (1.1)

$PhCH_2O_2^{\cdot} + PhCH_3 \rightarrow PhCH_2\,OOH + PhCH_2^{\cdot}$ (BH) (1.2)

$PhCH_2OOH + Ni(acac)_2\,L^2 \leftrightarrows K \rightarrow PhCH_2O^{\cdot} + Ni(acac)_2OH \cdot L^2$ (1.3)

$PhCH_2OOH + Ni(acac)_2OH\,L^2 \leftrightarrows K \rightarrow PhCH_2O_2^{\cdot} + Ni(acac)_2\,L^2 + H_2O$ (1.3')

$PhCH_2O^{\cdot} + PhCH_3 \rightarrow PhCH_2OH + PhCH_2$ (BZA) (1.4)

$PhCH_2O_2^{\cdot} + Ni(acac)_2 L^2 \rightarrow PhCHO + Ni(acac)_2OH{\cdot}L^2$, (BAL) (1.5)

*Scheme 2.*

$PhCH_2OH + Ni(acac)_2OH \leftrightarrows K \rightarrow PhC^{\cdot}HOH + H_2O + Ni(acac)_2$ (1.6)

$PhC^{\cdot}HOH + Ni(acac)_2OH \rightarrow PhCHO + Ni(acac)_2 + H_2O$ (BAL) (1.7)

$PhCHO + Ni(acac)_2OH \rightarrow PhCO^{\cdot} + Ni(acac)_2 + H_2O$ (1.8)

$PhCO^{\cdot} + Ni(acac)_2OH \rightarrow PhCO_2H + Ni(acac)_2$ (BA) (1.9)

$PhCO^{\cdot} + O_2 \rightarrow PhCOO_2$ (1.10)

$PhCOO_2^{\cdot} + PhCH_3 \rightarrow PhCO_3H$ (PBA) (1.11)

$PhCOO_2^{\cdot} + PhCHO \rightarrow PhCO_3H + PhCO^{\cdot}$ (PBA) (1.12)

$PhCO_3H + Ni(acac)_2L^2 \rightarrow PhCO_2^{\cdot} + Ni(acac)_2OH{\cdot}L^2$ (1.13)

$PhCO_3H + Ni_2(L^1)(L_{ox}^1)_3L^2 \rightarrow PhCO_2^{\cdot} + Ni_2(L^1)(L_{ox}^1)_3OH{\cdot}L^2$ (1.13')

$PhCO_2^{\cdot} + PhCH_3 \rightarrow PhCO_2H + PhCH_2^{\cdot}$ (BA) (1.14)

$PhCO_3H + PhCHO \rightarrow 2PhCO_2H$ (BA) (1.15)

Reaction steps (1.1), (1.3), (1.5) in mechanism for catalyzed oxidation are given by analogy with results of works [22, 27-29].

Additives of MP in the beginning stages of the oxidation of toluene in presence $Ni(acac)_2$ may accelerate the oxidation of BZA to BAL and BAL in BA, which proceeds on the nickel catalyst by reactions (1.6) - (1.7) and (1.8)-(1.9) [24, 25] (Scheme 2).

*The maximum selectivity in the oxidation of toluene to BH in presence of Ni(II)(acac)₂, without the addition of donor ligand MP $S_{BH}$ is about ~ 15 % at depth of reaction to ~ 2-3 % (Figure 2 (a)).*

*Additives of MP enhance the selectivity $S_{BH}$ of oxidation of toluene to BH, catalyzed with Ni(II)(acac)₂, to ~ 27 % at [MP] = $3.0{\cdot}10^{-2}$ mol/l* (Figure 2(a)) with the same extent of oxidation. Thereby on initial reaction steps the yield of BH significantly increases (~ in 4 times). With concentration of [MP] increase, when [MP] $\geq 7{\cdot}10^{-2}$ mol/l, the character of dependence $S_{BH}$ from degree of conversion $C$ is changed: the value of $S_{BH}$ is maximal on initial reaction steps: $S_{BH} \approx 30$-40% at 0.5 % of conversion $C$, and selectivity of BH formation is reduced at $C > 0.5$ %.

Introduction of MP in the reaction of toluene oxidation, catalyzed $Ni(II)(acac)_2$, results to significant synergic effect of increase of initial rate of toluene oxidation $w_0$, by more than a factor of 3 as the concentration [MP] is increased from 0 to 0.15 mol/l [17]. The dependence of $w_0$ on [MP] has maximum. It testifies in favour of formation catalytic coordinated unsaturated complexes $Ni(II)(acac)_2{\cdot}MP$ and coordinated saturated complexes $Ni(II)(acac)_2{\cdot}2MP$ [30]. The additives of MP in the absence of $Ni(II)(acac)_2$ does not practically influence on initial rate $w_0$ of reaction of toluene oxidation.

Similar dependences of initial rate on concentration of ligand-modifier at constant concentration of $Ni(II)(acac)_2$ that testified in favour of formation of catalytic active complexes $Ni(II)(acac)_2{\cdot}MP$ and $Ni(II)(acac)_2{\cdot}2MP$ [30] have been received by us in case of ethylbenzene oxidation in the presence of system {$Ni(II)(acac)_2$ ($3{\cdot}10^{-3}$ mol/l) + MP}[8, 39].

## Oxidation of Cumene, Catalyzed with System {Ni(II)(acac)$_2$ + MP}

The selectivity of cumene oxidation to cumyl hydroperoxide (CH) in the absence of catalyst at 90-100$^0$ is as a rule high, namely, $S_{CH}$ = 96-98 % [31-33].

The stability of CH, high values of rate constants for chain propagation and low yield of chain termination products due to low $k_5$ values facilitate the high concentrations of the hydroperoxide (Table 1).

High selectivity of cumene oxidation to hydroperoxide $S_{CH}$ can be connected with possibility of formation of cumyl hydroperoxide, CH, in *chain termination reaction* [33].

It is assumed, what recombination of cumyl peroxo radicals ROO• occurs through intermediate formation of tetroxide $RO_4R$, decomposition of which is realized on Russell mechanism [34], (through cyclic six-member transient state) and results to formation of decay products: CH, α-methyl styrene and $O_2$,on scheme:

$$2C_6H_5C(CH_3)O_2(CH_3)CC_6H_5 \rightarrow C_6H_5(CH_3)_2OOH{+}C_6H_5(CH_3){=}CH_2{+}O_2$$

However, the cumene oxidation reaction at 90-100$^0$ without catalyst, selective on hydroperoxide, is usually proceeds only to a low conversion due to self-inhition by the accumulated products of the extensive oxidation, - phenol and benzoquinone.

Thus, in the oxidation of cumene, we are faced with the problem of increasing the conversion and rate of cumene oxidation while retaining the high selectivity 97-99%.

The use of variable-valence metal compounds as catalysts, as a rule, is accompanied by a decrease in the selectivity of the oxidation of cumene to the hydroperoxide, mainly due to the formation of cumyl alcohol upon the decomposition of CH [35, 36]. An increase in the reaction rate with retention of high selectivity up to cumene conversion of 35-40 % is observed at oxidation of cumene when zinc and cadmium complexes of N-heterocyclic bases are used as catalysts [37].

These works appeared actually simultaneously with ours. In presence of heterogenised variable-valence metal compounds the values $S_{CH,max}$ and $C_{max}$ of cumene oxidations to hydroperoxide are lower. For example, at catalysis by silica – and polymer (poly-4-vinylpyridine) – supported $Cu(OAc)_2$, the increase in the rate of cumene oxidation by 68% was observed, and $S_{CH,max}$ was equal 92% at C = 22% [36].

Since the rate of cumene oxidation at 100$^0$ is rather high, on this system, on-visible, it is necessary to work with small concentrations of nickel catalyst, which can facilitate a high yield of the hydroperoxide [8, 39]. It is evident from analysis of scheme of catalyzed hydrocarbons oxidation, including participation of catalyst in chain initiation reaction under catalyst interaction with ROOH and also in chain propagation (Ct + $RO_2^{\bullet}\rightarrow$), that with decrease in $[Ct]_0$ the rate of reaction should be decreased, and $[ROOH]_{max}$ should be increased [8, 15]. It has appeared that together with increase in concentration $[ROOH]_{max}$ selectivity of ethylbenzene oxidation $S_{ROOH,max}$ increased also at decrease in $[Ni(II)(acac)_2]_0$ [8, 39].

Really, it was established by us earlier that selectivity of cumene oxidation to hydroperoxide CH increased from $S_{CH}$ = 95 % (in the uncatalyzed oxidation) up to = 97-98% in the presence of $6{\cdot}10^{-6}$ mol/l $Ni(II)(acac)_2$, while the conversion with the retention of $S_{CH,max}$ is increased from 10 to 15 % [17].

At this $Ni(II)(acac)_2$ catalyst does not affect the composition of the oxidation products, but alters their ratio, increasing the rate of accumulation of CH and its maximum concentration while virtually not affecting the kinetics of the accumulation of the other products of oxidation of cumene, namely, dimethyl phenylcarbinol (DMPC), acetophenone (AP) and phenol (PhOH). To that DMPC is formed in concentration, on order exceeding concentrations [AP] and [PhOH].

The value of $S_{CH,max}$ upon the introduction of MP to the cumene oxidation catalyzed by $Ni(II)(acac)_2$, is virtually unaltered (97-99%) but the cumene conversion is markedly increased to C~ 35% (and to C ≈ 40-45% at selectivity on the level not lower than $S_{CH,max}$ = 95%).

The addition of MP further accelerates the accumulation of CH and has a weak effect on the kinetics of the accumulation of other products [17].

In presence of $Ni(II)(acac)_2$ ($6{\cdot}10^{-6}$ mol/l) initial rate of oxidation of cumene increases by a factor of ~ 1.7 in comparison with the rate of uncatalyzed oxidation of cumene. Catalytic system $\{Ni(II)(acac)_2+MP\}$ increases initial rate of oxidation in 3 times in comparison with uncatalyzed oxidation, and in 1.6 time in comparison with the oxidation of cumene, catalyzed by $Ni(II)(acac)_2$.

These data indicate, what system $\{Ni(II)(acac)_2+L^2\}$ based on nickel complex catalyst, may serve as catalyst for the alkylarens oxidation: not only ethylbenzene but also toluene and cumene, - to the corresponding hydroperoxides.

*Thus the received kinetic laws of catalytic alkylarene oxidations testify in favour of a mechanism generality of catalysis of the investigated hydrocarbons by system* $\{Ni(II)(acac)_2+L^2\}$*:*

Dependence of the oxidation selectivity $S_{ROOH}$ on the conversion $C$ for the oxidation ethylbenzene (and of toluene also) at defined values $[L^2]$ ($L^2$=MP]) (Figure 2) traverse a maximum. The $S_{ROOH}$ value increases with increasing conversion, reaches a maximum value, and then decreases. Such character of dependence $S_{ROOH}$ on $C$, obviously, is connected with transformation of complexes $Ni(II)(acac)_2{\cdot}MP$ during reaction by effect of $O_2$ to a new selective catalyst, heteroligand macro complex $Ni_2(OAc)_3(acac){\cdot}MP$ ("A") [8, 39].

At MP concentrations, exceeding some defined value, the nature of dependence of $S_{ROOH}$ on $C$ changes: $S_{ROOH}$ is maximal in the beginning of reaction, and is then gradually reduced with increasing $C$ (Figure 2). Such dependence is characteristic for oxidation of cumene at all the MP concentrations examined.

Complexes $Ni(II)(acac)_2{\cdot}2MP$ may become responsible for high values of selectivity of alkylarenes oxidation to ROOH at sufficiently high MP concentration ([MP]). During alkylarenes oxidation complexes $Ni(II)(acac)_2{\cdot}2MP$, on-visible, are converted in coordinate unsaturated $Ni(II)(acac)_2{\cdot}MP$ (according to the reaction with $O_2$ or ROOH) [38], namely, the next reaction becomes possible:

$$Ni(acac)_2{\cdot}2MP+O_2 \rightarrow Ni(acac)_2{\cdot}MP{\cdot}O_2+MP \quad (**)$$

Complexes $Ni(acac)_2{\cdot}MP{\cdot}O_2$ are subsequently transformed in heteroligand macro complexes $Ni_2(OAc)_3(acac){\cdot}MP$ ("A").

So, the obtained data testifies to that that the next complexes formed *in situ* in the course of alkylarenes oxidations with dioxygen, catalyzed by system $\{Ni(II)(L^1)_2+L^2\}$, are responsible for observable kinetic laws: primary complexes $Ni(II)(L^1)_2{\cdot}L^2_n$ at I macro step of oxidation, and new catalytic heteroligand macro complexes of general formula $M^{II}_xL^1_y(L^1_{ox})_z(L^2)_n$ ($L^1_{ox}$=$(OAc)^-$) at II macro step of oxidation) [8, 39].

It was shown [39] that the selective catalyst heteroligand complex Ni(acac)(OAc)·$L^2$ was formed as a result of the ligand $L^2$-controlled regioselective addition of $O_2$ to a nucleophilic carbon atom (γ-C) of the $(acac)^-$ ligand. The coordination of the electron-donating axial ligand $L^2$ with the $Ni(acac)_2$ complex stabilized the intermediate zwitter-ion $[L^2(acac)Ni(acac)^+ \cdots O_2^-]$ and increased the probability of regioselective insertion of $O_2$ to the acetylacetonate ligand activated by coordination with the nickel(II) ion. Further incorporation of $O_2$ into the chelate acac-ring was accompanied by the proton transfer and the redistribution of bonds in the transition complex "A" leading to the scission of the cyclic system C(O)MeC(O)OCH(Me)O to form a chelate ligand $OAc^-$, acetaldehyde and CO (in the Criegee rearrangement) (Scheme 3).

The only known Ni(II)-containing dioxygenase ARD (acireducton dioxygenase), which catalyzes the oxidative decomposition of β-diketones, operates in the analogous way [40]. This applies to the functional enzyme models, namely, Cu(II)- and Fe(II)-containing quercetin 2,3-dioxygenases, which catalyze the decomposition of β-diketones in the enol form to carbonyl compounds with CO evolution [41-42]

*Scheme 3*

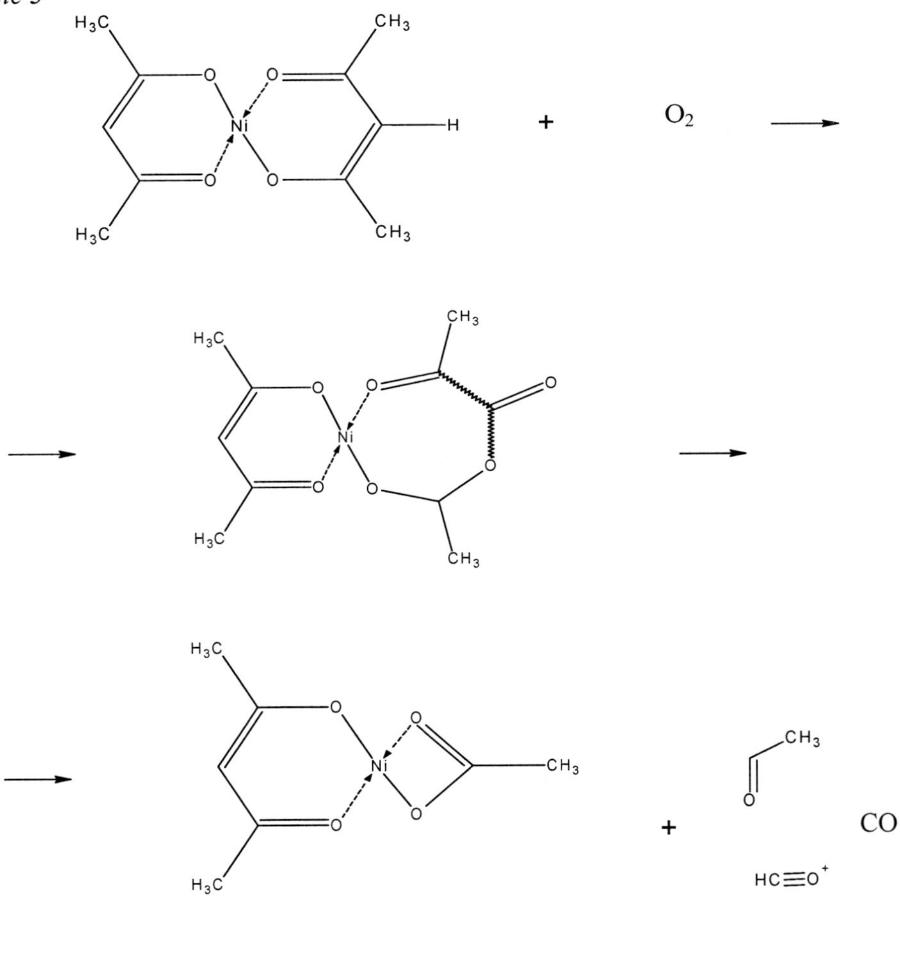

As a result of this process, reactive mono- and poly-nuclear heteroligand complexes with the general composition $Ni_x(acac)_y(AcO)_z(L^2)_n$ $L^2$ = HMPA, DMF, MP, M'St (M' = Na, Li, K), 18C6, $R_4NBr$ were formed [8].

$Ni(acac)_2\ L^2 + O_2 \longrightarrow [L^2(acac)Ni(acac)^+ \cdots O_2^-] \longrightarrow L^2(acac)Ni(AcO) + MeCHO + CO$,

$2Ni(acac)_2\ L^2 + O_2 \longrightarrow Ni_2(AcO)_3(acac)L^2 + 3MeCHO + 3CO + L^2$.

$L^2$ = HMPA, DMF, MP, M'St (M' = Na, Li, K).

## Mechanism of Formation of Products in Toluene, Cumene (and Ethylbenzene) Oxidations

For reactions of oxidation of toluene (and ethylbenzene), catalyzed $Ni(acac)_2$ changes in the mechanism of formation of the some products of the reaction it was observed in the presence of additives ligand MP, connected with changes of activity of various forms of the catalytic complexes, formed during oxidation process.

In the case of ethylbenzene oxidation in the beginning of reaction of oxidation in the presence of complexes $Ni(II)(acac)_2$ ($3.0\cdot10^{-3}$ mol/l) with monodentate ligand-modifier MP, by-products AP and MPC (P=AP, or MPC) are formed consistently, at decomposition of PEH, as well as at catalysis $Ni(II)(acac)_2$ without additives MP: $w_P/w_{PEH} \rightarrow 0$ at $t\rightarrow0$. However unlike uncatalyzed ethylbenzene oxidation and oxidation in the presence of $Ni(II)(acac)_2$, parallel formation AP and MPC is observed at catalysis with $Ni(II)(acac)_2\cdot MP$: $w_{AP}/w_{MPC} \neq 0$ at $t\rightarrow0$. In the developed reaction at catalysis with transformed form $Ni(II)(acac)_2\cdot MP$ (*$Ni_2(OAc)_3(acac)\cdot MP$ ("A")*) the direction of formation of products varies: AP and MPC are formed in parallel with PEH, and AP and MPC – also in parallel: $w_P/w_{PEH} \neq 0$ at $t\rightarrow0$, and, besides, $w_{AP}/w_{MPC} \neq 0$ at $t\rightarrow0$ in the chain propagation ($Ct + RO_2^{\bullet}\rightarrow$) and quadratic termination ($2RO_2^{\bullet}\rightarrow$) (as well as in the case of catalysis with complexes of $Ni(II)(acac)_2$ with macrocyclic polyethers and quaternary ammonium salts ($[Ni(II)(acac)_2]$ $=1.5\cdot10^{-4}$ mol/l) [8, 39].

Here it has been established by us for the first time that, *as well as in the case of ethylbenzene oxidation,* in toluene oxidation in the presence of system $\{Ni(II)(acac)_2+MP\}$ the route of formation major toluene oxidation products, benzaldehyde (BAL), benzyl alcohol (BZA) and benzoic acid (BA) is changed: *from successive* (at catalysis by complexes $Ni(II)(acac)_2\cdot MP$, formed *in situ* at the beginning of toluene oxidation, I macro step) *on parallel* (at catalysis by complexes $Ni_2(OAc)_3(acac)\cdot MP$ ("A"), formed during oxidation).

At catalysis with complexes $Ni_2(OAc)_3(acac)\cdot MP$ ("A"), benzaldehyde BAL and benzyl alcohol BZA are formed parallel each other ($w_{BAL}/w_{BZA} \neq 0$ at $t\rightarrow0$), apparently, in the steps of chain termination ($2PhCH_2O_2^{\bullet}\rightarrow(*)$ and reaction 1.5 (Scheme 1) (BAL)); BAL and BA – also are formed in parallel reactions ($w_{BA}/w_{BAL} \neq 0$ at $t\rightarrow0$) (Figure 3). At this BA is formed successively from PBA ($w_{BA}/w_{PBA}\rightarrow0$ at $t\rightarrow0$) (Scheme 2, reactions 1.13'-1.14). Unlike this, a successive formation BAL from BZA and BA from BAL (Figure 3) (and PBA)) occurs at catalysis with primary complexes $Ni(II)(acac)_2\cdot MP$ (Scheme 2).

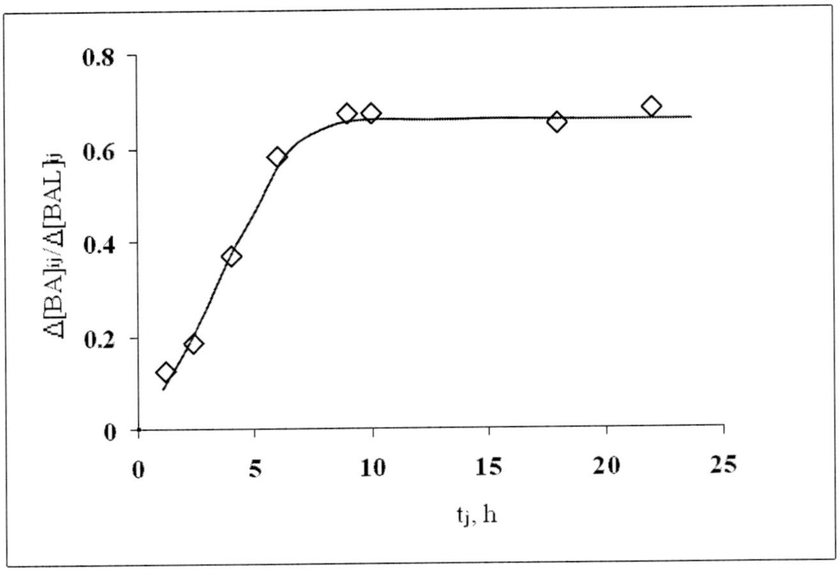


Figure 3. Dependence $\Delta[BA]_{ij}/\ \Delta[BAL]_{ij}$ on time $t_j$ in the course of toluene oxidation, catalyzed by {Ni(II)(acac)$_2$+MP}. [Ni(II)(acac)$_2$] =$3\cdot10^{-3}$ mol/l, [MP] =$3\cdot10^{-2}$ mol/l, 100°C.

The route of *benzaldehyde BAL and benzyl alcohol BZA* formation is changed from successive (*via benzyl hydroperoxide BH decomposition at catalysis with Ni(II)(acac)$_2$•MP* (Scheme 1, reactions 1.3-1.5) *on parallel with BH – at catalysis with Ni$_2$(OAc)$_3$(acac)·MP ("A") complexes, formed during oxidation.*

The interesting result was received in the case of research of cumene oxidation. At introduction of catalyst (with and without of ligand-modifier) in reaction of oxidation of cumene changes in the mechanism of formation of products it was not observed, as opposed to oxidation of ethylbenzene and toluene. So as well as in uncatalyzed oxidation or at catalysis by Ni(II)(acac)$_2$, and by system {Ni(II)(acac)$_2$ + MP} (complexes Ni(II)(acac)$_2$•2MP and *Ni$_2$(OAc)$_3$(acac)·MP*), *ketone AP and hydroperoxide CH are formed parallel* ($w_{AP}/w_{CH} \neq 0$ at $t\rightarrow0$), *but DMPC is product of CH decomposition* ($w_{DMPC}/w_{CH}\rightarrow0$) at $t\rightarrow0$) *during all reaction of oxidation* (Figure 4). At that *AP and DMPC are formed parallel each other* ($w_{AP}/w_{DMPC} \neq 0$ at $t\rightarrow0$)

In [43] mechanism of AP formation on reaction of quadratic chain termination of cumyl peroxo radicals ROO• is offered:

$$2PhC(CH_3)_2O_2^{\cdot}\rightarrow[PhC(CH_3)_2O_4(CH_3)_2CPh]\rightarrow2PhCOCH_3+CH_3O_2^{\cdot}+CH_3^{\cdot}$$

The growth of selectivity $S_{CH}$ from 95% (uncatalyzed reaction) to 99% (catalysis by complexes of nickel with MP) in this case seems to be connected with acceleration of formation ROOH for the account of participation of catalyst in a step of chain initiation ($O_2$ activation).

Here it has been established by us parallel formation of major products of ethylbenzene and toluene oxidation *at catalysis by coordinated saturated complexes Ni(II)(acac)$_2$·2MP as well as at catalysis with Ni$_2$(OAc)$_3$(acac)·MP ("A")*. So in the beginning of ethylbenzene oxidation, catalyzed with Ni(II)(acac)$_2$·2MP (Figure 2, b: kinetic curve 4), the oxidation products PEH, AP, and MPC are formed parallel, AP and MPC – also parallel each other. At that the value of $S_{PEH}$ is maximal on initial stages, and $S_{PEH,0} \sim 90\%$ at conversion $C \leq 8\%$. As it has been established by us earlier at catalysis with small concentrations of Ni(II)$(acac)_2$ ($1.5 \cdot 10^{-4}$ mol/l) without additives of ligands-modifiers and in the presence exo ligand $L^2$=18C6, namely at catalysis with Ni(II)$(acac)_2 \cdot L^2_n$ and heteroligand macro complexes of general formula $M^{II}_xL^1_y(L^1_{ox})_z(L^2)_n$ ($L^1_{ox}$=$(OAc)^-$), formed in the course of oxidation, products AP and MPC (P) are formed also not from ROOH (PEH) but in parallel with PEH, in steps of chain propagation (Ct + $RO_2^{\cdot}$→) and chain termination ($2RO_2^{\cdot}$→), $w_P/w_{PEH} \neq 0$ at t→0, and, besides, $w_{AP}/w_{MPC} \neq 0$ at t→0 [8, 39]. At that high values of selectivity $\boldsymbol{S}_{PEH,max}$= 90% (Ni(II)$(acac)_2$ ($1.5 \cdot 10^{-4}$ mol/l)), and $\boldsymbol{S}_{PEH,max}$ = 98% at catalysis with Ni(II)$(acac)_2 \cdot L^2_n$ are observed from the very beginning of reaction.

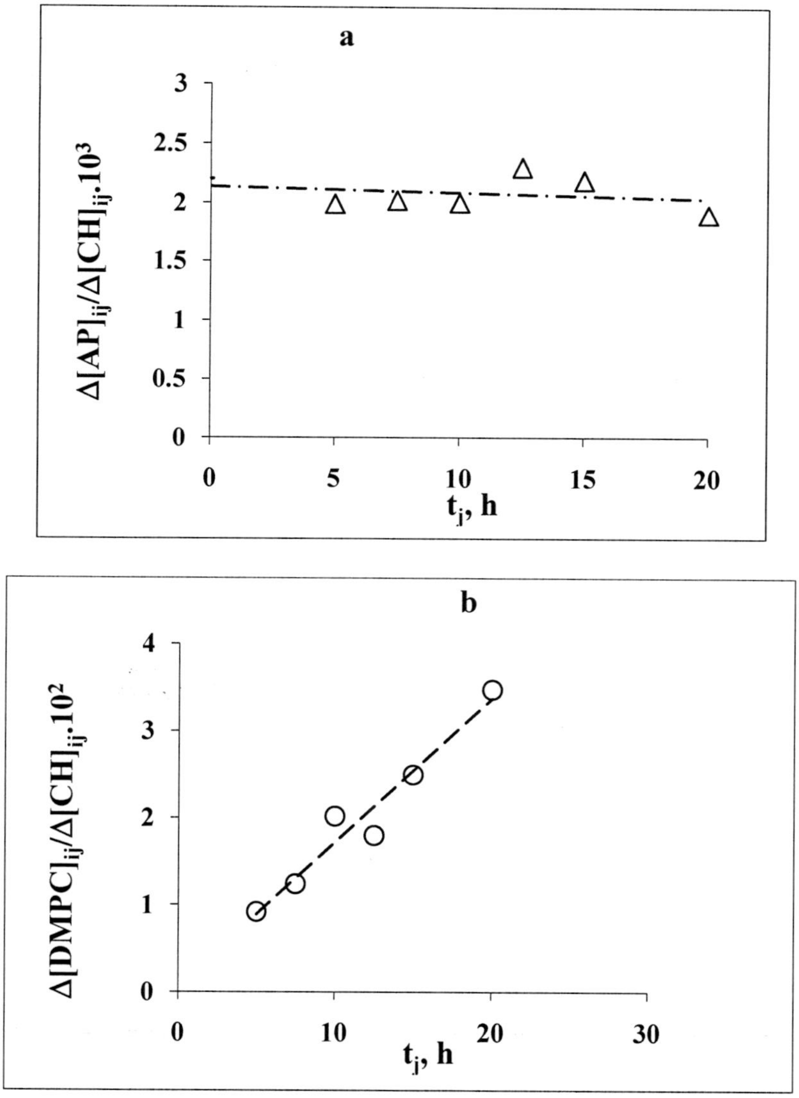


Figure 4. Dependences of product accumulation rations on time in the course of cumene oxidation, catalyzed by {Ni(II)$(acac)_2$+MP}: a. $\Delta[AP]_{ij}/\Delta[CH]_{ij}$ on $t_j$, b. $\Delta[DMPC]_{ij}/\Delta[CH]_{ij}$ on $t_j$. [Ni(II)$(acac)_2$] =$6 \cdot 10^{-6}$ mol/l, [MP] =$1.5 \cdot 10^{-4}$ mol/l, 100°C.

Similarly, benzaldehyde BAL and benzyl alcohol BZA are not products of decomposition of benzyl hydroperoxide BH, and formed parallel with BH at catalysis with coordinated saturated complexes $Ni(II)(acac)_2{\cdot}2MP$. BAL and BZA are formed in parallel reactions also. At that the value of $S_{BH}$ is maximal on initial stages. $S_{BH} \approx 30$-40% at 0.5 % of conversion $C$ (Figure 2, a: kinetic curves 3, 4).

As already it was mentioned above [21], thermal decay of μ-1,2-peroxodicopper(II) complex $[\{°LCu^{II}\}_2(O_2^{2-})]^{2+}$ (°L = an anisole-containing polypyridylamine potential tetra dentate ligand) in the presence of toluene or ethylbenzene leads to mainly benzaldehyde (40%) (toluene), or acetophenone/ 1-phenylethanol mixture (~1:1) (ethylbenzene), respectively formations. Very similar toluene oxygenation occurs with dicopper(III)-bis-μ-oxo species $[\{^{Bz}LCu^{III}\}_2(\mu\text{-}O^{2-})_2]^{2+}$ ($^{Bz}L$ is tridentate chelate, possessing the same moiety found in °L, but without the anisole O-atom donor). Authors suppose that alkylperoxo radical chemistry (e.g. the Russell mechanism, $RO_2\bullet$ radical disproportionation process) well explains the nearly equal yield of acetophenone and 1-phenylethanol produced from the ethylbenzene oxidation. This reaction mechanism was also reported by Que and co-workers to interpret nonheme diiron catalyzed alkyl benzene oxidation reactions [44].

It was established in [21] that $[\{°LCu^{II}\}_2(O_2^{2-})]^{2+}$ does not convert benzyl alcohol to benzaldehyde. Instead, the formation of benzaldehyde BAL and benzyl alcohol BZA occur at the same time during the decay process of $[\{°LCu^{II}\}_2(O_2^{2-})]^{2+}$

$$([\{°LCu^{II}\}_2(O_2^{2-})]^{2+} \rightarrow 2[°LCu^{I}]^{+} + O_2)$$ and toluene oxidation.

When the formation and decay of $[\{°LCu^{II}\}_2(O_2^{2-})]^{2+}$ was carried out in the presence of excess $O_2$, the yield of BAL increased to ~ 75%.

## CONCLUSION

One way of retention and increasing the activity of a homogeneous metal complex catalyst is the introduction of modifying ligands. The mechanism of action of the modifying ligands (monodentate, polydentate) in radical-chain ethylbenzene oxidation, catalyzed with transition metal (Ni(II), Fe(II,III), Co(II)) complexes are in detail investigated in our works [8, 39].

Research of the mechanism of formation of products of oxidation is a necessary condition of modeling effective catalytic systems for hydrocarbons oxidation in target products. In this work the routes of formation of the basic products of radical-chain toluene, cumene (and also ethylbenzene) oxidations by molecular oxygen, catalyzed with system $\{Ni(II)(acac)_2(3{\cdot}10^{-3}$ mol/l)$+L^2\}$ ($L^2$=electron donating exo ligand-modifier N-methylpirrolidon-2, MP) has been researched for the first time with kinetic method.

It has been established that the route of formation of products for toluene (and ethylbenzene} varies depending on a parity of components catalytic systems and is defined by the nature of the active complexes formed in situ during oxidation: active primary $Ni(II)(acac)_2{\cdot}nMP$ (n=1,2) and heteroligand complexes, $Ni_2(OAc)_3(acac){\cdot}MP$ ("A"), the intermediate products of oxygenation of $Ni(II)(acac)_2{\cdot}nMP$.

In nickel catalyzed toluene oxidation the route of formation major oxidation products, benzaldehyde (BAL), benzyl alcohol (BZA) and benzoic acid (BA) is changed: from *successive* (catalysis by coordinated unsaturated complexes Ni(II)$(acac)_2$·MP *on parallel* (catalysis by complexes $Ni_2(OAc)_3(acac)$·MP ("A").

The increase in selectivity of toluene (and ethylbenzene) oxidation to ROOH, catalyzed with complexes Ni(II)$(acac)_2$·2MP and $Ni_2(OAc)_3(acac)$·MP ("A"), as compared with catalysis by complexes Ni(II)$(acac)_2$·MP seems to be connected with as one of the reasons the change of a direction of formation of products BAL, BZA (toluene), AP, MPC (ethylbenzene), from consecutive *via* homolytic ROOH decomposition → the parallel, in steps of chain propagation with the assistance of the catalyst (Ct + $RO_2^{\cdot}$→), and chain termination, and also with acceleration of in chain initiation ($O_2$ activation).

In the case of cumene oxidation changes in the mechanism of formation of products it was not observed. So as well as in uncatalyzed oxidation or at catalysis by Ni(II)$(acac)_2$, and by complexes Ni(II)$(acac)_2$•2MP or *$Ni_2(OAc)_3(acac)$·MP, ketone AP and hydroperoxide CH are formed parallel each other, but alcohol DMPC is product of CH decomposition during of all reaction of oxidation.* The growth of selectivity $S_{CH}$ in this case seems to be connected with acceleration of formation ROOH for the account of participation of catalyst in a step of chain initiation ($O_2$ activation).

## References


[1] G.B. Shul'pin, Metal-catalyzed hydrocarbon oxygenations in solutions: the dramatic role of additives; a review, *J. Mol. Catal. A: Chem.*, 189, 39 (2002).

[2] *The Activation o/Dioxygen and homogeneous Catalyti'c Oxidation* (Eds. D. H. R. Barton, A. E. Martell, D. T. Sawyer) (New York: Plenum, 1993)

[3] D. Mansuy, New Model Systems for oxygenases, in *The Activation of Dioxygen and homogeneous Catalyti'c Oxidation* (Eds. D. H. R. Barton, A. E. Martell, D. T. Sawyer) (New York: Plenum, 1993) p. 347

[4] E. I. Karasevich, Liquid-phase oxidation of alkanes in the presence of metals compounds, in *Reactions and Properties of Monomers and Polymers (Eds. A. D'Amore, G. Zaikov)* (New York: Nova Science Publ., 2007) p. 43

[5] M. Szaleniec, C. Hagel, M. Menke, P. Novak, M. Witko, J. Heider, Kinetics and Mechanism of Oxygen-Independent Hydrocarbon Hydroxylation by Ethylbenzene *Biochemistry*, 46, 7637 (2007)

[6] K. Weissermel, H-J. Arpe, *Industrial Organic Chemistry* (3nd Ed).(New York: VCH, 1997)

[7] S. Menage, J.-B. Galey, J. Dumats, G.Hussler, M. Seite, I.G. Luneau, G. Chottard, and M. Fontecave, $O_2$ Activation and Aromatic Hydroxylation Performed by Diiron Complexes, *J. Am. Chem. Soc.*, 120, 13370-13382 (1998).

[8] L.I. Matienko, Solution of the problem of selective oxidation of alkylarenes by molecular oxygen to corresponding hydroperoxides. Catalysis initiated by Ni(II), Co(II), and Fe(III) complexes activated by additives of electron-donor mono- or multidentate extra-ligands, In: *Reactions and Properties of Monomers and Polymers* (Eds. A. D'Amore and G. Zaikov) (New York: Nova Sience Publ. Inc. 2007) p.21.

[9] L.A. Mosolova, L.I. Matienko, I.P. Skibida, Composition catalysts of ethylbenzene oxidation based on bis(acetylacetonato) nickel (II) and phase transfer catalysts as ligands. 1. Macrocyclic polyethers, *Izv. Akad. Nauk, Ser.Khim.*, 1406 (1994)[a]

[10] L.I. Matienko, L.A. Mosolova, The mechanism of selective ethylbenzene oxidation with molecular $O_2$, catalyzed by complexes of nickel (II), cobalt (II) and iron (III) with macrocyclic polyethers, in *Uspekhi v Oblasti Geterogennogo Kataliza i Geterotsiklov* (Progress in Heterogeneous Catalysis and Heterocycles) (Ed. D. L. Rakhmankulov) (Moscow: Khimiya, 2006) p. 235

[11] L.I. Matienko, L.A. Mosolova, I.P. Skibida, Composition catalysts of ethylbenzene oxidation based on bis(acetylacetonato) nickel (II) and phase transfer catalysts as ligands. 2. Quarternary ammonium salts, *Izv. Akad. Nauk, Ser. Khim.* 1412 (1994).

[12] L.I. Matienko, L.A. Mosolova, Selective oxidation of ethylbenzene with dioxygen into $\alpha$-phenylethylhydroperoxide. Modification of catalyst activity of Ni(II) and Fe(III) complexes upon addition of quaternary ammonium salts as exoligands, in *New Aspects of Biochemical Physics. Pure and Applied Sciences* (Eds. S. D. Varfolomeev, E. B. Burlakova, A. A. Popov, G. E. Zaikov) (New York: Nova Science Publ., 2007) p. 95

[13] L.I. Matienko, L.A. Mosolova. Selective Oxidation of Ethylbenzene with Dioxygen into $\alpha$-Phenylethylhydroperoxide. Modification of Catalyst Activity of Ni(II) and Fe(III) Complexes upon Addition of Quaternary Ammonium Salts as Exoligands, in: Biochemical Physics. Research Trends. (Ed. by Varfolomeev S.D., Burlakova E.B., Popov A.A., Zaikov G.E.) (New York: Nova Science Publ., Inc., 2009) p. 37.

[14] L.I. Matienko, L.A. Mosolova, Influence of Hexamethylphosphoric Triamide and Dimethylformamide Admixtures on the mechanism of Tris(acetylacetonato)iron(III)-Catalyzed Ethylbenzene Oxidation with Molecular Oxygen, *Neftekhimiya,* 47, 42 (2007) (*in Russian*).

[15] E.T. Denisov, N.M. Emanuel, *Usp. Khim.*, 29 1409 (1960) [*Russ. Chem. Rev.* 29 645 (1960)]

[16] N.M. Emanuel, D. Gall, *Okislenie Etilbenzola. Model'naya Reaktsiya* (0xidation of Ethylbenzene. Model Reaction) (Moscow: Nauka, 1984)

[17] L.A. Mosolova, L.I. Matienko, I.P. Skibida, Activation of Bis(acetylacetonato)nickel by N-methyl 2-pyrrolidone in the Oxidation of Monoalkyl-substituted Benzenes to Hydroperoxides, *Kinetika i kataliz*, 29, 1078-1083 (1988) (*in Russian*).

[18] S.J. Blanksby, G.B.Ellison, Bond Dissociation Energies of Organic Molecules, *Acc. Chem. Res.*, 36, 255 (2003).

[19] J.A. Howard, W.J. Schwalm, K.U. Ingold, Absolute Rate Constants for Hydrocarbons autoxidation. VII. The Reactivities of Peroxy Radicals towards Hydrocarbons and Hydroperoxides, in: *Preprints of the International Oxidation Simposium. San Francisco. California*. 1967, 1. p.3.

[20] W. Partenheimer, Metodology and scope of metal/bromide autoxidation of hydrocarbon, *Catalysis Today*, 23, 69 (1991).

[21] H.R. Lucas, L. Li, A.A.N. Sarjeant, M.A. Vance, E.I. Solomon, and K.D. Karlin, Toluene and Ethylbenzene Aliphatic C-H Bond Oxidations Initiated by a Dicopper(II)-$\mu$-1,2-Peroxo Complex, *J. Am. Chem. Soc.*, 131. 3230 (2009).

[22] J. Hanotier, M. Hanotier-Bridoux, P. de Radzitzky, Effect of Strong Acids of the Oxidation of Alkylarens by Manganic and Cobaltic Acetates in Acetic Acid, *J. Chem. Soc., Perk. Trans. II,* 381 (1973).

[23] P.G. Sergeev, V.V. Phedorova, The liquid oxidation of fat aromatic hydrocarbons with gaseous oxygen. The hydroperoxide of toluene, *Dokl. AN SSSR*, 109, 796 (1956) (*in Russian*).

[24] T.R. Reddy, V. Jagannadham, G.S.S, Oxidation of lactic and mandelic acids by nickel(III) ion in $H_2SO_4$ medium via addition/elimination, a kinetic study, *Orient. J. Chem.,* 2, 92 (1986).

[25] T.R. Reddy, G.S.S. Murthy, V. Jagannadham, Kinetiks and mechanism of oxidation of benzaldegyde and substituted benzaldegydes by Ni(III) ion in acedic acid – water mixtures, *Oxid. Commun*., 9, 83 (1986).

[26] G.A. Russel, The Rates of Oxidation of Aralkyl Hydrocarbons. Polar Effects in Free Radical Reaction, *J. Am. Chem. Soc*., 78, 1047 (1956).

[27] R.A. Sheldon, New catalytic methods for selective oxidation, *J. Mol. Catal*., 20, 1 (1983).

[28] T. Mlodnicka, Metalloporphyrin-Catalyzed Oxidation of Hydrocarbon with Dioxygen, in: *Metalloporphyrins in Catalytic Oxidation* (Ed. R.A. Sheldon) (New York, Basel, Hong Kong: Marcel Dekker, Inc. 1994) p.261.

[29] R. Boča, Molecular orbital study of coordinated dioxygen. V. Catalytic oxidation of toluel on cobalt- dioxygen complexes, *J. Mol. Catal*., 12, 351 (1981).

[30] V.A. Golodov, The Synergistic phenomena in catalysis, *Ross. Khim. Zh.*, 44, 45 (2000) (*in Russian*).

[31] V.L. Antonovsky, S.L. Khursan, *Fizicheskaja khimia organicheskich peroksidov* (Physic chemistry of organic peroxides) (Moscow: Academbook, 2003).

[32] Ja.K. Suyrkin, I.I. Moiseev, Mechanisms of some reactions with participation of peroxides, *Usp. Khim.*, 29, 425 (1960).

[33] C.B. Boozer, P.W. Ponder, J.C. Trisler, C.E. Wzightman, Deuterium isotope effects of the air oxidation of cumene, *J. Am. Chem. Soc*., 78, 1506 (1956).

[34] G.A. Russel, Deuterium-isotope effects in the autoxidation of aralkylhydrocarbons. Mechanism of the interaction of peroxy radicals, *J. Am. Chem. Soc*., 79, 3871 (1957).

[35] S. Weerarathas, M. Hronec, L. Malik, V. Vesely, Selectivity of metal polyphthalocyanine catalyzed oxidation of cumene, *React. Kinet. Catal. Lett*., 22, 7 (1983)

[36] Y.F. Hsu, M.H. Yen, Ch.P. Cheng, *J. Mol. Catal. A: Chem.,* Autoxidation of cumene catalyzed by transition metal compounds on polymeric supports. 105, 1377 (1996).

[37] S.K. Kozlov, Ph.I. Tovstokhat'ko, V.M. Potekhin, Isopropyl oxidation catalysts, activated with N-heterocyclic ligand – 1,10-phenantrolin, *Zh. Prikl. Khim*., 59, 217 (1986).

[38] R.D. Jones, D.A. Summerville, F. Basolo, Synthetic oxygen carriers related to biological systems, *Chem. Rev*., 79, 139 (1979).

[39] L.I. Matienko, L.A. Mosolova, G.E. Zaikov, The Modeling of Transition Metal Complex Catalysts in the Selective Alkylarens Oxidations with Dioxygen. The Role of Hydrogen – Bonding Interactions, *Oxid. Commun*., 32, 731 (2009).

[40] Y. Dai, Th. C. Pochapsky, R.H. Abeles, Mechanistic Studies of Two Dioxygenases in the Methionine Salvage Pathway of *Klebsiella pneumoniae*, *Biochemistry*, 40, 6379 (2001).

[41] B. Gopal, L.L. Madan, S.F. Betz, A.A. Kossiakoff, The Crystal Structure of a Quercetin 2,3-Dioxygenase from *Bacillus subtilis* Suggests Modulation of Enzyme Activity by a Change in the Metal Ion at the Active Site(s), *Biochemistry*, 44, 193 (2005).

[42] E. Balogh-Hergovich, J. Kaizer, G. Speier, Kinetics and mechanism of the Cu(I) and Cu(II) flavonolate-catalyzed oxygenation of flavonols, Functional quercetin 2,3-dioxygenase models, *J. Mol. Catal. A: Chem.*, 159, 215(2000).

[43] V.L. Antonovsky, B.I. Makalets, About sequence of formation of products at liquid cumene oxidation, *Dokl. AN SSSR*, 140, 1070 (1961) (*in Russian*).

[44] C. Kim, Y.H. Dong, L. Que, Modeling Nonheme Diiron Enzymes: Hydrocarbon Hydroxylation and Desaturation by a High-Valent $Fe_2O_2$ Diamond Core, *J. Am. Chem. Soc*. 119, 3635 (1997).

In: Chemical and Biochemical Reactions
Editors: G. E. Zaikov, L. Liu et al.

ISBN 978-1-61122-860-1


*Chapter 3*

# EFFECT OF TWO OXAZOLES ON MEMBRANE LIPID STRUCTURE OF LIVER MICROSOMES

***E. L. Maltseva**, *V. V. Belov* and *N. P. Palmina***
N. M. Emanuel Institute of Biochemical Physics Russian Academy of Sciemces, Kosygin str. 4, Moscow 119334, Russia

## ABSTRACT

The effect of 1,2-diphenyl-1,3-oxazole (DPhO) and its derivative – iod-methylate (IM-DPhO) in the concentration range of $10^{-2}$ -$10^{-13}$M were studied *in vitro* on the microviscosity (τ) and order parameter (S) of membrane lipids

of microsomes isolated from the liver cells of Balb mice. By EPR-technique using of spin probes 5-and 16-doxyl-stearic acids (5-and 16-DSA) localized in the surface and depth of membrane lipid regions respectively it was found that both oxazoles caused an increase in parameters τ and S, but the effect of DPhO was higher than its derivative. It was shown that the value of the effect depends on the concentrations of oxazoles by nonlinear and different manner for DPhO and IM-DPh-O. It was obtained a bimodal curve of change of parameter S depending on concentration of IM-DPhO with the maxima at $10^{-4}$ and $10^{-9}$ M The results obtained were discussed concerning radio-protecting properties of these oxazoles observed previously.

**Keywords:** oxazoles, spin-probes, microviscosity and order parameter of membrane lipids, microsomes

## INTRODUCTION

2,5-diphenyl-1,5-oxazole (DPhO) is well known component of scintillation liquid for measurement of radioactivity [1,2], fluorescence and chemiluminescence detection of

* E-mail: emal@sky.chph.ras.ru

oxazole-labeled amines and thiols [3,4]. DPhO was used as probe to determine a level of mono-oxygenation in liver as well [5]. Moreover, the oxazoles possess a number of biochemical and biophysical properties, for example, phototoxicity [6], inhibitory activity (phospholipase C) and others [7]. Additionally, it was found that DPhO and one of its derivative – iod-methylate (IM-DPhO) protected mice on irradiation, and this effect depends on both the dose of γ- radiation and concentration of these compounds [8,9].

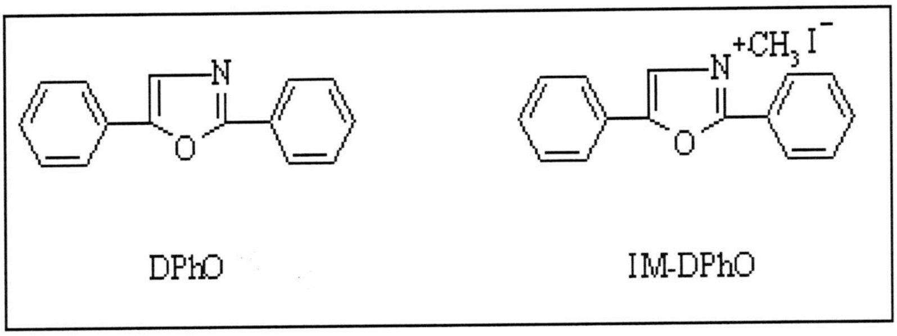


Due to the fact that radio-protective properties *in vivo* can be related with the change in lipid composition and microviscosity of cellular membranes, in particular [10,11], we proposed that these oxazoles can act on the membrane lipid structure. In general, a study of the effect of different compounds on the membranes *in vitro* is a good model as a first step to select the active one.

A purpose of the present work, therefore, was to study the effect of DPhO and IM-DPhO in a wide range of concentration ($10^{-2} – 10^{-14}$ mol/l) on the endoplasmic reticulum membranes (microsomes) isolated from Balb-line mice used in the experiments on irradiation. Electron paramagnetic resonance (EPR) technique and spin-probe method have been applied to estimate the microviscosity value (τ) and order parameter (S) of membrane lipids. The different effects on the membrane lipids of DPhO and IM – DPhO were suggested in a frame of "structure-activity" approach of drug action.

## EXPERIMENTAL

Balb-line mice available from the "Stolbovaya" nursery station were decapitated and used to obtain membrane preparations. Membranes of endoplasmic reticulum (microsomes) were isolated from liver cells of Balb mice as described in [12] with some modifications. Membrane isolation was produced at t = 0 - +4$^0$C. The steps were following: 1). 2.5 g of liver portion was replaced in 40 ml of medium containing 0.25 mg sucrose, 10mM Tris-HCl and 1mM EDTA (SET); 2). Liver washed and cut very small by scissors, then - in Potter homogenizer. 3). Centrifugation of homogenate for 15 min at 1500 g on centrifuge K-24 (Germany), repeated two times; 4). First supernatant centrifuged for 20 min at 2000g at the same centrifuge, the sediment thrown away (removal of heavy fractions); 5). Second supernatant dissolved in SET was centrifuged for 90 min at 10.5000 g on the centrifuge L-65 "Beckman" (Austria); 6). The last sediment containing microsomes was suspended in SET and poured in epindorphs.

**Table. The parameters of EPR spectra registration of doxyl spin-probes (5-and 16 – DSA) localized in the different lipid regions of microsomal membranes of liver cells**

| Parameters of EPR spectra | 5-DSA | 16-DSA |
|---|---|---|
| Central magnetic field | 3500 Gs | 3370 Gs |
| Assessment of magnetic field | 300 Gs | 200 Gs |
| Gain of magnetic field | $1^x10^5$ | $8{,}0^x10^4$ |
| Modulation of magnetic field | 5.0 Gs | 6.3 Gs |
| Frequency of UHF-radiation | 9,6 GHz | 9,6 GHz |
| Power of UHF-radiation | 20 MW | 20 MW |
| Time constant | 500 MS | 200 MS |
| Time of scanning | 500 s | 500 s |
| Temperature | $20^0$C | $20^0$C |

Protein level determined by Lowry method [13] using bovine serum albumin "Serva" (Germany) for calibration on the automatic spectrophotometer DU-50 "Beckman" (Austria) to measure an absorption. All membrane samples prepared finally contained 4 mg/ml of protein stored in refrigerator at t = -70$^0$C.

DPhO - from "Serva" (Germany) and its iod-methilate derivative IM-DPhO synthesized in the State Research Institute of Organic Chemistry and Technology (Moscow, Russia) were used as effectors.

The initial solution of DPhO – 1mol/l and IM-DPhO – $10^{-2}$ mol/l (maximal solubility) dissolved in ethanol and the concentrations ($10^{-(n+1)}$ M) were prepared by following: 20 μl ($10^{-n}$ mol/l of DPhO or IM-DPhO) + 180 μl of ethanol. These solutions of oxazoles (20 μl) added to the samples – suspension of membranes (200 μl) in a wide range of concentrations of oxazoles from $10^{-2}$ to $10^{-14}$ mol/l. The final acting concentrations of DPhO and IM-DPhO in the membranes were in 100 times less.

The lipid structure of membranes was studied by spin - probe method using two radicals – 5- and 16-doxyl-stearic acids (5-and 16-DSA) from "Sigma" (USA). The radicals prepared in concentration $10^{-2}$ mol/l in ethanol and added to membrane suspension in a final concentration $10^{-4}$ mol/l using Hamilton syringe: 200 μl membrane suspension + 2μl spin-probe, which was incubated for 30 min. The final concentration of ethanol in the membranes was less than 0.6%.

Electron paramagnetic resonance (EPR) – technique by use "Bruker-200D" (Germany) computerized spectrometer with special computer programs has been applied to registration the spectra. A basic EPR theory applicable to study a lipid structure described in [14, 15]. The parameters of registration of EPR-spectra are shown on the table.

A value of lipid microviscosity was estimated by a rotational correlation time (τ) of 16-DSA calculated from the spectra obtained using formula for fast motion range ($6^x10^{-11}<\tau<3^x10^{-9}$ s) of the nitroxyl radicals:

$$\tau = 6.65\, \Delta H_{+1} / [(I_{+1}/I_{-1})^{1/2} - 1]\, ^x10^{-10}\ s,$$

where $I_{+1}$ and $I_{-1}$ – the resonance heights of low- and high- field components of EPR spectra correspondingly, $\Delta H_{+1}$ – the resonance width of low-field component (Figure 1, right side).

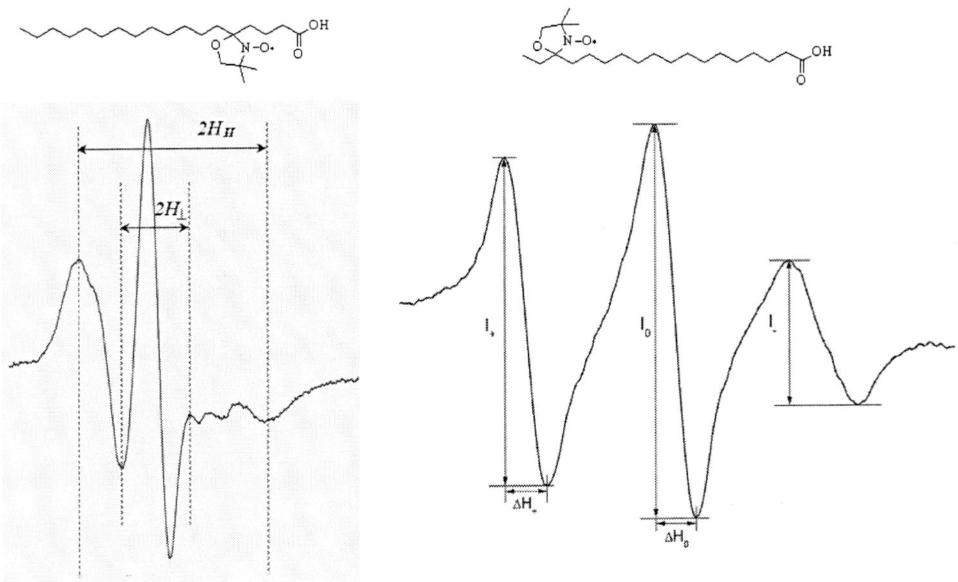


Figure 1. Structural formula of spin probes: 5-DSA (left side) and 16-DSA (right side) and EPR spectra observed in different regions of lipids of microsomal membranes.

An order parameter (S) related with the order degree of long axis of 5-DSA and molecular motion of lipid matrix has been calculated using formula:

$$S = 1.66\ (H_{//} - H_I) / (H_{//} + 2H_I),$$

where $2H_{//}$ – the resonance width between low- and high-field (external) extremes of EPR-spectra and $2H_I$ – between internal extremes (Figure 1, left side).

The values of S and τ were processed by methods of un-parametric statistics with 95% statistic reliability using special computer programs EXEL and STATISTICA. The data shown in the Figures obtained not less than in 3 independent experiments and 3 parallel measurements of membrane samples at each studied concentration of DPhO and IM-DPhO.

## RESULTS AND DISCUSION

The spin probes 5- and 16-DSA are localized in the different regions of membrane lipids: 5-DSA - in surface lipids at depth of ~ $8A^0$ and hydrophobic radical 16-DSA penetrated in lipid bilayer to the depth more than $20A^0$ [14, 16]. A motion of 16-DSA can be characterized by rotation correlation time (τ), by which a microviscosity value is estimated, anisotropic motion of 5-DSA - by order parameter (S).

The first step was to study the time-dependence of τ of 16-DSA after the effect of DPhO or IM-DPhO into the membrane samples. Figure 2 presents the changes of this parameter in the depth of lipid bilayer at the different concentration ($10^{-2}$ – $10^{-11}$ mol/l) of DPhO added to the microsomal membranes as compared with the control value (untreated membranes). It was shown that in spite of some differences the kinetic curves of these time-dependences changed extremely within 30 min and all concentrations of DPhO result in increase of microviscosity for the time period more than 60 min. We have compared these time-dependences with the effect of other biologically active substances on the microsomal membranes *in vitro*, for instance, phorbol ester (12-O-tetra-decanoyl-phorbol-13-acetate - TPA), antioxidants (α-tocopherol, ionol, phenozan) and protein and steroid hormones (insulin, glucagon, guanyl nucleotides) obtained previously [17, 18, 19]. It should be noticed that extremely changes of microviscosity value for first 30 min are the similar to the effect of TPA *in vitro*, but increased constant effect after that time – to antioxidants and hormones.

Maximum increase of τ value at each concentration of DPhO shown in Figure 2 used to obtain the concentration dependence, which is presented in Figure 3 (curve 1). A lipid microviscosity has been found to decrease un-monotonously while concentration of DPhO diminished from $10^{-2}$ M to $10^{-12}$ mol/l. The highest concentration – $10^{-2}$ mol/l is more effective and resuled to 25% increase of of τ 16-DSA, the second maximum (effect about of 15%) observed at $10^{-6}$ – $10^{-7}$ mol/l concentration of DPhO. It should be take into consideration that effective concentrations in membranes are usually in 100 times less than acting ones (added to the membranes).

The time-dependent effect of IM-DPhO was different: τ increased and reached approximately constant value after 15-20 min (data not shown). Otherwise, this kinetics is comparable with the effect of the antioxidants and hormones mentioned above. In this case the constant τ values at each concentration of IM-DPhO were used to obtain the "dose-effect" dependence, which is presented in Figure 3 (curve 2). The most increase of τ 16-DSA about 13% was shown at concentration $10^{-4}$ mol/l of IM-DPhO and the constant value of τ with the effect of 10% observed within concentration range $10^{-5}$ – $10^{-9}$ mol/l. A comparison of the effects of these oxazoles on the lipid microviscosity shows that the effect of DPhO is pronounced at high concentration - $10^{-2}$-$10^{-3}$ mol/l and more marked than IM-DPhO within concentration range $10^{-5}$ – $10^{-7}$ mol/l as well. But the effect of both oxazoles decreased to zero approximately the same concentrations ($10^{-12}$ mol/l –DPhO and $5^{x}10^{-11}$ mol/l – IM-DPhO).

Taking into consideration the cooperative structure of biological membranes 15-25% increase of lipid microviscosity can lead to the significantly changes, for instant, activity of membrane-bound enzymes and receptors *in vivo* and *in vitro* [20-23].

The order parameter – S of spin-probe – 5-DSA localized in the surface membrane lipids allows to estimate a rigidity of these regions. In our experiments both oxazoles shown the similar kinetic effect of S depending on time and reached maximum after 10 min kept it for 40 min of EPR spectra registration time (data not shown). The concentration dependences of this parameter are presented in Figure 4. In general, the quantitative changes of parameter S both *in vitro* and *in vivo* in 10 times less than τ. For example, the change by 1-3% of parameter S is physiologically important for membrane function [25-26].

Figure 4a shows that parameter S diminished from 3.5 % to 0 while concentration of DPhO decreased from $10^{-3}$ mol/l to $10^{-8}$ mol/l. A comparison of "dose-effect" dependences of τ and S on the concentration of DPhO shows their similarity and un-monotonous decreasing effect within the range from $10^{-3}$ to $10^{-8}$ mol/l with intermediate increase at $10^{-6}$ – $10^{-7}$ mol/l. But the structural changes in core (hydrophobic regions) of membrane lipids upon the action of DPhO are more prolonged on the concentration scale.

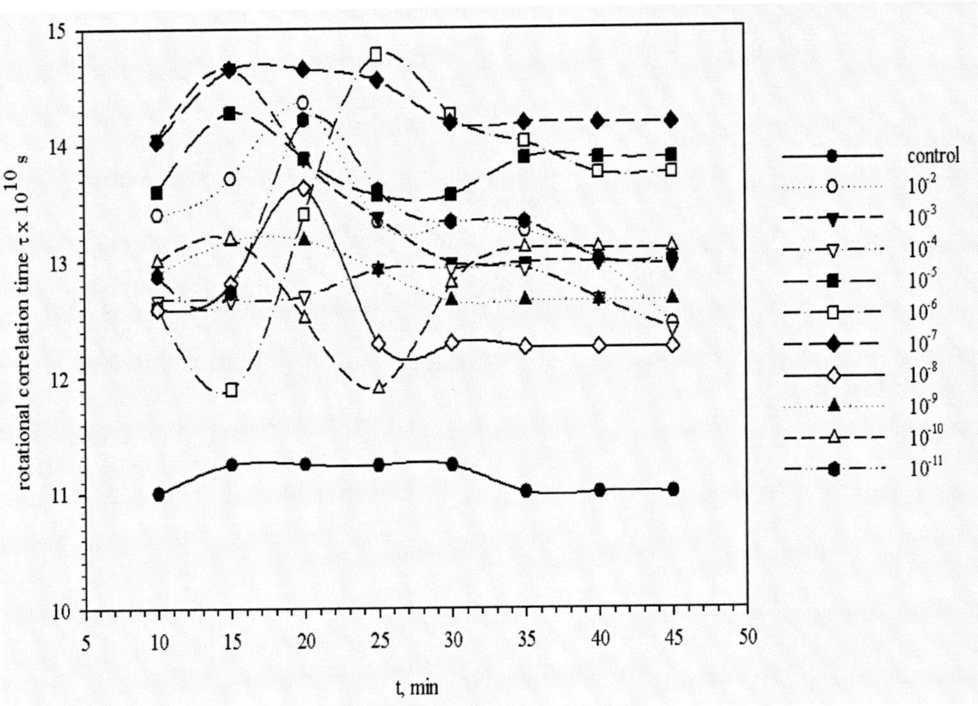


Figure 2. Time-dependences of microviscosity value (rotational correlation time-τ of 16-DSA) in the depth of membrane lipids at different concentration of DPhO ($10^{-2}$ – $10^{-11}$ M) *in vitro*.

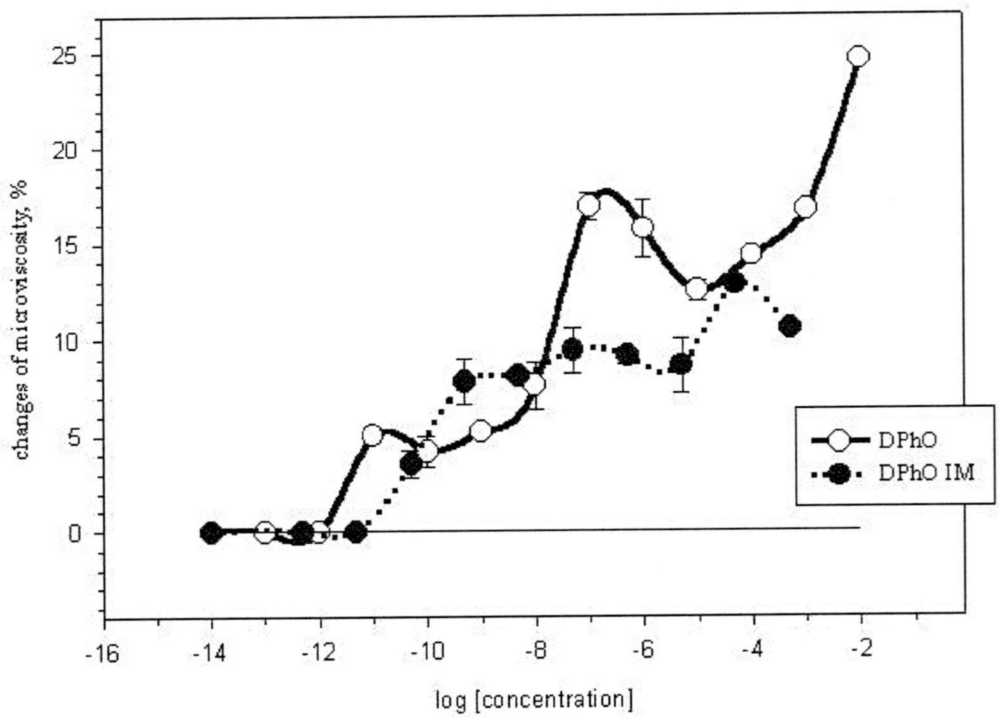


Figure 3. The effect (expressed in %) of the oxazoles on the lipid microviscosity (τ of 16-DSA) of microsomal membranes depending on the concentrations of DPhO (a) and IM-DPhO (b) *in vitro*.

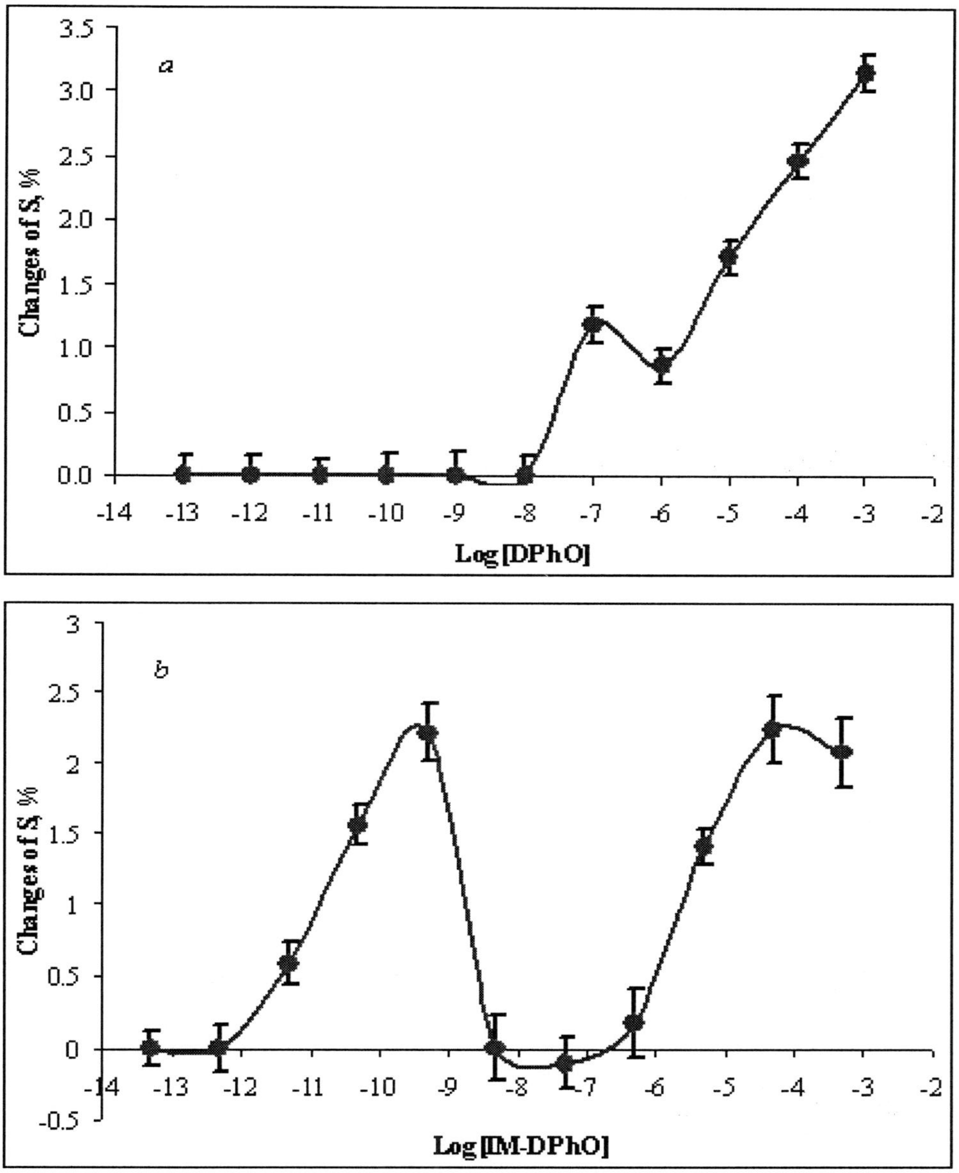


Figure 4. The effect (expressed in %) of the oxazoles on the order parameter (S of 5-DSA) depending on the concentrations of DPhO (a) and IM-DPhO (b) added to microsomal membranes.

Figure 4b illustrates a concentration dependence of the order parameter upon the effect of IM-DPhO. In this case we have obtained a bimodal curve with two maximum at $10^{-4}$ and $10^{-9}$ mol/l, where the effect is the same, and co-called "dead zone" at $5^{x}10^{-7}$ – $5^{x}10^{-9}$ mol/l, within the effect equals to zero. It should be noticed that such kind type of "dose-effect" dependence has been observed for a number of biological active substances acting on the membrane lipids and enzymes, in particular [19, 27]. Usually, the bimodal curve is characterized by second maximum at ultra-low doses - less than $10^{-11}$ – $10^{-12}$ mol/l But, for instance, in the case of low-affinity of the ligand-receptor interaction this bound can be at $10^{-9}$ -$10^{-10}$mol/l [28]. The second maximum of parameter S of the dependence on concentration of

IM-DPhO was observed at $10^{-9}$ mol/l, but as mentioned above the effective concentration of substances in membranes diminished in 100 times, therefore, IM-DPhO can be regarded as surface-active drug possessing maximum effect on the lipid rigidity at high and ultra low-bound concentrations ($10^{-11}$ mol/l).

The results obtained demonstrate that the effect of DPhO and its derivative IM-DPhO on lipid structure is differ in the depth of membrane lipids at concentration range of $10^{-2}$ -$10^{-7}$ mol/l and in the surface lipids at low concentration $10^{-9}$ -$10^{-11}$ mol/l. Thus, in the present paper it is evident that a modification of DPhO by iod-methyl group results in new properties of active oxazole to change the membrane lipid structure, in particular.

Regarding radio-protector properties of these compounds it can be pointed that, first of all, the survival curves of mice $\gamma$-irradiated by different doses following protected by IM-DPO were described by the bimodal type of "dose-effect" dependences [8,9], which can be related to the changes of lipid rigidity (parameter S) obtained by us.. The increase of microviscosity value ($\tau$) at $10^{-7}$ mol/l of DPhO and $10^{-5}$ mol/l of IM-DPhO can be related with this maximum on the survival curves of mice protected by comparable concentrations of these oxasoles *in vivo* at the average doses of irradiation (4 and 8 Gy).

## Conclusions

By spin probes 5- and 16-DSA localized in the different lipid regions of microsomal membranes it was found that DPhO and its iod-methylate derivate (IM-DPhO) increased the microviscosity ($\tau$) in the depth hydrophobic regions and rigidity (S) of surface membrane lipids depending on their concentration by nonlinear manner..

It was shown that IM-DPhO, but not DPhO changed the order parameter (S) of surface membrane lipids by bimodal type with two maxima ($10^{-4}$ and $10^{-9}$ mol/l) and "dead-zone" typically to the effect of substances in ultra-low doses.

It was proposed that the effects of DPhO and IM-DPhO on lipid structure of membranes can be related with the radio-protector properties of these oxazoles observed previously.

## Acknowledgments

This work was performed under Grant N1032 of International Science Technology Center (ISTC).

## References

[1] B. Chapman, A.J. Sutherland, *Chem. Commun.*, Vol.21, pp 84-85. 2003.

[2] S.J.Culliford, P. McCauley, A.J. Sutherland, M. McCair, J. Sutherland, J. Blackburn, R.Z. Kozlowski, *Biochem. Biophys. Res. Commun.*, Vol. 296, N4, pp. 857-863, 2002.

[3] N. DiCesare, J.R. Lakowicz, *Chem. Commun.*, Vol. 19, pp. 2022-2033, 2001.

[4] T. Toyo'oka, H.P. Chokshi, R.S. Givens, R.G. Carlson, S.M. Lunte, T. Kuwana, *Biomed. Chromatogr.*, Vol.4, pp. 208-216, 1993.

[5] E. Mutch, K.W. Woodhouse, F.M. Williams, D. Lambert, O.F. James, M.D. Rawlins, *Xenobiotica,* Vol.7, pp. 599-603, 1985.

[6] J. Kagan, C.P. Kolyvas, J.A. Jaworski, E.D. Kagan, I.A. Kagan, L.H. Zang, *Photochem. Photobiol*, Vol.4, pp. 479-483, 1984.

[7] A.Vogt, K.E. Pestell, B.W. Day, J.S. Lazo, P. Wipf, *Mol. Cancer. Ther.*, Vol.11, pp. 885-892, 2002.

[8] G.P,.Zhizhina, T.M. Zavarykina, E.B. Burlakova, V.G. Noskov, F.S. Dukhovich , *Radiats. Biol. Radioecol.* Vol. 45, N1, pp. 56-62, 2005.

[9] L.N. Shishkina, M.A. Smotryaeva, M.V. Kozlov, E.V. Kushnireva, F.S. Dukhovich, N.G. Noskov, *Radiats. Biol. Radioecol.*, Vol. 45, N5, pp. 610-615, 2005.

[10] E.B. Burlakova, G.V. Archipova, L.N. Shishkina, A.N. Goloshapov, Yu. A. Zaslavsky, *Studia Biophysica*, Vol. 53, pp. 63-65, 1975.

[11] E.B. Burlakova, A.N. Goloshapov "Spin probes in the study of the membranes of normal and tumor cells", pp. 212-225, in the book *"Method of spin probes and labels. Problems and Perspectives"* Ed. by N.M. Emmanuel, Moscow, *Nauka,* 1986.

[12] K.Y. Hostetler, D.B. Zenner, H.P. Morris, *Biochem. Biophys. Acta,* Vol. 441, pp. 231-235, 1976.

[13] O.Lowry, N. Rosebrough, A. Farr, R. Randall, *J. Biol. Chem.*, Vol. 193, pp. 265-269, 1951.

[14] O.H.Griffith, P.C. Jost "Lipid spin labels in biological membranes", pp.489-569, in the book *"Spin labeling theory and application"* Ed. by, L.J. Berliner, New York-San Francisco-London, *Acad. Press,* 1979.

[15] G.I. Likhtenshtein, *Method of spin labels in molecular biology*, Moscow, *Nauka,* 1974, p.255.

[16] A.V. Kuznetsov, "Method of Spin Probe", Moscow, *Nauka*, 1976, 210 pp.

[17] N.P. Palmina, N.G.Bogdanova, E.L. Maltseva, E.I. Pinzar, *Biol. Membr.,* Vol.9, N8, pp.810-820, 1999.

[18] I. Zlatanov, E.L. Maltseva, N.V. Borovok, V. Spassov, *Int.J. Biochem.* Vol.25, N7, pp.971-977, 1993.

[19] V.V. Belov, E.L. Mal'tseva, N.P. Pal'mina, *Biofizika*, Vol.52, N1, pp.75-83, 2007.

[20] M. Hershkowitz, .D Heron, D. Samuel, M. Shinitzky, *Prog. Brain. Res.,* Vol. 56, pp.419-422, 1982.

[21] S. Tsakiris, Naturforsh., Vol. 39, pp.1196-1202, 1984.

[22] E.L. Maltseva, O.S. Belokoneva, S.V. Zaitsev, S.D.Varfolomeev, *Biol. Membr.,*Vol. 8, N8, pp. 830-836, 1991.

[23] E.L. Maltseva, N.V. Borovok, I. Zlatanov, *Membr. Cell. Biol.,* Vol. 9, N6, pp. 621-630, 1996.

[24] P.G. Kury, P.W. Ramwell, H.M. McConnell, *Biochem. Biophys. Res. Commun.,* Vol.57, pp. 726- 730, 1974.

[25] W.H. Huetis, H.M. McConnell, Biochem. *Biohys. Res. Commun.*, Vol.57, pp. 726-730, 1974.

[26] A.S. Polezina, K.A. Anikienko, V.K. Kurochkin, *Russ. Chem J.*, Vol.18, N5, pp. 72-79, 1999.

[27] E.L. Maltseva, N.P. Palmina, E.B. Burlakova, *Membr. Cell. Biol.*,Vol. 12, N.2, pp. 251-258, 1998.
[28] E.B. Burlakova, *Russ. Chem.J.*, Vol.18, N5, pp.3-11, 1999.

In: Chemical and Biochemical Reactions
Editors: G. E. Zaikov, L. Liu et al.
ISBN 978-1-61122-860-1


*Chapter 4*

# The Characterization of Novel Biodegradable Blends Based on Polyhydroxybutyrate: The Role of Water Transport†*

***Yulia N. Pankova*[1], *Alexander N. Shchegolikhin*[2], *Alexey L. Iordanskii*[1], *Anna L. Zhulkina*[1], *Anatoly A. Ol'khov*[3] *and Gennady E. Zaikov*[2,†]**

[1]N.Semenov's Institute of Chemical Physics, Russian Academy of Sciences, Moscow, Russia Federation
[2]N.Emanuel's Institute of Biochemical Physics, Russian Academy of Sciences, Moscow, Russia Federation
[3]M.Lomonosov's Academy of Fine Chemical Technology, Moscow, Russia Federation

## Abstract

The present paper focuses on the study of novel blends based on poly(3-hydroxybutyrate) (PHB) and polymers with different hydrophilicity, namely low density polyethylene and polyamide (LDPE and PA), respectively. Polymer blends were produced from five ratios of PHB/LDPE in an effort to regulate the resistance to hydrolysis or (bio)degradation through the control of water permeability. The relation between the water transport and morphology (TEM data) shows the impact of polymer component ratio on the water flux regulation in the hydrophobic matrix. To elucidate the role of hydrophilicity of the second component present in the PHB-based blends, we studied the PHB/PA blends where PA is the polyamide resin composed of statistical copolymer of hexamethyleneadipinate and ε-caprolactam in the ratio 1:1. The use of

* Dedicated to Professor Victor Lobo on the occasion of his 70th birthday.

† Corresponding author: Prof. Gennady Zaikov. N. M. Emanuel Institute of Biochemical Physics, Russian Academy of Sciences, 4 Kosygin str., Moscow 119334, Russia. Phone: +7 (495) 939-73-20; Fax: +7 (495) 137-41-01; E-mail: mailto:chembio@sky.chph.ras.ru

techniques such as DSC and FTIR-imaging (for T-scale) demonstrates the interaction between PHB and PA in the temperature ranges of crystallization and melting. The general approach based on Flory-Huggins equation is presented as the way to choose the pairs of compatible or partly compatible polymers. Blending PHB with PA could be a simple and effective method to design new matrices for drug delivery in medicine, while the same procedure applied for PHB-LDPE system with better resistance against hydrolysis and lower price than the initial PHB, could be used as novel bioerodible packaging materials.

**Keywords:** polyhydroxybutyrate (PHB); water permeability; kinetics; diffusion; blends; biodegradation

## 1. INTRODUCTION

The families of biodegradable polymers, namely poly-α-hydroxyacides (PLA, PGA, and their copolymers, PLGA) as well as poly-β-hydroxyalkanoates (PHA): poly(3-hydroxybutyrate) (PHB) and its copolymers, especially with hydroxyvaleriate (PHBV), have very good performance characteristics. Their characteristics are sufficient to replace a significant portion of petroleum-based plastics used currently in medicine, packaging and food-service industry [1-3]. Along with biodegradability, the PHAs and their principal member PHB are biocompatible and environmental friendly [4-6]. In living organism and soil their macromolecules decompose to form the nontoxic intermediates and final products included in carbohydrate metabolism. Besides, bacterial PHB and its copolymers have also proved to be useful as model materials for investigating physical properties such as crystallization and diffusion.

However, the mechanical deficiency, a narrow processing window as well as the high cost of PHB/PHA, restricts its application as a friendly environmental biomedical material. For the solution of the above-mentioned three problems, we propose the specific blending of PHB, as a pivotal component, and a set of synthetic (LDPE and polyamide-66) polymers, with a different macromolecule polarity to design the (bio)degradable blends/compositions with improved transport and physical characteristics, which are useful for the development of packaging, biodegradation [7], and drug delivery [8].

## 2. EXPERIMENTAL

In the current study, we used PHB in the powder form which was kindly supplied by Biomer Co Krailing (Germany) Lot #216 with Mw = 206 kDa, density 1.248 g/cm$^3$. The blend films were produced using a) the low density polyethylene (LDPE), namely the commercial product (RF 15803-020 grade) in the form of granules with $M_\eta = 2.0 \times 10^5$ Da; and b) the polyamide resin composed of statistical copolymer of hexamethyleneadipinate and ε-caprolactam in the ratio 1:1 with $M_n = (1.94 \pm 0.06) \times 10^4$ Da and containing 38% of amide groups, with a content of terminal groups equals $(5.0 \pm 0.1) \times 10^{-2}$ mol/kg for —$NH_2$ groups and $(5.3 \pm 0.1) \times 10^{-2}$ mol/kg for —COOH groups. The blends were prepared by solvent casting from chloroform-ethanol solvent combination (both solvents of medical grade quality) and by melt/solid extrusion with a single-screw extruder. A series of LDPE–PHB blends was prepared with the following component ratios (w/w): 68:32, 84:16, 92:8, 96:4, and 98:2.

FTIR spectra of the blend films were recorded with the spectrometer "Bruker IFS-48" (Germany) at resolution 2 $cm^{-1}$ being the number of the scans about 70. Computer analysis of final spectra has been performed with the in-built programs in the range 500 – 3500 $cm^{-1}$. A Perkin-Elmer differential scanning calorimeter (Model DSC7), calibrated with indium, was applied for the study of the thermal behavior of PHB-PA blends, as well as for pure polymers (PHB and PA) separately. Specimens of about 5 mg were encapsulated in aluminium pans and heated/cooled at a constant rate 10 °C/min. The range of heating includes a temperature interval from room temperature to 182 °C. The enthalpies of fusion $\Delta H_{PHB}$ for PHB and $\Delta H_{PA}$ for PA were measured to calculate the crystallinity degrees $\alpha_{PHB}$ and $\alpha_{PA}$ using the following equations

$$\alpha_{PHB}\,(\%) = [\Delta H_{PHB}/\Delta H^{o}_{PHB}(1/w_{PHB})]\ 100 \quad (1)$$

$$\alpha_{PA}\,(\%) = [\Delta H_{PA}/\Delta H^{o}_{PA}(1/w_{PHA})]\ 100 \quad (2)$$

where $w_i$ (i=PHB or PA) is the weight fraction of each component in the blend, and $\Delta H^{o}_{PHB}$ and $\Delta H^{o}_{PA}$ are the enthalpies of fusion of completely crystalline polymers PHB (87 cal/g) and PA(54 cal/g), respectively [9].

# 3. Results and Discussion

## 3.1. General Approach in Framework of Flory Huggins Theory

The analysis of water solubilities in accessible intercrystalline area for about 50 polymers with different chemical structures (hydrophilicity) indicates that the classification of such polymers can be performed by a combination of the modified Flory-Huggins equation (at aw = 1) with an expression for the solubility parameter in the framework of the regular solution model [10] (eqs. 3 and 4, respectively)

$$\ln\varphi_w + \varphi_p + \chi_{wp}\varphi_p^2 = 0 \quad (3)$$

$$\chi_{WP} = \frac{\bar{V}_W}{RT}(\delta_W^2 - \delta_P^2) \quad (4)$$

Inserting expression of $\chi_{WP}$ in equation 3, the following final equation is achieved:

$$\ln\varphi_w + \varphi_p + \frac{\bar{V}_W}{RT}(\delta_W^2 - \delta_P^2)\varphi_p^2 = 0 \quad (5)$$

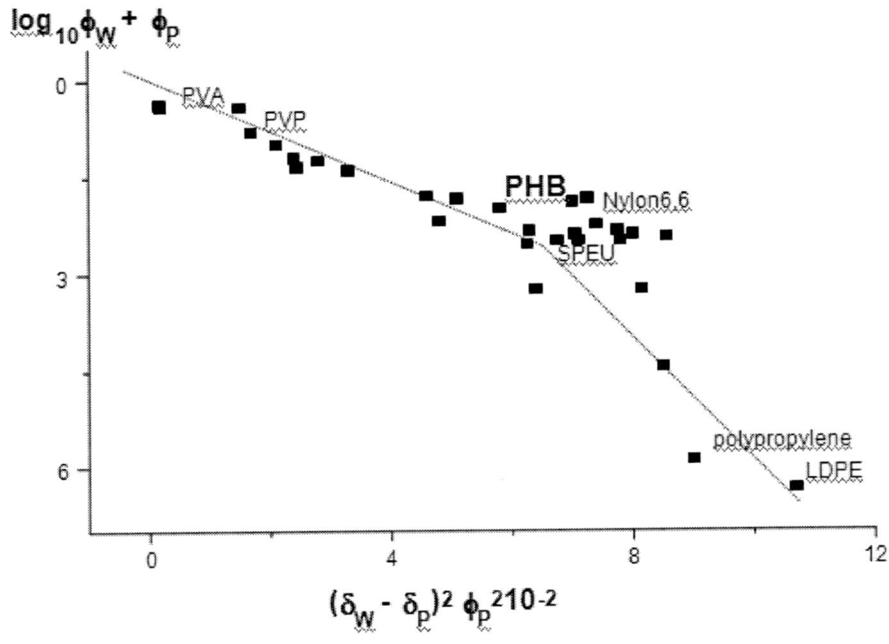


Figure 1. The pattern of water solubility in polymers with different hydrophilicity [11] in framework of Flory-Huggins equations.

where $\varphi_w$, $\varphi_p$ are the volume fractions of water and polymer, respectively, and $\delta_i$ are the solubility parameters of water (W) and polymer (P), $\chi_{WP}$ is the Flory interaction parameter, $\bar{V}_W$ is the molar volume of water, R is the gas constant and T is the absolute temperature.

The plot of Flory-Huggins equation (at water activity, aw = 1) using coordinates ($\ln\varphi_w + \varphi_p$) vs. $(\delta_W^2 - \delta_P^2)\,\varphi_p^2 \equiv Z$ is generally nonlinear and can be approximated by two intersectional lines in intervals: $Z \leq 600$ J/cm$^3$ for the hydrophilic polymers (poly(vinyl alcohol) - PVA, cellulose, chitosan and the like), and $Z \geq 800$ J/cm3 for the hydrophobic polymers (as for example, LDPE, high density polyethylene - HDPE, polypropylene, polystyrene and rubbers), see Figure 1. The points located in the vicinity of the intersection point of the above lines, at $600 < Z < 800$ J/cm3, belong to the moderately hydrophilic polymers [11].

In Figure 1 the water equilibrium sorption in the above systems of coordinates is presented. Two polymers (PHB and PHBV) from the family of polyhydroxyalkanoates with Hildebrand solubility parameters $\delta_p = 19.8$ and 19.4 J$^{1/2}$/cm $^{3/2}$ (Pa$^{1/2}$) [12], respectively, are situated in the vicinity of the intersection point. Data related to 55 polymers with different hydrophilicity have been collected by our group [13]. Variation of second polymer polarity in its blend/composite with the PHB (PHBV) regulates composite morphologies and, hence, both water and drug transport as well as thermophysical and mechanical behavior of polymer systems. Here we initiate our consideration with the description of an utmost situation where the PHB has been combined with typical hydrophobic polymer - polyethylene (LDPE).

Analyzing the micrographs obtained in a preliminary paper of our group [14], we can distinctly discern the porous space between continuous medium of the major phase of LDPE and the band-like fibrils of PHB. These materials have a microscopically complex structure with random distribution of pore sizes, ranging from a few micrometers to tens of micrometers. In packaging or pharmaceutical practice, in order to regulate the water permeability, the structure is slightly affected by an external force aroused during a film preparation or exploitation. Thus, controlling optimal values of the composite parameters such as porosity, permeability and hydrophilicity, are important issues in polyhydroxybutyrate -polyethylene research.

### 3.2. Transport Behavior of Water in Blend of PHB with Hydrophobic Polymer (PELD)

The total water permeability of the LDPE/PHB blends ($P_W$) can be described by the equation:

$$P_W = \alpha_S^{PE} F_{PE} P_W^{PE} + \alpha_S^{PHB} F_{PHB} P_W^{PHB} + \alpha_S^{Int} P_W^{H} \tag{6}$$

where $P_W^{PE}$, $P_W^{PHB}$ and $P_W^{H}$ are the water permeabilities in the phases of LDPE and PHB, and through the interphase, respectively; $\alpha_S^{PE}$, $\alpha_S^{PHB}$ and $\alpha_S^{int}$ ($\alpha_S^{int} = 1 - \alpha_S^{PE} - \alpha_S^{PHB}$) are the fractions of the film cross section possessed by the phase of each component and interphase; $F_{PE}$ and $F_{PHB}$ are the functions characterizing the orientation impact of the polymeric segments on the water permeability in the phases of LDPE and PHB respectively. In accordance with scanning electron microscopy results [14], the porous structure in the interphase is formed and, hence, $P_W^{Int}$ is equal to the hydraulic permeability [15]. Here, the porous medium is modeled as a continuum, composite material and its structure is characterized with a minimal set of properties, namely the porosity ($\alpha_S^{Int}$) and the hydraulic permeability ($P_W^{H}$), which might be functions of position (architecture) and structural element orientation (anisotropy) $F_{PE}$, $F_{PHB}$. From the local volume balance it follows that

$$\alpha_S^{PE} + \alpha_S^{PHB} + \alpha_S^{int} = 1 \tag{7}$$

The final term on the right hand side of Eq. (6) reflects the hydraulic movement of water through porous area, usually referred to as Darcy term [16]:

$$J_W^{Da} = -K_{Da} \frac{1}{\alpha_S \rho_w} \frac{d\mu}{dx} \tag{8}$$

where $J_W^{Da}$ is the water flux into a pore-containing area ($m^3$ [water] $m^{-2}$ [pore cross-sectional surface area] $s^{-1}$), $K_{Da}$ is the hydraulic Darcy constant ($m\ s^{-1}$) and $d\mu/dx$ is a water potential gradient ($m\ s^{-2}$) expressed through the water activity or the pressure of water vapor. The expression for water permeabilities through LDPE and PHB matrices has a classical form

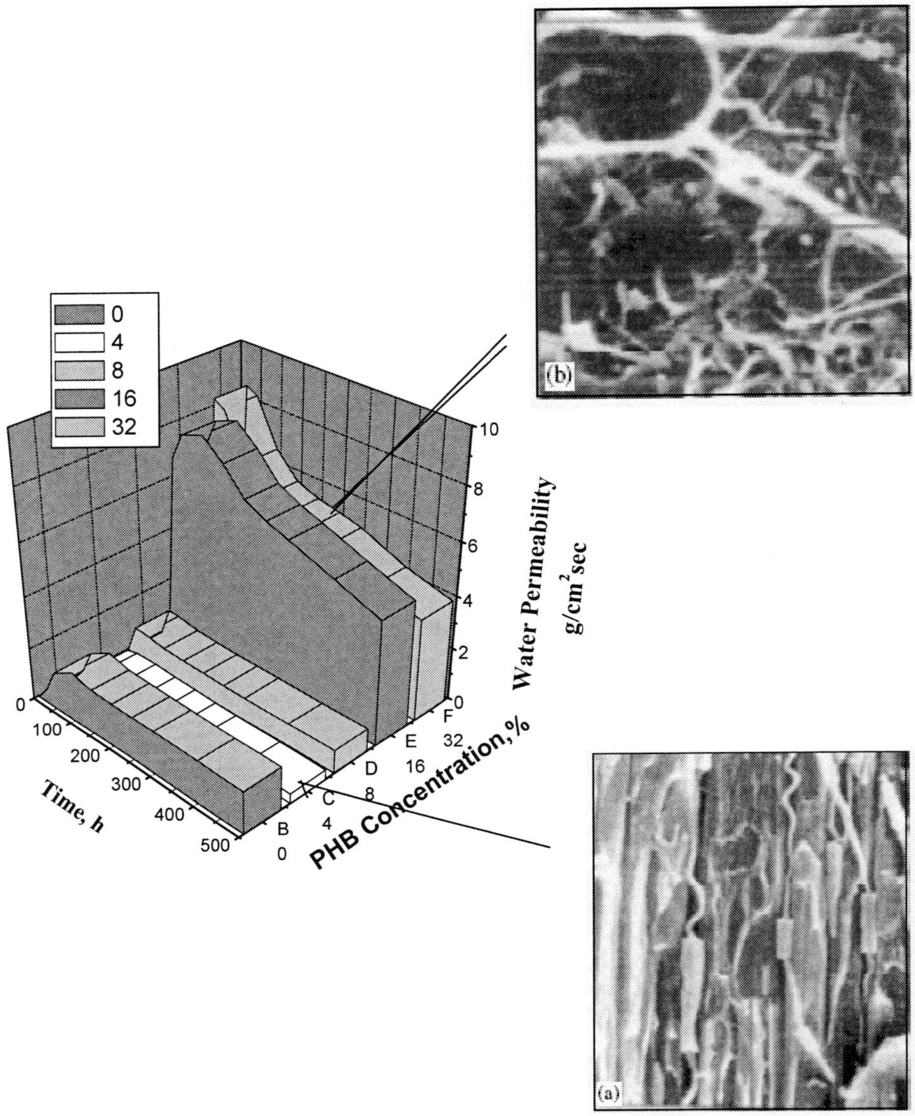


Figure 2. 3D-pattern of water permeability through the blend films prepared on the base of the different polymer fractions (shown on the top, wt%). The photo of anisotropic (a) and isotropic (b) morphologies of the blends. The lines show the correspondence between kinetics and TEM data (x3300).

$$J_W^i = -P_W^i \frac{dC_W}{dx} \tag{9}$$

where i - index used for PHB or LDPE, and $dC_W/dx$ is the gradient of water concentration. Figure 2 shows the total course of water permeability (fluxes) through the blend films as a function of PHB content. The sharp permeability jump from the matrices having the anisotropical orientation of the PHB bands in LDPE to matrices where the PHB fibrils formed the isotropic network is illustrated by corresponding photos [14] reflecting the two different types of morphology.

At the relatively low PHB concentrations (4-8 wt %), the water flux resistance is maximal and exceeds the resistance of the initial LDPE film, see Figure 3. Then, the resistance decreases with increasing PHB concentration. It is significant that those blends have both the same IR polarization ratio.

Note that in this case mechanical tensile strength dependence on the PHB concentration with minimum is situated at the same PHB/PE fraction, about 4- 8 wt % (here not presented, see [14]). It is apparent from the above data that the transport behavior of water in the blended films is substantially affected by the orientation of the polymer segments ($F_i$) in LDPE continuum phase and by the defects/pores formed in interphase areas ($\alpha_S^{Int}$ $P_W^{Int}$ in Eq.3) if one suggests that tensile strength was dominantly determined by defects (pores) of the heterogeneous matrix.

Analysis of extreme points for “transport coefficient - blend content” relationships combined with the morphology features controlled by FTIR and TEM techniques, and thermophysical testing (DSC), enables us to elucidate the impact of crystallinity and second component nature (macromolecular polarity) upon the mechanism of both controlled release under (bio)degradation.

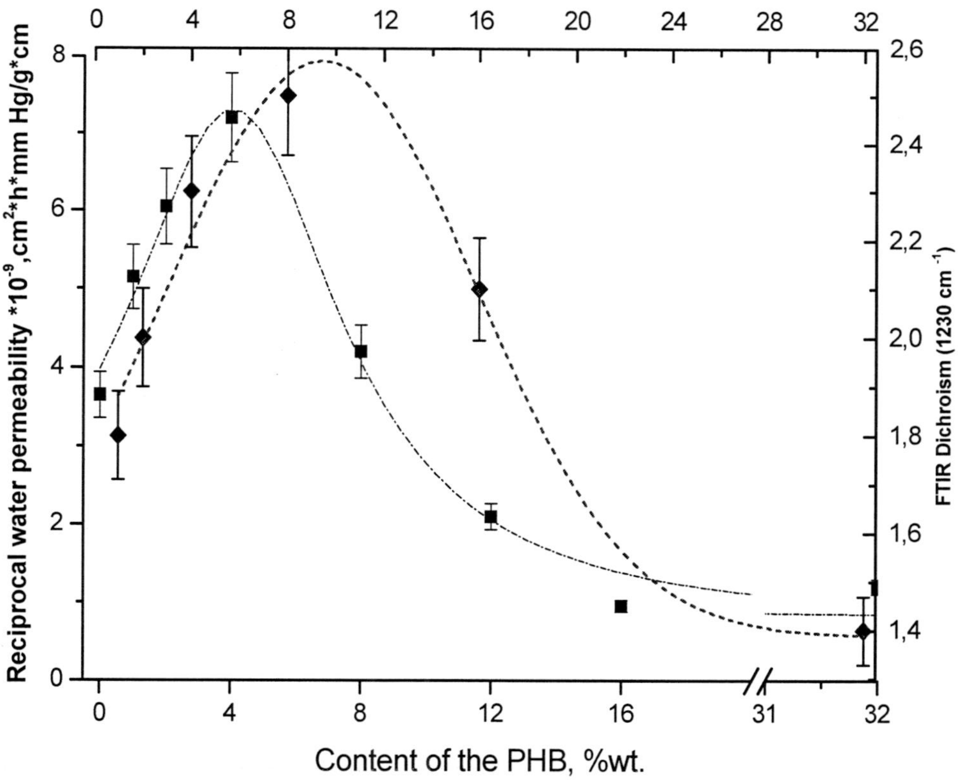


Figure 3. Comparison of reciprocal permeability of water (■), FTIR dichroism (♦), as functions of PHB content in PHB-LDPE composite films.

### *3.3. Phase Transitions in the Blend of PHB* with a Moderately Hydrophilic Polymer (PA)

For the preparation of a novel binary polymer composition where both components have moderate hydrophilicity (macromolecule polarity) we have selected poly(3-hydroxybutyrate) - PHB - and copolymer of polyamides PA54C characterized in the ref. [17] and hereafter designated by PA. One of the important reasons for such selection was the proximity of their melting points 175.7 °C (PHB) and 168.3 °C (PA), and hence the processing of their blends via the melt became much more efficacious. The other reason is the elasticity of PA, as low crystalline rubber-like copolymer (crystallinity degree ≈ 20 %). Therefore, we expected that addition of PA to high-crystalline PHB (with crystallinity degree ≈ 75 %) should decrease the crystallinity and improve viscoelasticity of the pure PHB.

The assay of repeated cycles 'melting-crystallization' provided by DSC for PHB-PA composition testifies a PHB crystallinity decrease as the result of their blending. The crystallinity degree was calculated through enthalpy melting on measuring endothermic peak areas, as shown in the Experimental section. Under blending with moderately hydrophilic polymer (PA), the crystallinity decline resembles qualitatively the course of decrease in crystallinity for the system PHB-PVA [18]. It is important to underline that the location of melting point for PHB on DSC curves displaced to low temperature area from the value 175.7 °C for pure PHB to 168.5 °C for PHB in the blend at 87.5% PA.

On the basis of FTIR spectroscopy data, we have shown recently [17] that in blend films of PHB-PA, hydrogen bonds between carbonyl group of PHB and amide one of PA are formed. Taking into account the accessibility of interacting groups in PHB and PA, the maximal interaction estimated as the shift of location for band maximum of carbonyl group happens at mole ratio PHB/PA = 1. In the context of FTIR data, we can suggest that in a chilled-down polymer composition the impact of the second component (PA) upon immobilization of PHB kinetic segments causes the decrease in their mobility and, hence, brings the drop of PHB segment fraction involved into crystallization.

The next step of blend crystallinity transfer study embraces the analysis of cooling thermograms from the melts PHB-PA. For all compositions (besides the blend with ratio PHB:PA = 1:9) there are two peaks with appropriate resolution on the temperature scale. Each peak belongs to the successive crystallization of individual components in the blend. It is interesting to note that the low-temperature maximum reflects the crystallization of the PHB melted at ~175 °C, while the high-temperature peak does the crystallization of the PA melted at lower temperature, about 165 °C. Consequently, the supercooling degree (designated as the difference in melting and crystallization points) is much higher for the high-crystalline PHB, relatively to PA.

For overcooled melts of the PHB blends, the addition of the second component gradually decreases the temperature of crystallization as a result of a solid crystalline phase formation in PA. The solidification of PA (its crystalline part) produces steric and kinetic hindrances effecting the PHB crystallization. Along with that, as mentioned above, blending PHB with PA decreases the melting point of the former. Generally, the locations of endothermic and exothermic peaks move on the temperature scale in such a way that the supercooling degree keeps nearly constant. In a number of works [e.g. 18] it has been shown that both the rate of crystallinity and the rate of heterogeneous nucleation increase with the value of supercooling degree ($\Delta T_{CM} = T_M - T_c$). Therefore, it is quite possible to assume that the formation of the crystalline phase in PHB must perform with constant rate, in such a way that the rate does not depend on the PHB-PA composition.

Acceleration or retardancy of segmental mobility for the blended components, reflecting the melting or crystallization, respectively, can be discovered by FTIR spectra obtained as the step-like series of temperatures. Figures 4 and 5 show a set of spectra recorded for heated and cooled samples of PHB-PA system, respectively.

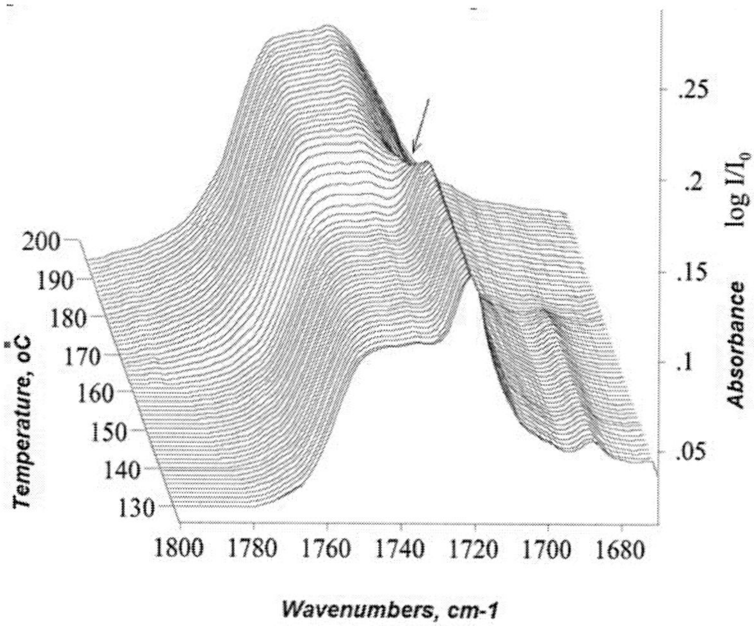


Figure 4. FTIR-imaging for melting in the blend PHA-PA in the area of C=O group vibration.

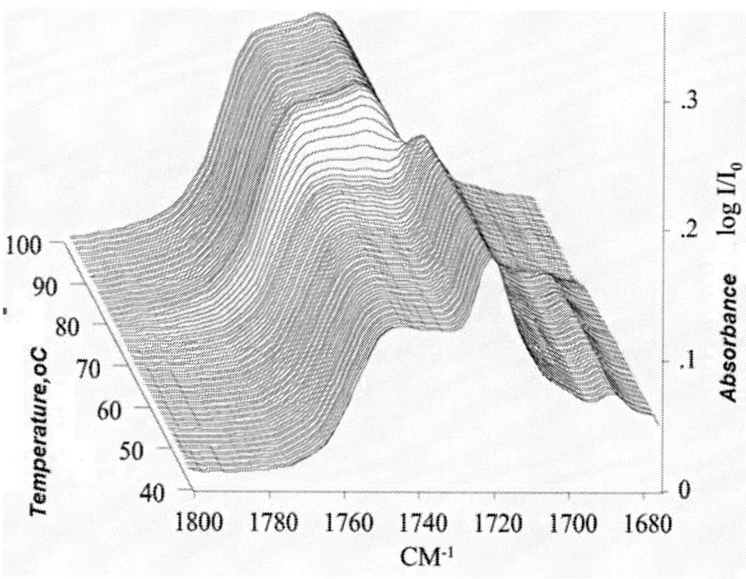


Figure 5. FTIR-imaging for crystallization in the blend PHA-PA in the area of C=O group vibration.

Here we show that stretching vibration bands of the carbonyl group of the ester entity of PHB are very sensitive to "defrosting" in the mobility of polymer molecules. Such jump of segment mobility is the result of PHB crystal fusion. A low-frequency shoulder of the multicomposed band of >C=O group vibration (at 1724 $cm^{-1}$) disappears in the vicinity of the melting temperature of PHB in the PHB/PA (1:1) blend and, for the blend melt above 172 $^{o}C$, there is only one diffuse band with the maximum at 1740 $cm^{-1}$. The extended maximum in this area most likely reflects the existence of a wide distribution of mobile conformers responsible for the vibration of the ester groups in the melt. We emphasize that temperature conforming to mobility deceleration (FTIR data) is in good agreement with melting point (DSC data) for PHB in the blend.

The parallel temperature series has been performed under cooling of the melt blend (Figure 5). The successive record of the FTIR spectra, at different temperatures, clearly demonstrates the transfer from the liquid state in the melt to crystalline state. In the vicinity of 65 $^{o}C$, this process is characterized by the sharp appearance of low-frequency mode related with immobilized carbonyl groups. Under blend cooling we observe also the coincidence of this temperature with the temperature interval of phase transition, measured by a DSC procedure. Hence, both methods, independently applied to the polymer blend systems, show a first-kind phase transition - crystallization of PHB. It is important that both thermophysical results being related with a heat effects and spectroscopic data being related with dynamic of a bond vibration are in agreement with each other on the temperature scale. Analogous result has been obtained during the study of melting for the same polymer blend.

Summarizing the section devoted to PHB and moderately hydrophilic PA blends, we have shown that phase transition occurs as a thermo-induced process when each polymer component has the proper structure evolution and both components interact each other. The displacements of endothermal ($T_m$) and exothermal ($T_c$) peaks on the temperature scale as a function of blend composition give also evidences of intermolecular interactions between the polymers forming the novel blend composition. Combination of thermophysical behavior and FTIR-imaging could help us to design not only a novel bioerodible barrier material, but a novel therapeutic system for controlled drug delivery.

## 4. Conclusions

We propose a simplified approach (relative to the 3D Hansen model [19]) for the preliminary selection of polymer components to produce novel biodegradable blends. The approach is based on modified Flory-Huggins equation and uses the equilibrium water sorption in polymers with different segmental polarity (hydrophilicity). As an example of an immiscible pair forming the blend, we studied the PHB/LDPE composition. The morphological features of the blends demonstrate the self-reinforced matrix where the minor component (PHB) forms the like-band fibrils embedded in the LDPE matrix. Above 16 % of PHB, the blend system undergoes the morphology transfer from oriented/anisotropic PHB structure to isotropic one where the PHB fibrils are transformed into a network. The distinction between two morphologies reflects different values of water permeability through the blend films.

To elucidate the role of hydrophilicity of other component present in the PHB blends, we studied the PHB/PA blends where PA is the polyamide resin composed of statistical copolymer of hexamethyleneadipinate and ε-caprolactam in the ratio 1:1. Several techniques, including DCS and FTIR-imaging (for T-scale), show the interaction between PHB and PA in the temperature ranges of crystallization and melting. The general approach based on Flory-Huggins equation is presented as the way to choose the pairs of compatible, noncompatible or partly compatible polymers in blends. Blending PHB with PA could be an effective method to design new matrices for drug delivery in medicine, while the same procedure applied for PHB-LDPE system with better resistance again hydrolysis and lower cost than the initial PHB could be used as novel bioerodible packaging materials.

## Acknowledgments

This work was financially supported by a special grant from the Presidium of the Russian Academy of Sciences (2010 (21P) "Academic Sciences to Medicine" and partially by RFBR grant 08-03-00929. The authors would like also to thank the Biomer Co (Krailing, Germany) and personally Dr. U.Hanggi for its valuable supply of PHB.

## References

[1] G.Q. Chen, *A Polyhydroxyalkanoates Based Bio- and Materials Industry*, Chem. Soc. Rev. 38 (2009), 2434-2446.

[2] V. Siracusa, P. Rocculi, S. Romani, M. Dalla Rosa, *Biodegradable polymers for food packaging: a review,* Trends in Food Science & Technology, Vol. 19, Issue 12, pp. 634-643.

[3] B.P. Mooney, *The second green revolution? Production of plant-based biodegradable plastics.* Biochem. J. 418 (2009) 219-232.

[4] R.A. Gross, B. Kalra, *Biodegradable Polymers for the Environment*, Science, 297 (2002) 803-806.

[5] Z.W. Zhong, B. Song, C.X. Huang, *Environmental impacts of three polyhydroxyalkanoate (PHA) manufacturing processes*, Mater. Manuf. Process. 24 (5) (2009) 519-523.

[6] M. Akiyama, T. Tsuge, Y. Doi, *Environmental life cycle comparison of polyhydroxyalkanoates produced from renewable carbon resources by bacterial fermentation*, Polym. Degrad. Stab. 80 (2003) 183-194.

[7] E. Chiellini, A. Corti, S. D'Antone, R. Solaro, *Biodegradation of poly(vinyl alcohol) based materials*, Prog. Polym. Sci. 28 (2003) 963-1014.

[8] R. Gabathuler, *Approaches to transport therapeutic drugs across the blood-brain barrier to treat brain diseases*, Neurobiol. Dis. 37 (1) (2010) 48-57.

[9] A. Gonzalez-Montiel, H. Keskkula, D.R. Paul, *Morphology of nylon6/polypropylene blends compatibilized with maleated polypropylene*, J. Polym. Sci.: Pt. B: Polym. Phys. 33 (1995) 1751-1767.

[10] Y. Hu, R.H. Hurt, *Thermodynamics of Carbonaceous Mesophase II: General Theory for Nonideal Solutions,* Carbon 39 (2001) 887-896.

[11] A.L. Iordanskii, *Impact of structure and morphology upon water transport in polymers with moderate hydrophilicity. From traditional to novel environmentally friendly polymers.* Nova Science Publishers, Inc, Hauppauge, 2003.

[12] M. Terada, R.H. Marchessault, *Determination of solubility parameters for poly(3-hydroxyalkanoates)*, Int. J. Biol. Macromol. 25 (1-3) (1999) 207-215.

[13] A.L. Iordanskii, T.E. Rudakova, G.E. Zaikov, *Interaction of Polymers with Bioactive and Corrosive Media*, VSP, Tokyo, 1994, p. 298.

[14] A.A. Ol'khov, C.V. Vlasov, L.S. Shibryaeva, I.A. Litvinov, N.A. Tarasova, R.Y. Kosenko, A.L. Iordanskii, *Structural features of LDPE-poly(3-hydroxybutyrate) blends, Polymer Science*, Series A, 42 (4) (2000) 447-452.

[15] C.G. Williams, J.V. Macpherson, P.R. Unwin, C. Parkinson, *Laser scanning confocal microscopy coupled with hydraulic permeability measurements for elucidating fluid flow across porous materials: application to human dentine,* Anal. Sci. 24 (4) (2008) 437-442.

[16] S. Furmaniak, A.P. Terzyk, P.A. Gauden, G. Rychlicki, *Applicability of the generalised D'Arcy and Watt model to description of water sorption on pineapple and other foodstuffs,* J. Food Eng. 79 (2) (2007) 718-723.

[17] Y.N. Pankova, Ph.D Dissertation, Moscow State Academy of Fine Chemical Technology, 2007.

[18] A.A. Olkhov, S.V. Vlasov, A.L. Iordanskii, G.E. Zaikov, V.M.M. Lobo, Water transport, structure features and mechanical behavior of biodegradable PHB/PVA blends. J. Appl. Polym. Sci. 90 (6) (2003) 1471-1476.

[19] C.M. Hansen, Hansen Solubility Parameters: A User's Handbook. Second edn. CRC Press, Hoersholm, 2007.

In: Chemical and Biochemical Reactions
Editors: G. E. Zaikov, L. Liu et al.
ISBN 978-1-61122-860-1


*Chapter 5*

# EFFECT OF NEW HYBRID ANTIOXIDANTS—ICHPHANS—ON THE SURFACE ARCHITECTONICS OF ERYTHROCYTES

***E. Yu. Parshina*[1], *L. Ya. Gendel'*[2] *and A. B. Rubin*[1]**
[1]Biological Faculty, Moscow State University, Moscow, 119899 Russia
[2]Emanuel Institute of Biochemical Physics, Russian Academy of Sciences, Moscow, 117977 Russia

## ABSTRACT

Scanning electron microscopy was used to study how the surface architectonics of erythrocytes is affected by new-generation drugs, Ichphans, which have a combined antioxidant and anticholinesterase effect. It was found that intercalation of each of the studied compounds, which have a positive charge on a quaternary nitrogen atom, in the erythrocyte membrane and its distribution in the intramembrane space are accompanied by manifestations of the echinocytogenic, stomatocytogenic, and compensatory effect on the erythrocyte shape. Time-dependent morphological transformations of erythrocytes are likely to be caused by changes in the preferential distribution of the compounds in the outer, inner, or both monolayers of the erythrocyte membrane, respectively. Differences in the transforming effect of the drugs with different hydrophobic properties were revealed.

**Keywords:** hybrid compounds, synthetic antioxidants, anticholinesterase drugs, membrane transport, erythrocyte morphology

## INTRODUCTION

One of the lines of search for new efficient drugs is to create drugs producing a combined effect.

Recently, the biological activity of Ichphans, a new class of hybrid compounds containing groups with antioxidant and anticholinesterase effects, has been intensely studied [1]. The original chemical design of Ichphans makes them promising drugs for treating Alzheimer's disease and other age-related pathologies [2-4].

Previously [1-4], a high antioxidant activity of Ichphans was revealed, their inhibitory effect on soluble and membrane-bound human erythrocyte acetylcholine esterase was established, and differences in efficiency were shown between their derivatives containing nitrogen-bound alkyl substituents differing in hydrophobic properties.

The erythrocyte shape and its changes are known to be among the most important indices of the functional state of these cells [5, 6]. Xenobiotics, in particular, synthetic antioxidants, can affect the erythrocyte morphology [7]. According to the concepts developed by Sheetz and Singer [8-10], the changes in the erythrocyte shape that are induced by membrane-intercalating chemical agents characterize specific features of the distribution of these agents in the intramembrane space. Such information is important to obtain to understand the mechanisms of action of biologically active substances.

In this context, the purpose of this work was to perform a scanning electron microscopy study of the effect of Ichphan derivatives differing in hydrophobic properties on the erythrocyte morphology.

Ichphans

| Structural formula | R | Notation |
|---|---|---|
| 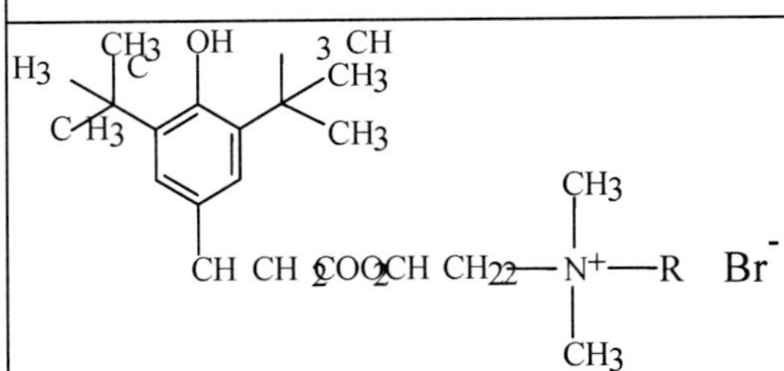 | $-C_{10}H_{21}$ | I(C-10) |
| | $-C_{12}H_{25}$ | I(C-12) |
| | $-C_{16}H_{33}$ | I(C-16) |

## EXPERIMENTAL

The object of investigation was erythrocytes from the blood of mongrel albino rats (with heparin as an anticoagulant). Erythrocytes were isolated according to a published procedure [7]. In the experiments, erythrocytes were suspended in buffer A (145 mM NaCl, 5 mM KCl, 4 mM $Na_2HPO_4$, 1 mM $NaH_2PO_4$, 1 mM $MgSO_4$, 1 mM $CaCl_2$, 10 mM glucose, pH 7.4). The cell concentration in the suspension was (8-10)$\times 10^6$ cells per microliter. The erythrocyte mass obtained was stored at 4°C and used within 8 h.

In this work, we used Ichphan derivatives containing different aliphatic substituents at the quaternary nitrogen atom, which were synthesized at Institute of Biochemical Physics, Russian Academy of Sciences, Moscow, Russia [11]. The table presents their structural formulas. Ichphans were used as ethanol solutions, the ethanol concentration in samples did not exceed 0.8%.

In studying the effect of Ichphans on the erythrocyte morphology, the cells were incubated for different times from 2 to 120 min (the incubation time is indicated in the figure captions) at (18 ± 2)°C with a certain Ichphan derivative. Samples incubated under similar conditions in the absence of the compound under investigation served as reference samples.

The samples were fixed in 1% solution of glutaraldehyde (Serva, Germany) in buffer A for 3 h, and then the cells were washed according to published recommendations [6].

The samples were coated with a platinum-palladium layer in an EICO IB-3 ion coater. Microscopic examination was carried out with a CamScan scanning electron microscope (UK). The magnifications are indicated in the figure captions.

Statistical processing was performed using the Student's-test.

## Results and Discussion

The table presents the structural formulas of Ichphans derivatives studied in this work. In the left part of the structure of all the derivatives is a shielded phenol, which ensures their antioxidant properties, and in the right part are a group determining the anticholinesterase activity and also an alkyl sub-stituent bound to a quaternary nitrogen atom, whose aliphatic chain contains 10, 12, or 16 carbon atoms in the case of I(C-10), I(C-12), and I(C-16), respectively. In this series of compounds, the longer the aliphatic chain of the substituent, the higher the hydrophobicity of the derivative. The quaternary nitrogen atom in the structure of Ichphans imparts a positive charge and properties of organic cations to them.

Figure la presents a typical electron micrograph of reference samples of rat erythrocytes. It is seen that most of erythrocytes are discocytes and the other morphological forms are present in small amounts.

To reveal characteristic features of the effect of Ichphans derivatives on the erythrocyte shape, we studied the kinetics of change in the concentration β, % of various morphological forms of erythrocytes after adding I(C-10), I(C-12), or I(C-16) to a concentration of $10^{-4}$ M. Figures lb-ld and 2a show that incubation of an erythrocyte suspension with I(C-10) leads to time-dependent morphological transformations of a part of the cells. Early in the incubation, the concentration of normal discoid forms of the cells decreased and the concentration of echinocytes increased. Later, the pattern became opposite: the concentration of echinocytes decreased and the concentration of discocytes increased; in addition, stomatocytes emerged and their concentration increased. I(C-12), a derivative with longer aliphatic chain, induced similar morphological changes in erythrocytes (Figure 2b). I(C-16) also caused the above transformations of the erythrocyte shape, but the changes in the concentrations of different morphological forms of the cells were much slower and less pronounced (Figure 2c).

Earlier, by the example of the homologous series of spin-labeled nonelectrolytes (benzo-γ-carboline derivatives), it was shown that derivatives with longer aliphatic side substituents (which increase the hydrophobicity of the compound) are more slowly intercalated in the erythrocyte membrane and more slowly reach their distribution regions [12]. One can assume that certain differences in the effect of the amphipathic substances I(C-10), I(C-12), and I(C-16) on the erythrocyte morphology are also caused by differences in kinetics of intercalation in the membrane and distribution in the intramembrane space between these compounds, which differ in hydrophobic properties and molecular volume. Note that the enlargement of the hydrophobic substituent by two carbon atoms causes no significant changes in the efficiency of the substance: the morphological pattern in the case of the effect of I(C-12) on erythrocytes is close to that in the case of I(C-10). However, I(C-16) differs considerably from less hydrophobic members of the homologous series.

The differences in modifying effect of I(C-10), I(C-12), and I(C-16) on the cell shape can be caused by the fact that the erythrocyte membrane capacity for an entering substance is limited; therefore, larger homologues are intercalated in the membrane in smaller amounts [7] and, hence, have a weaker effect.

With allowance for the concepts developed by Sheetz and Singer and also other researchers [8-10, 13] about specific features of the erythrocyte membrane distribution of exogenous chemical substances that have an echinocytogenic or stomatocytogenic effect on the erythrocyte morphology, one can probably represent the erythrocyte plasma membrane distribution of I(C-10), I(C-12), and I(C-16) as follows.

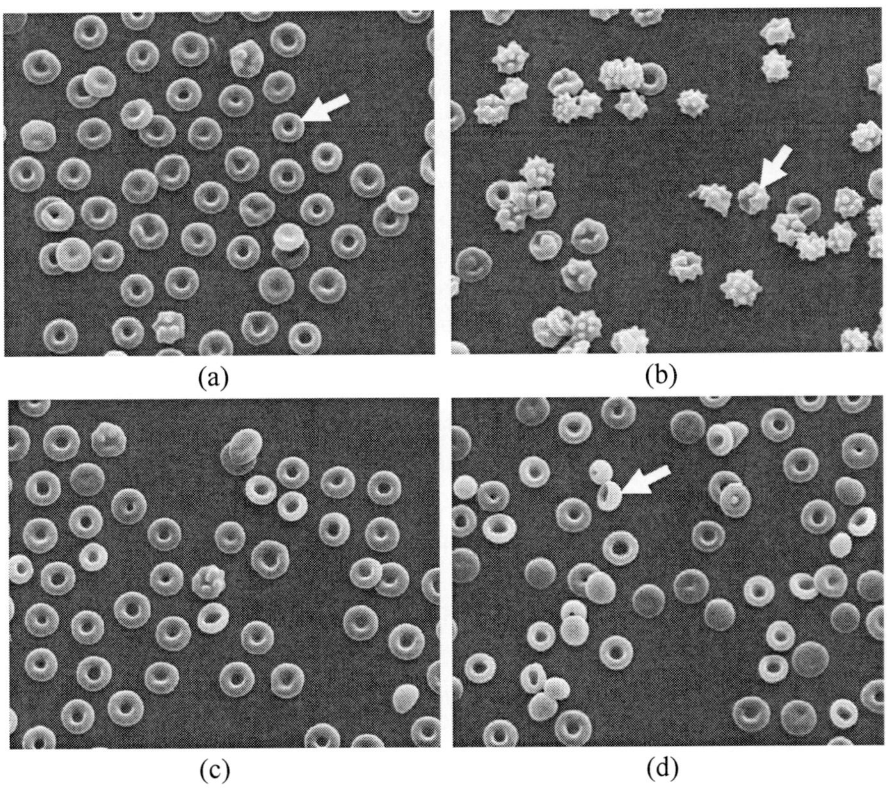


Figure 1. Effect of I(C-10) on the rat erythrocyte morphology in (a) reference sample and in (b) 2, (c) 30 and (d) 90 min after introducing the drug into the erythrocyte incubation medium (×1000). The arrows indicate various morphological forms: (a) discocyte, (b) echinocyte, and (d) stomatocyte

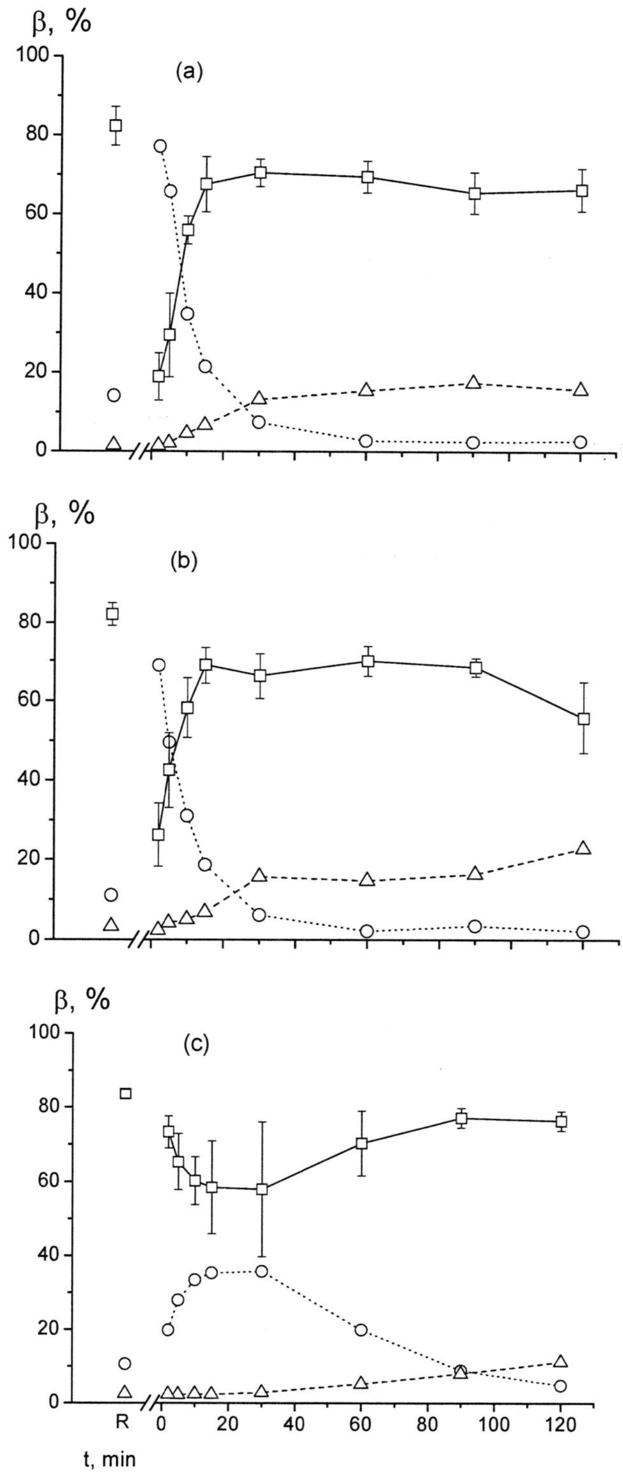


Figure 2. Kinetic curves of change in the concentrations (β (%) of discocytes (squares), echinocytes (circles), and stomatocytes (triangles) in an erythrocyte suspension after adding (a) I(C-10), (b) I(C-12), and (c) I(C-16) to a concentration of $1\times10^{-4}$ M. The isolated points represent the concentrations of different forms of erythrocytes in a reference sample R (in the presence of 0.8 vol % ethanol).

These compounds have a positive charge. In the erythrocyte membrane, the inner monolayer because of the preferential location of phosphatidyl serine has a negative charge [8]. Therefore, I(C-10), I(C-12), and I(C-16) propagating in the intramembrane space tend to be distributed in the inner monolayer. However, the initial step of membrane transport of each of these compounds is its intercalation in the outer monolayer of the erythrocyte membrane. This increases the ratio of the area of the outer monolayer to the area of the inner monolayer. This area disbalance leads to transformation of discocytes into echinocytes (Figure lb). Further ingress of the substance into the inner monolayer gradually increases its area, thus relieving the area disbalance; and once a compensatory effect is attained, the cells assume a discoid shape (Figure 1c). Further preferential ingress of the agent into the inner monolayer causes a decrease in the ratio between the areas of the outer and inner monolayers, which leads to transformation of discocytes into stomatocytes, caliciform cells (Figure 1d).

As Figure 1 shows, the shape transformation involves only a part of the cells in the erythrocyte suspension. Probably, it is in these cells that Ichphans are intercalated. The fact that the concentration of cells transformed into stomatocytes is low in comparison with the concentration of cells transformed into echinocytes is likely to suggest that, under equilibrium conditions, the distribution of Ichphans in the membranes of a significant part of erythrocytes ensures a compensatory effect. Note that the morphological changes in erythrocytes were observed at high concentrations of Ichphans ($10^{-4}$ M and more), which should be taken into account in analyzing their biological activity spectrum.

Previously [1, 2, 14], a high inhibitory effect of Ichphans on erythrocyte acetylcholine esterase was demonstrated. Our data suggest that one of the factors providing for the anticholinesterase effect of the drugs is the structural changes they induce in the erythrocyte membrane. Such rearrangements can affect both the character of the binding of the enzyme to the membrane and the conformational properties determining its catalytic activity; moreover, direct interaction of the hydrophobic substances not only with the active site of the enzyme but also with the apoenzyme is also possible.

The revealed character of the morphological changes induced by the membrane transport of I(C-10), I(C-12), and I(C-16) indicates that the antioxidant effect of these compounds can take place in the inner and outer monolayers of the erythrocyte membrane.

## CONCLUSION

Scanning electron microscopy revealed morphological transformations of erythrocytes that are induced by the fact that three Ichphans derivatives differing in hydrophobic properties are intercalated in the membrane and distributed in the intramembrane space. The changes in the surface architectonics of the erythrocyte membrane can probably have a certain effect on the membrane-bound acetylcholine esterase activity, which was detected in a number of studies of the biological effect of Ichphans. The character of the morphological changes suggests that the antioxidant effect of the Ichphans studied can take place in both monolayers of the erythrocyte membrane.

## Acknowledgments

We thank G.A. Nikiforov for providing Ichphan preparations.

## References

[1] Burlakova E., Molochkina E., Nikiforov G. *Chemisry & chemical technology*. 2008. Vol. 2. № 3. P. 163-173.

[2] Braginskaya, F.I., Molochkina, E.M., Zorina, O.M., Ozerova, I.B., and Burlakova, E.B., *Progress in Alzheimer's Disease and Parkinson Disease,* Fisher, A. and Hanin,L, Eds., N.Y.: Plenum Press, 1998, pp. 177-182.

[3] Molotchkina, E.M., Ozerova, LB., Burlakova, E.B., *Free Radical Biology and Medicine,* 2002, vol. 33, Issue 2S1, no. 610, pp. S229-S230.

[4] Burlakova, E. B.; Molochkina, E. M.; Zorina, O. M.; Fatkullina, L. D.; Goloschapov, A. N; Nikifirov, G. A. *Journal of Neurochemistry*. 2005. № 94. Suppl. 2. P. 42.

[5] Bessis, M., *Corpuscles: Atlas of red blood cell shape,* Berlin-Heidelberg-N.Y.: Springer Verlag, 1974.

[6] Kozinets, G.I. and Simovart, Yu.A., *Poverkhnostnaya arkhitektonika kletok perifericheskoi krovi v norme ipri zabolevaniyakh sistemy krovi* (Surface Architectonics of Peripheral Blood Cells in Normalcy and in Pathologies of Blood System), Tallinn: Valgus, 1984.

[7] Luneva, O.G., Gendel', L.Ya., and Kruglyakova, K.E., *Biophysics,* 2002, vol. 47, №. 1, pp. 38-44.

[8] Sheetz, M.P. and Singer, S.J., *Proc. Natl. Acad. Sci. USA,* 1974, vol. 71, pp. 4457^461.

[9] Sheetz, M.P. and Singer, S.J., *J. Cell Biol.,* 1976, vol. 70, pp. 247-251.

[10] Sheetz, M.P. and Singer, S.J., J. *Cell Biol.,* 1976, vol. 70, pp. 193-203.

[11] Nikiforov, G.A., Belostotskaya, I.S., Vol'eva, V.B., Komissarova, N.L., and Gorbunov, D.B., *Nauchn. vestn. Tyumenskoi meditsinskoi akademii, spetsial'nyi vypusk "Biooksidanty"* (Sci. Herald of Tyumen Medical Academy: Special Issue "Biooxidants"), 2003, pp. 50-51.

[12] Gendel', L.Ya. and Kruglyakova, K.E., in *Metod spinovykh metok i zondov* (Spin Labeling and Probing Method), Moscow: Nauka, 1986, pp. 163-194.

[13] Isomaa, B., Hagerstrand, H., and Paatero, G., *Biochim. Biophys. Acta,* 1987, vol. 899, pp. 93-103.

[14] Ozerova, L.B., New Antioxidants—Screened Phenols—as Modulators of Acetylcholine Esterase Activity *in vivo* and *in vitro, Cand. Sci. (Biol.) Dissertation,* Moscow: Emanuel Inst. Biochem. Phys., Russ. Acad. Sci., 2000, P. 1-25.

In: Chemical and Biochemical Reactions
Editors: G. E. Zaikov, L. Liu et al.
ISBN 978-1-61122-860-1


*Chapter 6*

# ATOMIC HYDROGEN TRAPPING AS A PROBE OF INTRAMOLECULAR PHOTOELECTRON TRANSFER MEDIATED BY BOUND ACIDIC PROTON IN OXYACID $AO^{\delta-}$—$H^{\delta+}$

***O. S. Nedelina*[*], *O. N. Brzevskaya and E. N. Degtyarev***
Emanuel Institute of Biochemical Physics, Russian Academy of Sciences, ul. Kosygina 4, Moscow, 119334 Russia

## ABSTRACT

pH-dependent detection of the EPR spectra of atomic hydrogen $H^{\bullet}$ ($g$ = 2.0023, $A$ = 50.6 mT) in frozen aqueous solutions of oxyacids AOH initiated by photoinjected low energy electrons upon energy impact below the excitation thresholds of water or oxyacids (photon energy of not more than 4.5 eV) was crucial for identification of the dissociative electron capture by bound acidic proton. The data on direct electron attachment to an oxyacid that can create an acidic bound [H+] network under certain conditions allows one to consider direct participation of the acidic bound proton in the intramolecular ET as one electron relay being highly effective in mediating electron transport in acid catalysis of redox reactions.

**Keywords:** photoinjected electrons, photoinduced intramolecular electron transfer, one electron relay, acid catalysis, EPR studies, atomic hydrogen, highly reactive intermediates, chemical trapping, low-energy electrons, dissociative electron attachment, polarized oxyacids,bound acidic proton

Artificial photoredox systems involving an electron donor and an electron acceptor are potentially applicable for converting solar energy into chemical. Many different host systems

[*] E-mail olga.nedelina@gmail.com

have been tried to improve the efficiency of the energy storage by preventing the rapid back electron transfer reaction.

The most prevalent approach to photocatalysis is the construction of three-component systems comprising a sensitizer to absorb light, a proton reduction catalyst, and an electron relay to shuttle reducing equivalents from the sensitizer to the catalyst [1-3].

The rates of electron transfer (ET) can be finely controlled by the fundamental electron transfer properties of electron donor and acceptors (the redox potentials and the reorganization energy) and also by the electron donor-acceptor distance and the environment surrounding them. Various lines of experimental evidence have led to the conclusion that both thermal and photoinduced' electron transfer may occur between species separated by a distance significantly exceeding the sum of their van der Waals radii. In fact many biological electron transport processes including the primary steps of photosynthesis involve such long-range electron-transfer events.

Electron transfer rate constant falls off expotentially with increasing distance between the donor and acceptor molecules [4]. In contrast introduction of an appropriate third component acting as an effective catalyst or relay that control the direction and dynamics of ET by through-bond-coupling effects has been a central issue in designing modern chemical reactions which involve cleavage and formation of chemical bonds. ET over long distances doesn't proceed in a single step, but in multistep process reaction occurs in several faster steps over shorter distances each of them is a superexchange reaction. A condition for such a mechanism is that certain subunits or functional groups can play the role of relay stations where the charge resides for a short time before the ET continious by next hopping step. This configuration enables the photounduced ET process to take place in a sequence of a short rapid "hops" between adjacent relays the final result being efficient formation of a long-lived giant charge-separated state $^{+}D \rightarrow [relay] \rightarrow A^{-}$ [5-7].

The data presented repeatedly provide unambiguous evidence for the important influence of the intervening medium on the rate of electron transfer. Solvent molecules may play an important role as they actively take part in the reaction, being reactant, relay or product.

Here we report a simple approach to the problem of electron-transfer reaction distances based on EPR studies of photoinduced electron transfer in a condensed frozen solution (scheme 1), that would allow the underlying chemical and photochemical events to be probed with greater precision. This is the reductive electron transfer mechanism where an excited sensitizer or fluorophore acts as an electron-injecting species.

The classical photochemical system (Scheme 1) we used in combination with control of the stepwise transformation of photoinjected electrons by EPR, nanosecond laser photolysis, and fluorescence quenching can serve as an adequate method for identification of one electron relays in the intramolecular electron transfer [8-10].

A discussion of photoionization in condensed rigid media must involve some consideration of the photoelectron transport and trapping processes.

It has been known for some time that a number of specific photochemical processes occur in frozen rigid solutions [11]. In this project the light energy is used to overcome the thermodynamic barrier associated with electron removal or attachment. Photon energy range corresponding to the near ultraviolet that is from 3 to 4,5 eV. The photon range is limited at low energies by the absorption threshold of the aromatic solute molecule (fluorophore) and at high energies by the water absorption.

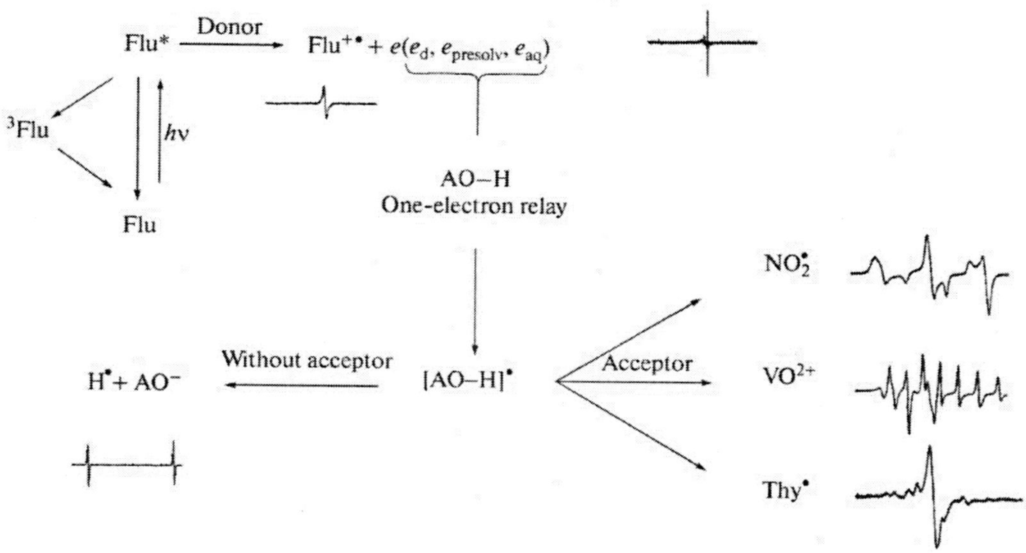


Scheme1. Capture of photoinjected low-energy electrons by bound acidic proton of oxyacids AOH as a a model of elementary steps of donor(e)-[H+]-acceptor(R) intramolecular transport.

Depending on the nature of the solute and the solvent, the photochemical events that can occur subsequent to illumination of such a rigid solutions are a) solute (or fluorophore) photoionization that lead to the formation of an aromatic radical cation which is physically or chemically trapped in the solvent. The product obtained upon irradiation of fluorophore is identical to that produced by chemical oxidation. The details of these processes are not our major concern however and they will be only considered as factors which in many instances influence the primary processes of photoelectron transport.

Studies in frozen solvents are particularly amenable to systematic investigations of the electronic, structural and distance factors. In frozen solvents at liquid-nitrogen temperature, excited-state ET reactions must be highly optimized in order to compete with other radiative and nonradiative decay processes.

Condensed rigid systems related to those studied here appear to be of great value in such studies not only providing more insight into the thermodynamics of ET and its solvent dependence but especially also into the role of the nature of the coupling of donor and acceptor. Frozen matrices are known to be so rigid that highly reactive intermediates such as dry or solvated electrons and radicals may be trapped indefinitely, the generation of an initially random distribution of electrons and radicals alters to the steady state concentration gradients when the electrons diffuse to the scavenger ions.

Generally two types of trapping can be distinguished: a) physical trapping: the electron is trapped in a “self-created hole”. The trapped or solvated electron is a specific entity with characteristic optical and ESR absorption b) chemical trapping: in the presence of an electron acceptor or scavenger or relay the electron attaches to a molecule or some other species and it is absorbed into the acceptor orbitals, which can in certain cases result in chemical change such as dissociation-fragmentation [11]. The trapping of electrons in frozen water is an extremely broad problem and it is currently the subject of extensive studies. This discussion is not intended to examine the details of electron trapping but rather to consider how it can be linked to photoelectron long-range transfer events between donor and distant acceptor in low concentrations.

The ability to unravel the more intricate details of electron transfer (ET) reactions has greatly benefited from the widespread interest that the chemistry of reactive intermediates including radicals, radical ions and metastable or resonant states has attracted in recent decades.

The interaction of a molecule with an electron can provide, apart from its activation in chemical reaction or molecule fragmentation, also intermolecular electron transport from donor to acceptor, thus giving rise to intermediate electron relays. The considerable amount of experimental and theoretical data concerning these reactive intermediates that has become available has allowed a more accurate analysis of the dynamics of such ET reactions. Investigations in this area aim at an understanding of elementary reactions involved in such ET process.

Electron-induced chemistry at subionization energies can proceed either *via* electronic excitation of the target or, particularly at energies too low for electronic excitation, by way of dissociative electron attachment (DEA). The low-energy electrons can resonantly attach themselves to molecular targets and yield formation of temporary negative ions (TNI) [12]. The energized TNI in turn can either evolve by autodetaching the extra electron or could dispose of the excess energy within its molecular framework a path which may ultimately lead to a rupture of a chemical bond within the target: this process yielding a stable anion and a radical fragment, is known as dissociative electron attachment (DEA). In modern terms, this reaction proceeds via the formation of short_lived electron adducts (negative ions, metastable anions, resonants) followed by energy and electron density transfer toward the weakened bond being cleaved to give stable products (free radical and anion): $e$ + M–ABC → (M—ABC)$^-$ → M$^-$ + ABC$^\bullet$ [13-14], in particular, $e$ + MH → (MH)$^\bullet$ → (M–H)$^-$ + H$^\bullet$. Moreover, in the dissociating transient negative ion any translational excess energy will nearly exclusively be transferred to the light hydrogen atom (via the principle of linear momentum conservation more than 99%) thus facilitating the removal of the hydrogen radical in this DEA reaction. Most often the anion M–H detected by mass spectroscopy serves as the marker of this reaction (light H$^\bullet$ and D$^\bullet$ atoms are undetectable by this method). The formation of the (M–H)$^-$ anion, which has given off atomic hydrogen, was demonstrated experimentally by resonance electron capture mass spectroscopy for many compounds containing OH groups [15-18].

The energy of the electron at which the resonance increase in the capture is observed is different for different substances. For example, for compounds that easily dissociate and have high electron affinity, the probability of capture is the highest for nearly zero electron energy. If the affinity of the electron for an atom incorporated into a polyatomic molecule exceeds the energy for abstraction of this atom from the molecule, the electron attachment to this atom in a polyatomic molecule may occur at indefinitely low energy of the electron even at thermal energy. Functional group dependence is observed in the dissociative electron attachment (DEA) to various organic molecules in which the DEA features seen in the precursor molecules of the groups are retained in the bigger molecules. This functional group dependence is seen to lead to site-selective fragmentation of these molecules at the hydrogen sites [19].

Here we show that the DEA process can also be used to probe the distance dependence of the electronic coupling in frozen solutions of oxyacids AOH. We have studied the photooxidation of different fluorophores and photoreduction of some biological acceptors in frozen oxyacids (AOH) diluted solution. It was found that the AOH has a strong effect on both the photooxidation of fluorophores and photoreduction of remote acceptors. Detection of atomic hydrogen as a probe of low energy electron attachment to an oxyacid provides grounds for considering direct participation of the bound acidic proton of AOH as a one_electron relay in the intramolecular ET donor → [AOH] → acceptor [8-10].

Acid catalysis in redox reactions has been described once and again [20-22].

Quite a few facts have been accumulated indicating capture of electrons by Brønsted acids HX, which is detected by appearance of atomic hydrogen [23]. It has been suggested that acids HX can add an additional electron to give $HX^-$ as the primary product: $e_{aq} + HX \rightarrow HX^- \rightarrow H + X^-$.This was shown experimentally (NMR) for the radiation hydroxonium: $e_{aq} + H_3O^+ \rightarrow H_3O \rightarrow H^{\bullet} + H_2O$ [24]. In radiation experiments, the irradiation energies are very high and the determined atomic hydrogen is the overall product of several reactions including decomposition of water ($H_2O^* \rightarrow H^{\bullet} + OH^{\bullet}$ or $H^{\bullet} + O^{\bullet} + H^{\bullet}$, the threshold energy is not lower than 5.5 eV), homolytic bond cleavage X–H → X• + H•,and dissociative electron capture $e + HX \rightarrow [HX]^- \rightarrow X^- + H^{\bullet}$ [24]. Detection of atomic hydrogen occurring in subionization energy range is unambiguous evidence for interaction of the low energy electron with oxyacids AOH in the low energy dissociative electron capture, $e + AOH \rightarrow [AO\text{–}H]^{\bullet} \rightarrow AO^- + H^{\bullet}$. DEA is in fact the only mechanism to induce a bond cleavage at such low electron energies.

Investigation of low energy electron addition and transfer to acids AOH containing a polarized $O^{\delta-}$—$H\delta^+$ bond, that is if there is a large accumulation of positive charge on the H atoms in a bond OH (charge delocalization usually determines the degree of acidity p*K*α) can be useful for elucidation of mechanisms of acid catalysis. The p*K*a value of an oxyacid is ultimately determined by the difference in the free energy between the dissociated and undissociated forms of the acid in the aqueous medium, characterized by the dissociation of the OH covalent bond. As a result, the acidity of an oxyacid is directly reflected in the OH bond strength, exhibited by the equilibrium OH bond length and stretching frequency υ(OH). This explains the origin of correlations between the molecular properties (OH bond length, stretching frequency and positive charge on the bound proton) and the acidity. The larger the intrinsic acidity is of the proton concerned, the larger the polarization of the OH bond and the resulting OH dipole moment [25-26]. When acidity is considered as being related to the elongation of the O–H bond, prior to transfer to a base, we find that the relative acid strength are related to the vibrational properties ν of the acid site. A vibrational fine structure in the effective ion yield curves at threshold energies for formation of $(M\text{-}H)^-$ ions through OH-bond cleavage, has been assigned to the υ(OH) stretching mode. It has been supposed that the formation of these ions occurs predissociation of the vibrationally excited state of the parent anion [18] being due to electron capture, followed by direct dissociation.

In this context, particular attention has been focused on pH dependence of atomic hydrogen detection as a probe of dissociative attachment of low energy electron to polarized oxyacids AOH.

The main principle for instrumental identification of a one electron relay in the donor → relay → acceptor chain is detecting the product of its interaction with the mobile electron in the absence and in the presence of the remote acceptor. The photochemical system (Scheme 1) with appropriate fluorophores donating electrons with specified energy (Flu (fluorophore) + $h\nu \rightarrow (Flu)^* \rightarrow Flu^{\bullet +} + e^-$) to aqueous solutions of oxyacids makes it possible, by varying the composition of the medium, to elucidate the sequence and the nature of electron traps on the way of the photoinjected electron.

Electron migrates a considerable distance from the fluorophore cation radical prior to recombination during which time it is free to react with accessible scavengers or relays. The distance between the donor and relay as an acceptor is obviously very important. The pH-dependent survival probability of the donor mobile electron depends on the concentration of relay, i.e. acidic bound proton of hydroxyl groups which can serve as the electron captor to form hydrogen atoms as shown by the appearance of the hydrogen atom ESR signal. In the present work, we have varied the distance between the probe and relay by changing the relay (captor) concentration values.

The pH-dependent dissociative capture of a low energy electron by oxyacids AOH tracked by the behavior of the EPR spectrum of atomic hydrogen was observed in frozen (77 K) dilute (0.1 to 0.5 M) aqueous solutions of inorganic oxyacids AOH, in particular, HCl ($H_3O^+$), $H_2SO_4$ ($H_3O^+$, $HSO_4^-$), $H_2PO_4^-$ , $HPO_4^{2-}$, $H_3BO_3$ on excitation of different fluorophores (tryptophan, NADH, phenothiazine, tetrasulfopyrene, quercetin), donating photoinjected electrons in their absorption band by exposure to near_UV radiation ($240 \leq \lambda \leq 400$ nm) whose energy is below the excitation threshold of both water and acceptor acids.

Thus, samples were irradiated at 77 K and examined by ESR at 77 K to detect hydrogen atoms. Figure 1 does show hydrogen atom spectra at 77 K characterized by two peaks separated by 506 G with g = 2.014.

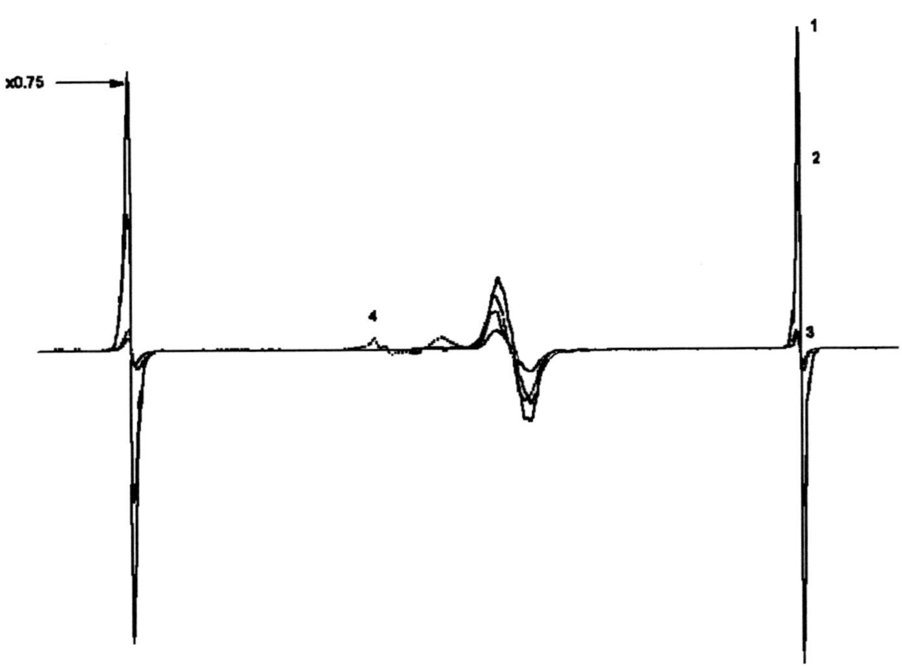


Figure 1. EPR spectra of atomic hydrogen obtained upon UV irradiation ($\lambda \geq 240$ nm) of frozen solutions of inorganic oxyacids in the presence of tryptophan (Trp) as the donor of photoinjected electrons at 77 K and a 2 μW microwave irradiation: (*1*) 0.5 M $HSO_4^-$ (pH =0.8, p*K*α = 1.96); (*2*) 0.5 M $H_2PO_4^-$ (pH = 4.1, p*K*α = 7.2); (*3*) 0.5 M $H_3BO_3$ (pH = 9.0, p*K*α = 9.2); (*4*) 0.5 M HCl (pH = 0.6).

Proton spinflip satellite lines are also seen on each side of the main hydrogen atom lines. The pattern of the EPR spectrum of atomic hydrogen, characterizing its proton environment was the same for oxyacids AOH with different p$K\alpha$ . For deuterated samples, the intensity of the EPR signals of atomic hydrogen sharply decreased in parallel with appearance of the deuterium spectrum, the EPR spectrum being narrowed down (Figure 2). H-D exchange occurs more easily on terminal OH group which brings about the highest OH vibrational band at around 3700cm-1 and the lowest value in proton signal in NMR spectrum.

The H/D isotope effect for the formation of H atoms is more than 7 for the phosphate monoanion and more than 30 for boric acid. This high H/D effect can be interpreted only assuming that the reaction first gives the intermediate electron adduct $[H_2PO_4\ ]^{\bullet}$ and/or $[D_2PO_4]^{\bullet}$, which then decomposes to give H or D.

It was found that attachment of low energy electrons with oxyacids, which we observed, is a pH dependent reaction; i.e. the intensity of the EPR spectrum of atomic hydrogen is proportional to the acidic bound proton concentration in AOH, and the highest amount of detectable atomic hydrogen is accumulated around the highest molar concentration of the acid in the predissociated state with the polarized $AO^{\sigma-}$–$H^{\sigma+}$ bond near its p$K\alpha$. Indeed, for oxyacids with p$K\alpha$ in acidic (sulfuric acid, –2.8 and 1.92), neutral ($H_2PO_4^-$, 7.21), or alkaline (boric acid, 9.24) regions, the hydrogen atom is detected mainly in exactly these pH ranges, which is in line with electron capture by the acidic proton of the AOH group: $e + AOH \rightarrow [AOH]^{\bullet} \rightarrow H^{\bullet} + AO^{-}$ . In the absence of distant acceptor R, the intensity of atomic hydrogen signals obtained for different oxyacids correlated with their acidity constants p$K\alpha$, which confirms the presence of a linear relationship between the p$K\alpha$ of acids and the logarithm of the rate constants for their reactions with electron [24] (Figure 3).

Therefore, it was assumed that the acidic (electrophilic) proton is the attachment site for the low energy electron, and this allows detecting the product as the characteristic EPR spectrum of atomic hydrogen (the left hand part of Scheme 1).

According to Brønsted and Lowry, a stronger acid has a smaller p$Ka$ value whereas a stronger Lewis acid has a stronger capability to accept a pair of electrons. The findings reveal that a stronger Brønsted acid in most cases behaves as a stronger Lewis acid as well. An acid with a lower p$Ka$ value is expected to have a higher electron affinity, electronegativity and electrophilicity compared to other acids of similar type if the two theories need to correlate. Proton acidity should greatly influence direct electron attachment to the OH group, as well as acting as an efficient intermediary in transferring electrons from one portion of the molecular to another [ 27].

It was suggested that an electron acceptor relay process in the presence of the remote acceptor R in a low concentration occurs from site to site in the acids framework. Apparently, this relay process does not generate hydrogen atoms, but remote acceptors radical $R^{\bullet}$ (Scheme1).

In a typical experiment, a small concentration of photoexcited fluorophore (<$10^{-3}$M) as a source of donor electron, is embedded in a frozen solvent amid a higher concentration (>0,1M) of randomly distributed primary acceptors AOH molecules as a relay (phosphate, sulfate,.borate) and a low concentration (<0,01M) of distant terminal acceptors R (nitrate, thymine, vanadate). If the weaker primary electron captor AOH is present in large excess, acidic bound proton can be reduced (chemical trapping) and, the excess electron later transfers to the remote acceptor of greater electron affinity.

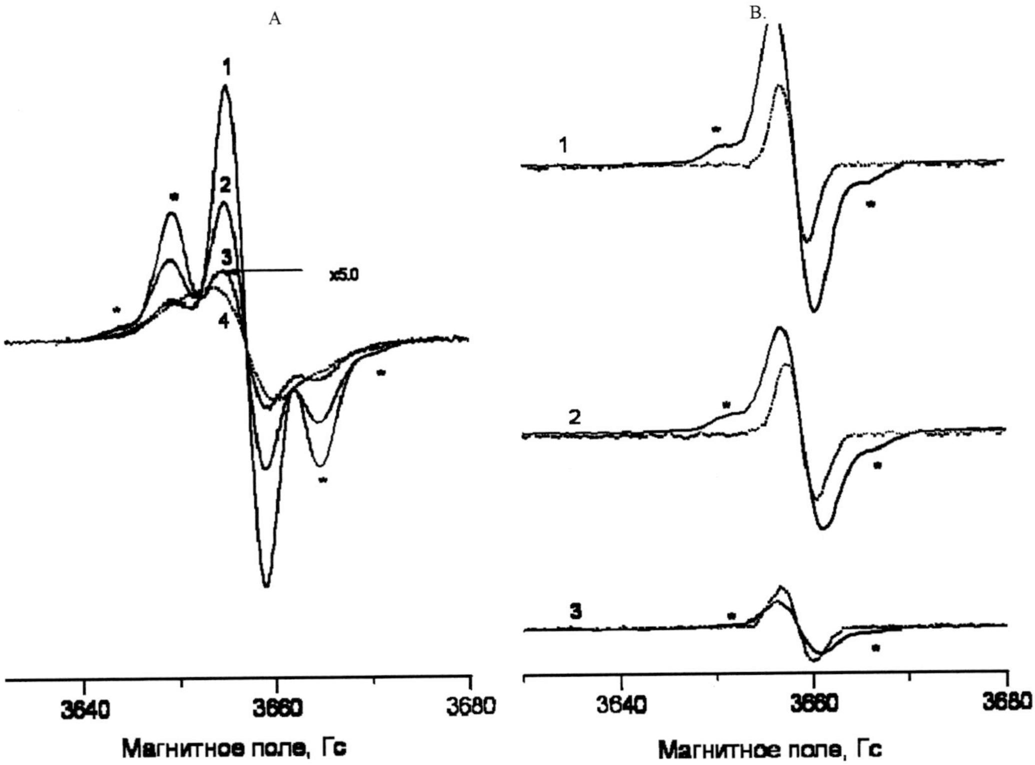


Figure 2. High_field component of the EPR spectrum of H atoms obtained upon UV irradiation ($\lambda \geq$ 240 nm) of frozen aqueous solutions of inorganic oxyacids in the presence of tryptophan as a sensitizer at 77 K: (1) 0.5 M $HSO_4^-$ (pH =0.8, p*K*α = 1.96); (*2*) 0.5 M $H_2PO_4^-$ (pH = 4.1, p*K*α = 7.2); (*3*) 0.5 M $H_3BO_3$ (pH = 9.0, p*K*α = 9.2); (*4*) 0.5 M HCl (pH = 0.6). The continuous curves on the right are satellite lines arising upon simultaneous flip_flop change in the spin state of adjacent protons (microwave radiation power 200 and 2 μW). The dotted lines on the right show the spectral components of the H atom in solutions of the corresponding acids in $D_2O$.

The oxyacids system is the ET reaction we are aware of for which the spectra of all species involved in the reaction – the reactants, intermediates relays, and products R·– can be detected and well characterized (Scheme1).

In our experiments we obtained evidence for the fact that in the presence of remote acceptors R in a low concentration, the products of electron capture $[AO–H]^•$ can transmit the electron to remote electron acceptors R (nitrate, vanadate, thymine), which is detected in the experiment as a decrease in the intensity of the EPR spectrum of hydrogen and appearance of the EPR spectra of R• (the right_hand part of Scheme 1). This effect is observed at about the highest concentration of the bound acidic proton having a high electron affinity (Figure 4, 5), which suggests that intramolecular transport occurs via successive attachment of the electron to the acidic proton as a relay removing the distance limitations of one electron transport. The effect was attributed to the enhancement of charge separation of the photoproduced primary ion pair due electron injection from the photoreaction centers into the electron relay system and the succeeding electron migration.

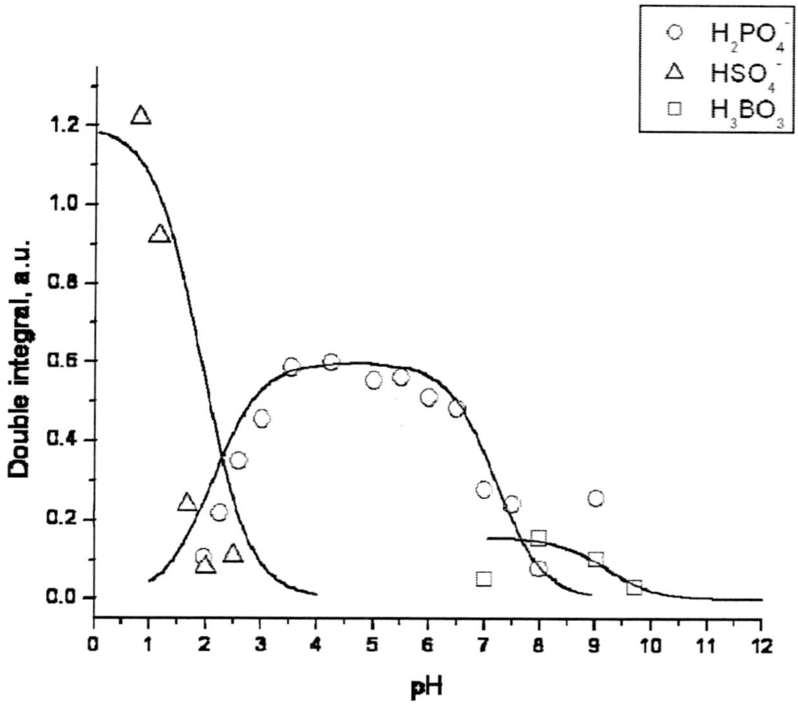


Figure 3. pH dependence of atomic hydrogen intensity for oxiacids with different pK$\alpha$. The curves are calculated pH dependences of mole fractions of AO-H (*1triangles*) $HSO_4^-$, (*2circles*) $H_2PO_4^-$ and (*3squares*) $H_3BO_3$ ions.

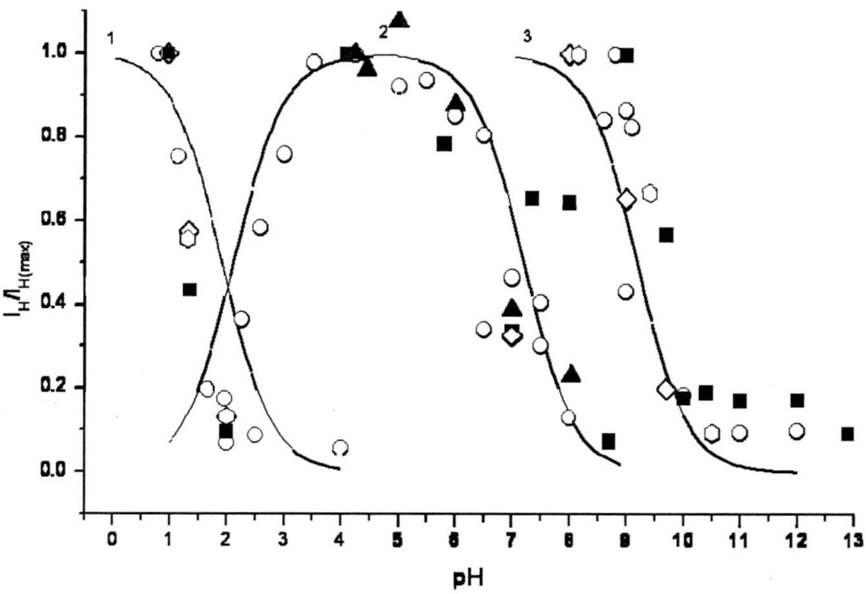


Figure 4. pH dependence of the yield of H atoms (circles) in the absence of acceptors R and corresponding yield of acceptors $R^{\cdot}$ for solutions of oxyacids with different p*K*$\alpha$ at 77 K( squares – $NO_3^{2-}$, triangles $VO^{2+}$). The curves are calculated pH dependences of mole fractions of AO-H (*1*) $HSO_4^-$, (*2*) $H_2PO_4^-$ and (*3*) $H_3BO_3$ ions.

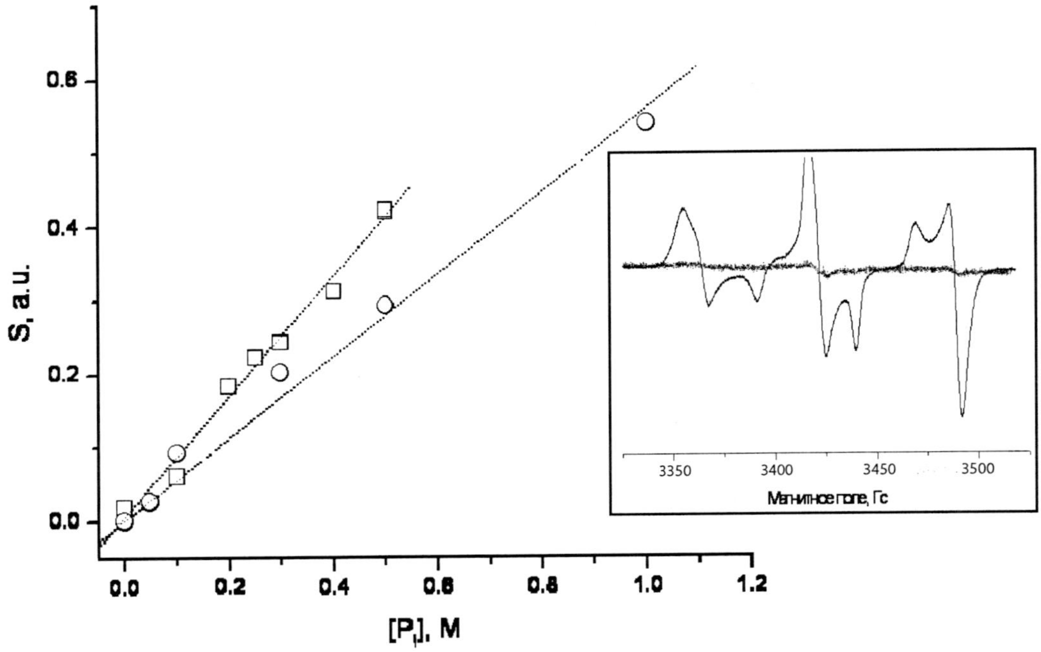


Figure 5. Yields of atoms (*1 circles*) H and (*2 squares*) $NO_3^{\ 2-}$ in Trp + $H_2PO_4^-$ systems without electron acceptor $KNO_3$ and Trp + $H_2PO_4^-$ + $KNO_3$ systems with electron acceptor vs. $H_2PO_4^-$ concentration on exposure to UV radiation at $\lambda \geq 240$ nm (the *y* axis shows the double integrals of EPR spectra of H atoms and radicals $NO_3^{\ 2-}$ reduced to the same microwave power). Reagent concentrations: Trp, $5 \cdot 10^{-4}$; $H_2PO_4^-$ 0.5 M; $KNO_3$ 0.01 M (*3, 4*) Inset: EPR spectra of $NO_3^{\ 2-}$ at pH 5.0 recorded on UV irradiation of an aqueous solution of Trp and $KNO_3$ only in the presence of phosphate.

In this study, detection of the EPR spectra of atomic hydrogen $H^{\bullet}$ ($g = 2.0023$, $A = 50.6$ mT) in frozen aqueous solutions initiated by low energy photoinjected electrons was crucial for identification of the electron capture by oxyacids AOH. The detection of the hydrogen atom upon energy impact below the excitation thresholds of water or oxyacids (photon energy of not more than 4.5 eV, Figure 1) in a frozen aqueous medium at different pH values serves as unambiguous proof for dissociative capture of the low_energy photoinjected electron by AOH acids with different p*K*α [6-8].

Thus our studies of low energy electron attachment to the bound acidic proton of oxyacid $AO^{\delta-}$–$H^{\delta+}$ containing easily controllable polarized bond OH following distant acceptor reduction can be useful for more precise elucidation of the acid catalysis mechanism. The data on direct electron attachment to an oxyacid that creates an acidic bound [H+] network under certain conditions [28] allows one to consider direct participation of the acidic bound proton in the intermolecular ET as one electron relay.

In our opinion, the mechanism of acid catalysis of redox reactions is based on activation of resonance capture of low energy (≤4 eV) electrons by the acidic proton of XH molecules (or their functional groups OH, NH, CH); the acidity (electrophilicity) of the bound proton controlled by specific catalysis determines the rate of ET reactions.

One of the most remarkable features of our results (anticipated from theoretical and experimental studies on the atomic hydrogen dynamics in pH-dependent photoelectron capture) is the observation that a bound acidic proton $AO^{\delta-}$ - $H^{\delta+}$ can be highly effective in mediating intramolecular electron transport.

## ACKNOWLEDGMENTS

We are grateful to Academician A.E. Shilov for permanent attention to the work and fruitful discussion of the results.

This work was supported by the Program of the Division of Chemistry and Material Science of the

RAS "Theoretical and Experimental Studies of the Nature of the Chemical Bond and Mechanisms of Chemical Reactions and Processes."

## REFERENCES

[1] Bolton J.R. (1996) Solar photoproduction of hydrogen: a review. *Solar energy* 57, 37-50.

[2] Barbara P. F. , Meyer T. J., Ratner M. A. (1996) Contemporary Issues in Electron Transfer Research. *J. Phys. Chem.* 100, 13148–13168.

[3] Esswein A. J. and Nocera D.G. (2007) Hydrogen Production by Molecular Photocatalysis. *Chem. Rev.* 107, 4022–4047.

[4] Closs G.L., Miller J.R. (1988) Intramolecular Long-Distance Electron Transfer in Organic Molecules. *Science* 240, 440 – 447.

[5] Paddon-Row M.N . (1994) Investigating Long-Range Electron-Transfer Processes with Rigid, Covalently Linked Donor-( Norbornylogous Bridge)-Acceptor Systems. *Acc. Chem. Res.* 27, 18-25.

[6] Gray H. B. and Winkler J. R. (2005) Long-range electron transfer. *Proc. Natl. Acad. Sci. U S A.* 102, 3534–3539.

[7] Rice S. A. (1980) Do Excess Electrons Migrate from Trap to Trap in Low Temperature Molecular Glasses? Some Energetic and Kinetic Considerations. *J. Phys. Chem.* 84, 1280-1286.

[8] Brzhevskaya O. N., Degtyarev E. N., Levin P. P., Lozinova T. A., and Nedelina O. S. (2005)

[9] Hydrogen Atoms in Frozen Phosphate-Containing Aqueous Media Photosensitized by Biological Cofactors. *Doklady Biochemistry and Biophysics* 405, 395–399. *Translated from Doklady Akademii Nauk,* 405, 259-263.

[10] Degtyarev E. N., Brzhevskaya O. N. and O. S. Nedelina (2006) Detection of the Products of Acceptor Interaction between Orthophosphate and Photoejected Electron by EPR Spectroscopy. *Doklady Biochemistry and Biophysics* 410, 286–291. *Original Russian Text © E.N. Degtyarev, O.N. Brzhevskaya, O.S. Nedelin, published in Doklady Akademii Nauk,* 2006, 410, 542–547.

[11] Degtyarev E. N., Brzhevskaya O. N. and O. S. Nedelina (2009) Inorganic Oxyacids $AO^{\delta-}$-$H^{\delta+}$ as One_Electron Relay in Acid Catalysis of Redox Reactions *Doklady Chemistry,* 428, 233–237. *Original Russian Text © O.S. Nedelina, O.N. Brzhevskaya,*

*E.N. Degtyarev, 2009, published in Doklady Akademii Nauk, 2009,* Vol. 428, No. 4, pp. 474–479.

[12] Lesclaux R.and Joussot-Dubien J. (1973), Electron photoinjection from aromatic molecules in condensed media. In *Organic Molecular Photophysics* edited by J.B.Birks (Wiley, London) Vol.1, p.457-487.

[13] Massey H. S. W., Negative Ions (Cambridge Univ. Press, Cambridge, England, 1976), 3rd ed.

[14] Ingólfsson O., Weik F. and Illenberger E. (1996) The reactivity of slow electrons with molecules at different degrees of aggregation: gas phase, clusters and condensed phase *Int.J.Mass Spectrom.* 155, 1-68.

[15] Simons J. (2005) Response of a molecule to adding or removing an electron. *Adv.Quantum Chem.* 50, 213-233.

[16] Liu B, Hvelplund P., Nielsen S.B., (2004) Hydrogen loss from nucleobase nitrogens upon electron attachment to isolated DNA and RNA nucleotide anions. *J. Chem. Phys.* **121**, 4175-4179.

[17] Burrow P.D., Gallup G.A. , Modelli A. (2008) Are there π*Shape resonance in electron scattering from phosphate group?*J.Phys.Chem.A* 112, 4106-4113.

[18] Pelc A., Sailer W., Scheier P., Probst M., .Mason N.J., Illenberger E., Mark T.D. (2002).

[19] Dissociative electron attachment to formic acid (HCOOH) *Chemical Physics Letters.* 361, 277-284.

[20] Vasil'ev Y.V, Muftakhov M.V., Tuimedov G.M., Khatymov R.V., AbzalimovR.R., Mazunov V.A., Drewello T. (2001) Specific formation of $(M\text{-}H)^-$ ions from OH-group containing molecules. *Int.J.Mass Spectrom.* 205, 119-135.

[21] Prabhudesai V.S., Kelkar A. H., Nandi D, and Krishnakumar E. (2005) Functional Group Dependent Site Specific Fragmentation of Molecules by Low Energy Electrons. PRL *Phys.Rev.Lett.* 95, 143202 (1-4).

[22] Khrushch A.P., Shilov A.E., Vorontzova T.A. (1974) Role of acids in reduction of acetylene catalyzed by molybdenum-thiol complexes. *J.Amer.Chem.Soc.* 96. 4987-4989.

[23] Fukuzumi S., Ishikawa K., Hironaka K., and Tanaka T. (1987) Acid Catalysis in Thermal and Photoinduced Electron-Transfer Reactions. *J. Chem. Soc., Perkin Trans. 2*, 751-760.

[24] Irebo T., Reece S.Y., Sjodin M., Nocera D.G., Yammarstrom L. (2007) Proton-coupled Electron Transfer of Tyrosine Oxidation: Buffer Dependence and Parallel Mechanisms. *J.Amer.Chem.Soc.* 129, 15462-15464.

[25] Kevan L., Fine C. (1966) Trapped hydrogen yields and precursors in frozen solution radiolysis. *J. Amer. Chem. Soc.* 88, 896–872.

[26] Hart, E.J. and Anbar, M., *Hydrated Electron*, New York: Wiley, 1970. Translated under the title *Gidratirovannyi elektron*, Moscow: Atomizdat, 1973.

[27] Rustad J.R., Dixon D. A., Kubicki J. D. and Felmy A. R. (2000) Gas-Phase Acidities of Tetrahedral Oxyacids from ab Initio Electronic Structure Theory. *J. Phys. Chem. A*, 104, 4051–4057.

[28] Smiechowski M., Gojło E. and Stangret J. (2009) Systematic Study of Hydration Patterns of Phosphoric(V) Acid and Its Mono-, Di-, and Tripotassium Salts in Aqueous Solution *J. Phys. Chem. B*, 113, 7650–7661.

[29] Gupta K, D.R. Roy D.R., Subramanian V. and Chattaraj P.K. (2007) Are strong Brønsted acids necessarily strong Lewis acids? Journal of Molecular Structure: THEOCHEM, 13-24.

[30] Reece S.Y., Hodgkiss J.M., Stubbe J.A., Nocera D.A. (2006) Proton-coupled electron transfer: the mechanistic underpinning for radical transport and catalysis in biology, *Phil.Trans.R.Soc.B.* 361, 1351-1364.

In: Chemical and Biochemical Reactions
Editors: G. E. Zaikov, L. Liu et al.
ISBN 978-1-61122-860-1


*Chapter 7*

# QUANTUM-CHEMICAL RESEARCH OF THE MECHANISM OF SYNTHESIS OF 5-ACETYLOXYMETHYL-2-CHLORO-5-ETHYL-1,2,3-DIOXAPHOSPHORYNANE

***V. A. Babkin*[1],**, *V. U. Dmitriev*[1], *G. A. Savin*[2],*†, *E. S. Titova*[3] *and G. E. Zaikov*[4]*‡

[1]Volgograd State Architect-Build University Sebrykov Departament, 403300 Michurin Street 21, Michailovka, Volgograd region, Russia
[2]Volgograd State Pedagogical University, Volgograd; Russia
[3]Volgograd State Technical University, Volgograd; Russia
[4]Institute of Biochemical Physics Russian Academy of Sciences; 117977 Kosigin Street 4, Moscow, Russia

## ABSTRACT

Quantum-chemical research of the mechanism of synthesis of 5-acetyloxymethyl-2-chloro-5-ethyl-1,2,3-dioxaphosphorynane for the first time is executed by classical method MNDO. Synthesis of this compound is the result of interaction of acetyl chloride and 4-ethyl-2,6,7-trioxa-1-phosphabicyclo[2.2.2]octane. Interaction proceeds on the mechanism of bimolecular nucleophylic substitution $S_N2$. Reaction is endothermic and has barrier character. The size of a power barrier makes 178 kJ/mol.

**Keywords:** quantum-chemical research, the mechanism of synthesis, 5-acetyloxymethyl-2-chloro-5-ethyl-1,2,3-dioxaphosphorynane, 4-ethyl-2,6,7-trioxa-1-phosphabicyclo[2.2.2] octane, acetyl chloride, method MNDO

---

* E-mail: sfi@reg.avtlg.ru
† E-mail: gasavin@mail.ru
‡ E-mail: chembio@sky.chph.ras.ru

## AIMS AND BACKGROUNDS

5-Acetyloxymethyl-2-chloro-5-ethyl-1,2,3-dioxaphosphorynane is synthesized by direct acylation of 4-ethyl-2,6,7-trioxa-1-phosphabicyclo[2.2.2]octane with acetyl chloride for the next scheme:

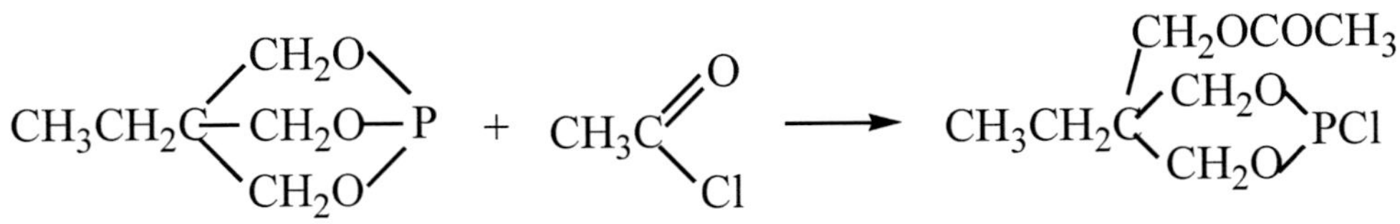

This synthesis can be rather perspective for reception of various new chemical compounds and, in particular, medical products with useful and probably unique properties. The mechanism of synthesis of 5-acetyloxymethyl-2-chloro-5-ethyl-1,2,3-dioxa-phosphorynane now is not studied. Research of the mechanism of synthesis of this compound was the purpose of the present work. Research was spent by one of classical quantum-chemical methods.

## METHODICAL PART

The Quantum-chemical semi empirical method MNDO with optimization of geometry on all parameters by the standard gradient method which has been built in PC GAMESS has been chosen for research of the mechanism of synthesis of 5-acetyloxymethyl-2-chloro-5-ethyl-1,2,3-dioxaphosphorynane [2]. This method well enough reproduces power characteristics and stability of chemical compounds, including the substances containing multiple bonds [1]. Calculations were carried out in approach to the isolated molecule to a gas phase. The Program MacMolPlt was used for visual representation initial, intermediate and final models [3].

The Mechanism of synthesis of studied compound was investigated by method MNDO. Initial components of synthesis – 2,6,7-trioxa-4-ethyl-1-phosphabicyclo[2.2.2]octane and acetyl chloride – settled down on distance 2.8-3.0Å from each other. On such distance any interactions between them practically are absent. As coordinate of reaction distance $R_{O7C22}$ (figure 1) has been chosen. This distance is most energetically a favorable direction of interaction of initial components by the maximal values of charges on atoms of molecular system $P_{10}$,$O_7$,$Cl_{25}$ and $C_{22}$ directly participating in reaction. Optimization on all parameters of initial components at $R_{O7C22}$ = 2.8Å was carried out. Values of lengths of bonds and valent corners, values $E_0$ (the general energy of system) and $q_H$-a charge on atoms along coordinate of reaction $R_{O7C22}$ were fixed and brought in Tables 1-3 after optimization. The coordinate of reaction changed with 2.8Å up to 1.2Å at each step of optimization. The step on coordinate of reaction $R_{O7C22}$ has made 0.2Å.

**Table 1. Change of energy along coordinate of reaction $R_{O7C22}$**

| $R_{O7C22}$,Å | $E_0$, kJ/mol |
|---|---|
| 2,8 | -291546 |
| 2,6 | -291530 |
| 2,4 | -291507 |
| 2,2 | -291475 |
| 2,0 | -291433 |
| 1,8 | -291394 |
| 1,6 | -291365 |
| 1,4 | -291544 |
| 1,3 | -291533 |
| 1,2 | -291444 |

**Table 2. Change of lengths of bonds along coordinate of reaction $R_{O7C22}$**

| $R_{C22-O7}$,Å | $R_{C1-C2}$ | $R_{C2-C3}$ | $R_{C3-C4}$ | $R_{C3-C5}$ | $R_{C3-C6}$ | $R_{C4-O7}$ | $R_{C5-O8}$ | $R_{C6-O9}$ | $R_{P10-O8}$ | $R_{P10-O7}$ |
|---|---|---|---|---|---|---|---|---|---|---|
| 2.8 | 1.53 | 1.55 | 1.58 | 1.58 | 1.58 | 1.40 | 1.39 | 1.39 | 1.60 | 1.61 |
| 2.6 | 1.53 | 1.55 | 1.58 | 1.58 | 1.58 | 1.40 | 1.39 | 1.39 | 1.60 | 1.62 |
| 2.4 | 1.53 | 1.55 | 1.58 | 1.58 | 1.58 | 1.40 | 1.39 | 1.39 | 1.60 | 1.62 |
| 2.2 | 1.53 | 1.55 | 1.58 | 1.58 | 1.58 | 1.40 | 1.39 | 1.39 | 1.60 | 1.63 |
| 2.0 | 1.53 | 1.55 | 1.58 | 1.58 | 1.58 | 1.41 | 1.39 | 1.39 | 1.60 | 1.64 |
| 1.8 | 1.53 | 1.55 | 1.57 | 1.58 | 1.58 | 1.42 | 1.39 | 1.39 | 1.60 | 1.65 |
| 1.6 | 1.53 | 1.55 | 1.57 | 1.58 | 1.58 | 1.43 | 1.39 | 1.39 | 1.59 | 1.69 |
| 1.4 | 1.53 | 1.57 | 1.57 | 1.58 | 1.57 | 1.40 | 1.38 | 1.38 | 1.58 | 3.62 |
| 1.3 | 1.53 | 1.57 | 1.57 | 1.58 | 1.57 | 1.41 | 1.38 | 1.38 | 1.59 | 3.66 |
| 1.2 | 1.53 | 1.57 | 1.57 | 1.58 | 1.57 | 1.42 | 1.38 | 1.38 | 1.59 | 3.71 |

| $R_{C22-O7}$,Å | $R_{P10-Cl25}$ | $R_{H11-C1}$ | $R_{H12-C1}$ | $R_{H13-C1}$ | $R_{H14-C2}$ | $R_{H15-C2}$ | $R_{H16-C4}$ | $R_{H17-C4}$ |
|---|---|---|---|---|---|---|---|---|
| 2.8 | 4.43 | 1.11 | 1.10 | 1.10 | 1.11 | 1.11 | 1.11 | 1.11 |
| 2.6 | 3.59 | 1.11 | 1.10 | 1.10 | 1.11 | 1.11 | 1.11 | 1.11 |
| 2.4 | 3.39 | 1.11 | 1.10 | 1.10 | 1.11 | 1.11 | 1.11 | 1.11 |
| 2.2 | 3.19 | 1.11 | 1.10 | 1.10 | 1.11 | 1.11 | 1.11 | 1.11 |
| 2.0 | 3.03 | 1.11 | 1.10 | 1.10 | 1.11 | 1.11 | 1.11 | 1.11 |
| 1.8 | 2.86 | 1.11 | 1.10 | 1.10 | 1.11 | 1.11 | 1.11 | 1.11 |
| 1.6 | 2.63 | 1.11 | 1.10 | 1.10 | 1.11 | 1.11 | 1.11 | 1.11 |
| 1.4 | 2.02 | 1.11 | 1.10 | 1.10 | 1.11 | 1.11 | 1.12 | 1.12 |
| 1.3 | 2.02 | 1.11 | 1.10 | 1.10 | 1.11 | 1.11 | 1.12 | 1.12 |
| 1.2 | 2.02 | 1.11 | 1.10 | 1.10 | 1.11 | 1.11 | 1.11 | 1.12 |

**Table 2. Continued**

| $R_{C22-O7}$,Å | $R_{H18-C5}$ | $R_{H19-C5}$ | $R_{H20-C6}$ | $R_{H21-C6}$ | $R_{C23-C22}$ | $R_{O24-C22}$ | $R_{Cl25-C22}$ | $R_{H26-C23}$ | $R_{H27-C23}$ | $R_{H28-C23}$ |
|---|---|---|---|---|---|---|---|---|---|---|
| 2.8 | 1.11 | 1.11 | 1.11 | 1.11 | 1.51 | 1.20 | 1.80 | 1.10 | 1.10 | 1.10 |
| 2.6 | 1.11 | 1.11 | 1.11 | 1.11 | 1.51 | 1.20 | 1.80 | 1.10 | 1.11 | 1.10 |
| 2.4 | 1.11 | 1.11 | 1.11 | 1.11 | 1.51 | 1.20 | 1.81 | 1.10 | 1.11 | 1.10 |
| 2.2 | 1.11 | 1.11 | 1.11 | 1.11 | 1.52 | 1.20 | 1.82 | 1.10 | 1.11 | 1.10 |
| 2.0 | 1.11 | 1.11 | 1.11 | 1.11 | 1.52 | 1.21 | 1.84 | 1.10 | 1.11 | 1.10 |
| 1.8 | 1.11 | 1.11 | 1.11 | 1.11 | 1.53 | 1.21 | 1.87 | 1.10 | 1.11 | 1.10 |
| 1.6 | 1.11 | 1.11 | 1.11 | 1.11 | 1.54 | 1.22 | 1.95 | 1.10 | 1.10 | 1.10 |
| 1.4 | 1.12 | 1.12 | 1.12 | 1.12 | 1.51 | 1.22 | 4.42 | 1.10 | 1.10 | 1.10 |
| 1.3 | 1.12 | 1.12 | 1.12 | 1.12 | 1.53 | 1.23 | 4.43 | 1.10 | 1.10 | 1.10 |
| 1.2 | 1.12 | 1.12 | 1.12 | 1.12 | 1.54 | 1.24 | 4.45 | 1.10 | 1.10 | 1.10 |

**Table 3. Change of charges of system along coordinate of reaction $R_{O7C22}$**

| $R_{C22-O7}$, Å | 2,8 | 2,6 | 2,4 | 2,2 | 2,0 | 1,8 | 1,6 | 1,4 | 1,3 | 1,2 |
|---|---|---|---|---|---|---|---|---|---|---|
| C1 | 0.02 | 0.02 | 0.02 | 0.02 | 0.02 | 0.02 | 0.02 | 0.02 | 0.02 | 0.02 |
| C2 | 0.02 | 0.02 | 0.02 | 0.02 | 0.02 | 0.02 | 0.01 | 0.00 | 0.00 | 0.00 |
| C3 | -0.23 | -0.23 | -0.23 | -0.23 | -0.23 | -0.23 | -0.22 | -0.17 | -0.17 | -0.17 |
| C4 | 0.20 | 0.20 | 0.20 | 0.19 | 0.19 | 0.19 | 0.19 | 0.21 | 0.22 | 0.23 |
| C5 | 0.20 | 0.20 | 0.20 | 0.20 | 0.20 | 0.20 | 0.20 | 0.23 | 0.23 | 0.22 |
| C6 | 0.20 | 0.20 | 0.20 | 0.20 | 0.20 | 0.20 | 0.20 | 0.23 | 0.23 | 0.23 |
| O7 | -0.52 | -0.53 | -0.54 | -0.53 | -0.53 | -0.51 | -0.45 | -0.35 | -0.32 | -0.28 |
| O8 | -0.50 | -0.50 | -0.50 | -0.50 | -0.50 | -0.51 | -0.51 | -0.51 | -0.51 | -0.51 |
| O9 | -0.51 | -0.51 | -0.52 | -0.52 | -0.51 | -0.52 | -0.53 | -0.52 | -0.52 | -0.51 |
| P10 | 0.98 | 1.00 | 1.00 | 1.02 | 1.03 | 1.07 | 1.14 | 0.98 | 0.98 | 0.97 |
| H11 | 0.00 | 0.00 | 0.00 | 0.00 | 0.00 | 0.01 | 0.01 | 0.00 | 0.00 | 0.01 |
| H12 | 0.00 | 0.00 | 0.00 | 0.00 | 0.00 | 0.00 | 0.00 | 0.00 | 0.00 | 0.00 |
| H13 | 0.00 | 0.00 | 0.00 | 0.00 | 0.00 | 0.00 | 0.00 | 0.00 | 0.00 | 0.00 |
| H14 | 0.01 | 0.01 | 0.01 | 0.01 | 0.01 | 0.01 | 0.02 | 0.01 | 0.01 | 0.01 |
| H15 | 0.01 | 0.01 | 0.01 | 0.01 | 0.01 | 0.01 | 0.02 | 0.01 | 0.01 | 0.01 |
| H16 | 0.03 | 0.02 | 0.02 | 0.01 | 0.02 | 0.03 | 0.03 | 0.01 | 0.02 | 0.03 |
| H17 | 0.03 | 0.03 | 0.03 | 0.03 | 0.04 | 0.05 | 0.04 | 0.01 | 0.02 | 0.03 |
| H18 | 0.02 | 0.02 | 0.02 | 0.02 | 0.03 | 0.03 | 0.03 | 0.03 | 0.03 | 0.04 |
| H19 | 0.02 | 0.02 | 0.02 | 0.02 | 0.02 | 0.02 | 0.03 | -0.01 | 0.00 | 0.00 |
| H20 | 0.02 | 0.02 | 0.02 | 0.02 | 0.02 | 0.02 | 0.03 | 0.00 | 0.00 | 0.00 |
| H21 | 0.02 | 0.02 | 0.02 | 0.02 | 0.02 | 0.02 | 0.02 | 0.03 | 0.04 | 0.04 |
| C22 | 0.34 | 0.35 | 0.36 | 0.38 | 0.4 | 0.43 | 0.45 | 0.36 | 0.35 | 0.35 |
| C23 | 0.02 | 0.03 | 0.03 | 0.03 | 0.03 | 0.03 | 0.03 | 0.05 | 0.04 | 0.04 |
| O24 | -0.23 | -0.23 | -0.24 | -0.25 | -0.28 | -0.33 | -0.39 | -0.35 | -0.39 | -0.44 |
| Cl25 | -0.24 | -0.24 | -0.25 | -0.27 | -0.29 | -0.34 | -0.44 | -0.38 | -0.37 | -0.36 |
| H26 | 0.03 | 0.03 | 0.03 | 0.03 | 0.02 | 0.02 | 0.01 | 0.02 | 0.02 | 0.03 |
| H27 | 0.03 | 0.03 | 0.03 | 0.03 | 0.03 | 0.03 | 0.04 | 0.04 | 0.03 | 0.02 |
| H28 | 0.04 | 0.04 | 0.04 | 0.03 | 0.03 | 0.03 | 0.02 | 0.04 | 0.03 | 0.03 |

**Table 4. Change of valence corners along coordinate of reaction $R_{O7C22}$**

| **Coordinate of reaction $R_{O7C22}$,Å** | **2.8** | **2.6** | **2.4** | **2.2** | **2.0** | **1.8** | **1.6** | **1.4** | **1.3** | **1.2** |
|---|---|---|---|---|---|---|---|---|---|---|
| C(3)C(2)C(1) | 119 | 119 | 119 | 119 | 119 | 119 | 119 | 119 | 119 | 119 |
| C(4)C(3)C(2) | 113 | 113 | 113 | 112 | 112 | 112 | 112 | 108 | 107 | 107 |
| C(5)C(3)C(2) | 109 | 109 | 109 | 109 | 109 | 109 | 108 | 106 | 106 | 106 |
| C(6)C(3)C(2) | 113 | 113 | 113 | 113 | 113 | 112 | 112 | 109 | 109 | 109 |
| O(7)C(4)C(3) | 109 | 109 | 110 | 110 | 110 | 110 | 109 | 111 | 111 | 111 |
| O(8)C(5)C(3) | 109 | 109 | 109 | 109 | 109 | 109 | 109 | 115 | 115 | 114 |
| O(9)C(6)C(3) | 109 | 109 | 109 | 109 | 109 | 108 | 108 | 115 | 114 | 114 |
| P(10)O(8)C(5) | 120 | 120 | 120 | 121 | 121 | 121 | 122 | 127 | 126 | 126 |
| H(11)C(1)C(2) | 109 | 109 | 109 | 109 | 109 | 109 | 109 | 109 | 109 | 109 |
| H(12)C(1)H(11) | 107 | 107 | 107 | 107 | 107 | 107 | 107 | 106 | 106 | 107 |
| H(13)C(1)C(11) | 107 | 107 | 107 | 107 | 107 | 107 | 107 | 107 | 107 | 107 |
| H(14)C(2)C(1) | 108 | 108 | 108 | 108 | 108 | 108 | 108 | 108 | 108 | 108 |
| H(15)C(2)H(14) | 105 | 105 | 105 | 105 | 105 | 105 | 105 | 105 | 105 | 105 |
| H(16)C(4)O(7) | 108 | 108 | 108 | 108 | 108 | 108 | 108 | 110 | 109 | 108 |
| H(17)C(4)H(16) | 106 | 106 | 106 | 106 | 107 | 107 | 107 | 106 | 107 | 107 |
| H(18)C(5)O(8) | 109 | 109 | 109 | 109 | 109 | 109 | 109 | 107 | 107 | 107 |
| H(19)C(5)H(18) | 106 | 106 | 106 | 106 | 106 | 106 | 106 | 106 | 106 | 106 |
| H(20)C(6)O(9) | 109 | 109 | 109 | 109 | 109 | 109 | 109 | 108 | 108 | 108 |
| H(21)C(6)H(20) | 106 | 106 | 106 | 106 | 106 | 106 | 106 | 106 | 106 | 106 |
| C(22)O(7)C(4) | 117 | 122 | 123 | 123 | 123 | 123 | 122 | 123 | 125 | 129 |
| C(23)C(22)O(7) | 96 | 95 | 96 | 97 | 99 | 103 | 108 | 113 | 114 | 116 |
| O(24)C(22)O(7) | 128 | 128 | 128 | 128 | 128 | 126 | 125 | 127 | 124 | 120 |
| Cl(25)C(22)O(7) | 112 | 112 | 112 | 111 | 110 | 108 | 104 | 59 | 59 | 58 |
| H(26)C(23)C(22) | 111 | 111 | 111 | 111 | 111 | 111 | 110 | 110 | 110 | 110 |
| H(27)C(23)H(26) | 108 | 108 | 108 | 108 | 108 | 108 | 108 | 108 | 108 | 108 |
| H(27)C(23)H(26) | 108 | 108 | 108 | 108 | 108 | 108 | 108 | 108 | 108 | 108 |

## RESULTS AND DISCUSSION

Optimized geometrical parameters of initial compounds, an intermediate condition of system (in the moment of separation of atom $Cl_{25}$) and the final generated condition of a molecule of 5-acetyloxymethyl-2-chloro-5-ethyl-1,2,3-dioxaphosphorynane are presented to the moment in Figures 1-3. Change of lengths of bonds, valence angles, the general energy of all molecular system and charges on atoms along coordinate of reaction $R_{O7C22}$ with step 0.2Å are shown in Tables 1-3 and on diagrams Figures 4-8. The specified parameters testify that essential changes with system of initial components except for mutual orientation from each other at steps 1-6 ($R_{O7C22}$ changes from 2.8Å up to 1.8Å) not occur. In this connection we shall define this stage as a stage of coordination. At II stage of interaction (7-8 steps, $R_{O7C22}$ change from 1.8Å up to 1.4Å) there is a simultaneous and almost full break of two bonds $C_{22}$-$Cl_{25}$ and $P_{10}$-$O_7$. Lengths of bonds accordingly change from 1.95Å up to 4.42Å and from 1.6Å up to 3,62Å accordingly (table 2). We shall define this stage, as a stage of break of bonds.

Thus atom $Cl_{25}$ of acetyl chloride attacks atom $P_{10}$ of 4-ethyl-2,6,7-trioxa-1-phosphabicyclo[2.2.2]octane, forming covalent bond $P_{10}$-$Cl_{25}$ (2.02Å). Atom $O_7$ and $C_{22}$ form covalent bond $O_7C_{22}$ ($R_{O7C22}$ = 1.4Å). Formation of a required product comes to an end at last III stage. We shall name it's the final stage of synthesis of 5-acetyloxymethyl-2-chloro-5-ethyl-1,2,3-dioxaphosphorynane.

It was important to Us to track change of charges on atoms directly participating in reaction $P_{10}$, $O_7$, $C_{22}$ and $Cl_{25}$, and also change $E_0$ (the general energy of molecular system) along coordinate of reaction $R_{O7C22}$ (Figures 4-8 and Tables 1).

Reaction has barrier character (Figure 4). The size of a barrier makes 178 kJ/mol. Thus positive charges on atoms $P_{10}$ and $C_{22}$ (Figures 5 and 7) change direct proportion to change $E_0$. Charges reach the maximal values during the moment of break of bonds C22-C125 and P10-O7. Negative charges on atoms $O_7$ and $Cl_{25}$ change inversely to change $E_0$. Negative charges also reach the maximal values (on the module) during the moment of break of the same bonds. We analyze all stated above: behavior of atoms directly participating in studied synthesis $P_{10}$, $O_7$, $C_{22}$ and $Cl_{25}$, change of charges on these atoms and power of reaction - it is easy to notice, that the mechanism of synthesis of 5-acetyloxymethyl-2-chlorine-5-ethyl-1,2,3-dioxaphosphorynan represents the coordinated process with simultaneous break of bonds $P_{10}$-$O_7$ and $C_{22}$-$Cl_{25}$ and formation of new bonds $P_{10}$-$Cl_{25}$ and $C_{22}$-$O_7$ and carrying all features of $S_N2$ nucleophylic substitution.

5-Acetyloxymethyl-2-chloro-5-ethyl-1,2,3-dioxaphosphorynane is synthesized by Us for the first time and the mechanism of its synthesis is studied by quantum-chemical semi empirical method MNDO. By us it is shown, that synthesis of this compound is the result of the coordinated interactions of acetyl chloride and 4-ethyl-2,6,7-trioxa-1-phosphabicyclo [2.2.2]octane under scheme $S_N2$ nucleophylic substitution. It is established, that this reaction is endothermic. It will qualitatively be coordinated with experiment and has barrier character. Size of a power barrier of studied reaction makes 178 kJ/mol.

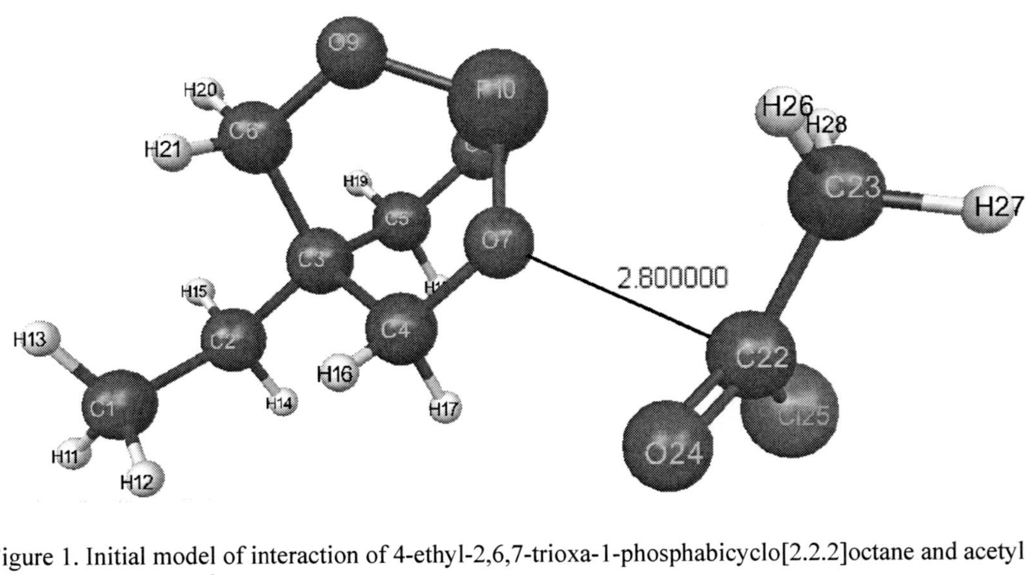


Figure 1. Initial model of interaction of 4-ethyl-2,6,7-trioxa-1-phosphabicyclo[2.2.2]octane and acetyl chloride. $R_{O7C22}$ = 2.8Å.

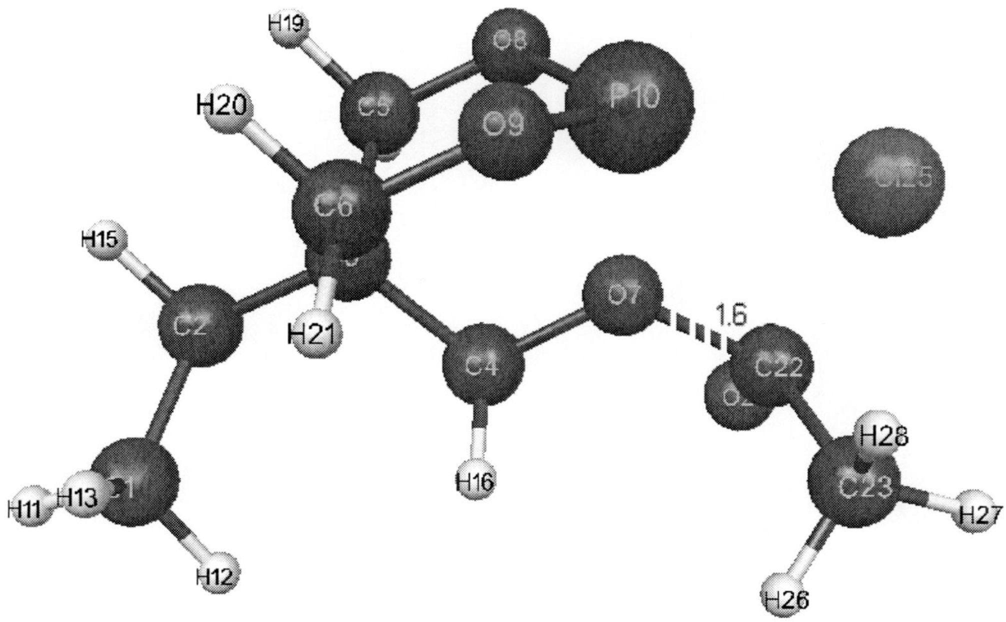


Figure 2. Model of a stage of break of bonds (a transitive condition). $R_{O7C22} = 1.6Å$.

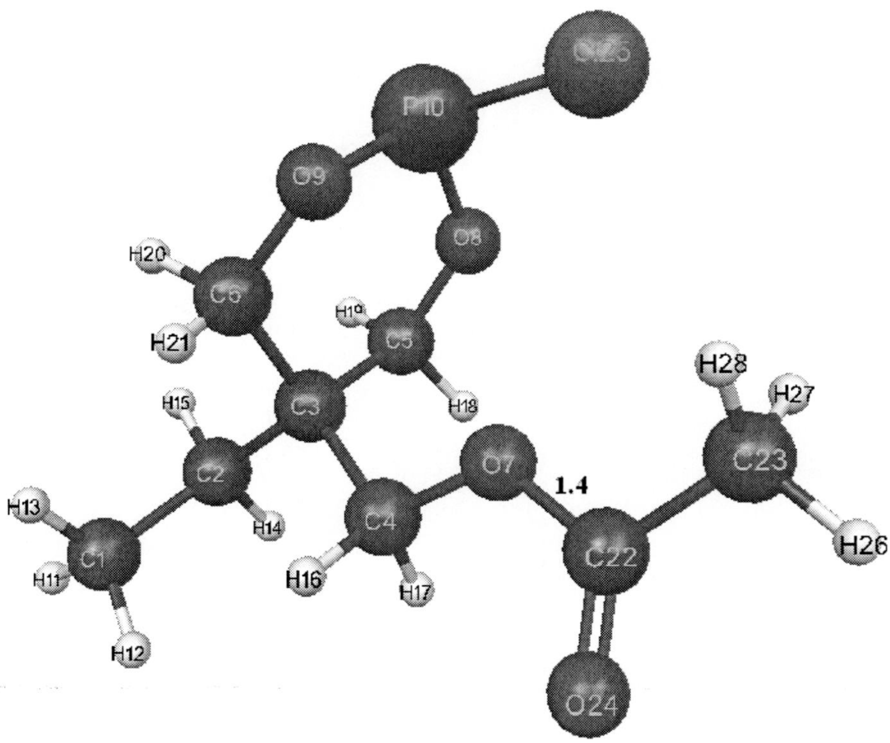


Figure 3. Final model - formation of 5-acetyloxymethyl-2-chloro-5-ethyl-1,2,3-dioxaphosphorynane. $R_{O7C22} = 1.4Å$.

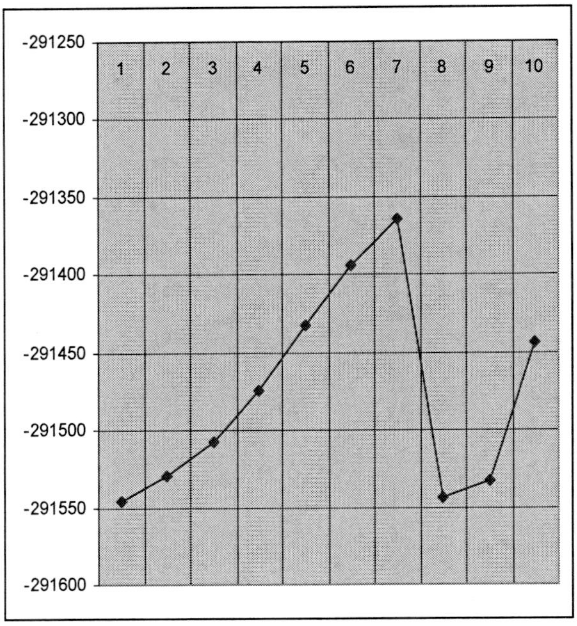


| № Step | $R_{O7C22}$, Å | $E_0$, kJ/mol |
|---|---|---|
| 1 | 2,8 | -291546,1384 |
| 2 | 2,6 | -291530,4007 |
| 3 | 2,4 | -291506,79415 |
| 4 | 2,2 | -291475,31875 |
| 5 | 2,0 | -291433,35155 |
| 6 | 1,8 | -291394,0073 |
| 7 | 1,6 | -291365,15485 |
| 8 | 1,4 | -291543,51545 |
| 9 | 1,3 | -291533,02365 |
| 10 | 1,2 | -291443,84335 |

Figure 4. Change of the general energy of system along coordinate of reaction $R_{O7C22}$.

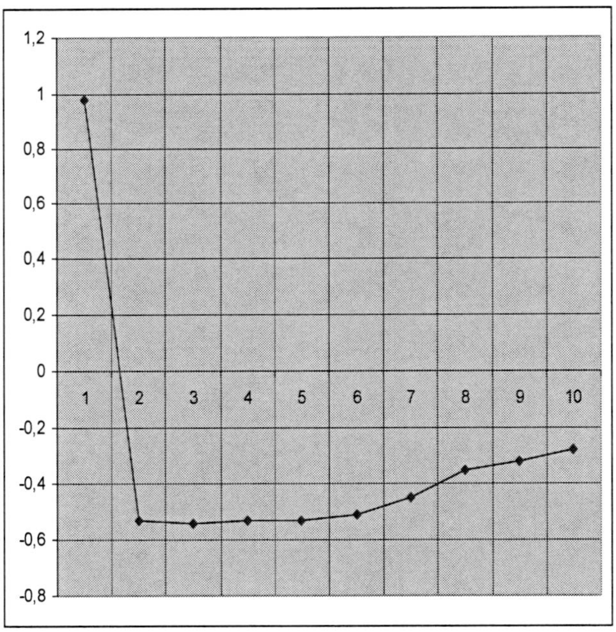


| № Step | $R_{O7C22}$,Å | q |
|---|---|---|
| 1 | 2,8 | 0,98 |
| 2 | 2,6 | -0,53 |
| 3 | 2,4 | -0,54 |
| 4 | 2,2 | -0,53 |
| 5 | 2,0 | -0,53 |
| 6 | 1,8 | -0,51 |
| 7 | 1,6 | -0,45 |
| 8 | 1,4 | -0,35 |
| 9 | 1,3 | -0,32 |
| 10 | 1.2 | -0,28 |

Figure 5. Change of a positive charge on atom of phosphorus $P_{10}$ along coordinate of reaction $R_{O7C22}$.

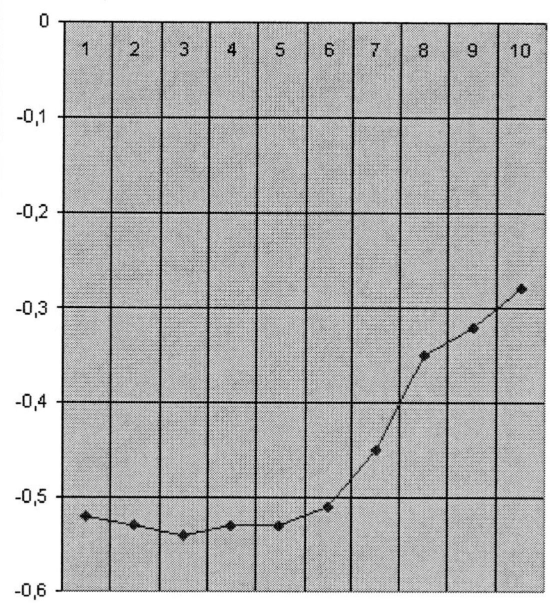


| № Step | $R_{O7C22}$, Å | q |
|---|---|---|
| 1 | 2,8 | -0.52 |
| 2 | 2,6 | -0.53 |
| 3 | 2,4 | -0.54 |
| 4 | 2,2 | -0.53 |
| 5 | 2,0 | -0.53 |
| 6 | 1,8 | -0.51 |
| 7 | 1,6 | -0.45 |
| 8 | 1,4 | -0.35 |
| 9 | 1,3 | -0.32 |
| 10 | 1.2 | -0.28 |

Figure 6. Change of a negative charge on atom $O_7$ along coordinate of reaction $R_{O7C22}$.

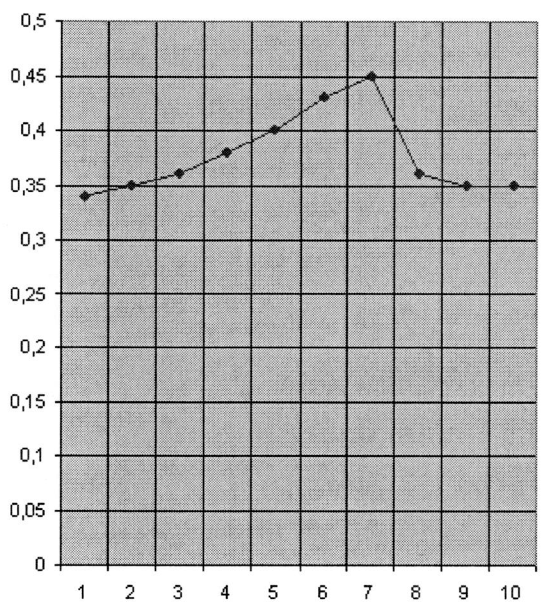


| № Step | $R_{O7C22}$,Å | q |
|---|---|---|
| 1 | 2,8 | 0.34 |
| 2 | 2,6 | 0.35 |
| 3 | 2,4 | 0.36 |
| 4 | 2,2 | 0.38 |
| 5 | 2,0 | 0.4 |
| 6 | 1,8 | 0.43 |
| 7 | 1,6 | 0.45 |
| 8 | 1,4 | 0.36 |
| 9 | 1,3 | 0.35 |
| 10 | 1.2 | 0.35 |

Figure 7. Change of a positive charge on atom of carbon $C_{22}$ along coordinate of reaction $R_{O7C22}$.

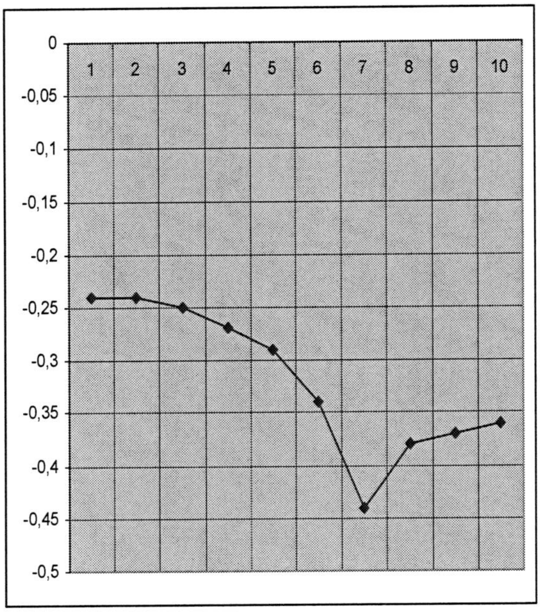


| № Step | $R_{O7C22}$,Å | q |
|---|---|---|
| 1 | 2,8 | -0.24 |
| 2 | 2,6 | -0.24 |
| 3 | 2,4 | -0.25 |
| 4 | 2,2 | -0.27 |
| 5 | 2,0 | -0.29 |
| 6 | 1,8 | -0.34 |
| 7 | 1,6 | -0.44 |
| 8 | 1,4 | -0.38 |
| 9 | 1,3 | -0.37 |
| 10 | 1.2 | -0.36 |

Figure 8. Change of a negative charge on atom Cl, along coordinate of reaction $R_{O7C22}$.

## REFERENCES

[1] Clark T. *The Computer chemistry*. - M.: Mir. 1976. 512 p.

[2] Schmidt M.W., Baldrosge K.K., Elbert J.A. et. al. // *J. Computer Chem.* 1993. Vol. 14. P. 1347-1363.

[3] Bode B.M. and Gordon M.S. // *J. Mol. Graphics Mod.* 1998. Vol. 16. P. 133-138.

In: Chemical and Biochemical Reactions
Editors: G. E. Zaikov, L. Liu et al.

ISBN 978-1-61122-860-1


*Chapter 8*

# ESTIMATION OF ACID FORCE OF COMPONENTS OF SYNTHESIS OF 5-ACETYLOXYMETHYL-2-CHLORO-5-ETHYL-1,2,3-DIOXAPHOSPHORYNANE

***V. A. Babkin*[1]*,*, *V. U. Dmitriev*[1]*, *G. A. Savin*[2]*,†,*
***E. S. Titova*[3] *and G. E. Zaikov*[4]*‡***

[1]Volgograd State Architect-Build University Sebrykov Departament, 403300 Michurin Street 21, Michailovka, Volgograd region, Russia
[2]Volgograd State Pedagogical University, Volgograd; Russia
[3]Volgograd State Technical University, Volgograd; Russia
[4]Institute of Biochemical Physics Russian Academy of Sciences; 117977 Kosigin Street 4, Moscow, Russia

## ABSTRACT

Quantum-chemical calculation of molecule of 5-acetyloxymethyl-2-chloro-5-ethyl-1,2,3-dioxaphosphorynane (I) and components of its synthesis: acetyl chloride (II) и 4-ethyl-2,6,7-trioxa-1-phosphabicyclo[2.2.2]octane (II) for the first time is executed by method AB INITIO in basis 6-311G**. The optimized geometrical and electronic structures of these compounds are received. Their acid force is theorized. All these compounds concern to a class very weak C-H-acids (pKa > 14, pKa- the universal parameter of acidity). Dependence between acid force of components pKa (II) and pKa (III) of synthesis and acid force pKa (I) a required product is positioned: pKa (I) = pKa (II) - $\sqrt{pKa(III) + 14.7}$

* E-mail: sfi@reg.avtlg.ru
† E-mail: gasavin@mail.ru
‡ E-mail: chembio@sky.chph.ras.ru

**Keywords:** Quantum-chemical calculation, method AB INITIO, acidic force, 5-acetyloxymethyl-2-chloro-5-ethyl-1,2,3-dioxaphosphorynane, acetyl chloride, 4-ethyl-2,6,7-trioxa-1-phosphabicyclo[2.2.2] octane

## AIMS AND BACKGROUNDS

5-Acetyloxymethyl-2-chloro-5-ethyl-1,2,3-dioxaphosphorynane (I) can be rather perspective start substance for reception of various chemical compounds, and in particular, the medical products possessing useful and probably unique properties.

This compound is received by interaction of acetyl chloride (III) and 4-ethyl-2,6,7-trioxa-1-phosphabicyclo[2.2.2] octane (II) in a gas phase:

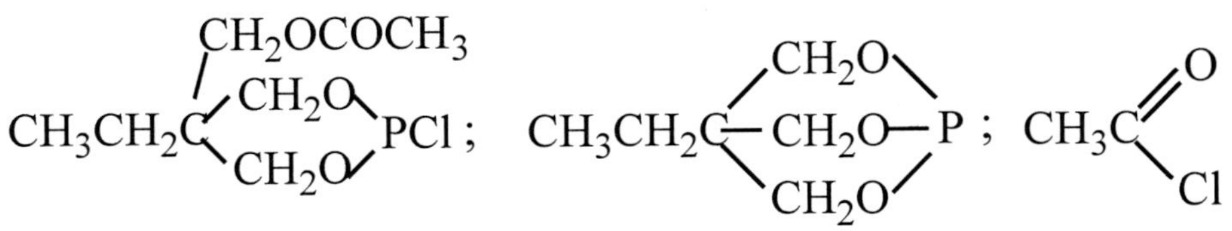

**I** **II** **III**

The mechanism of reaction of direct acylation of bicyclophosphites by chlorine anhydride of carboxylic acids now is not investigated. One of the first investigation phases of the mechanism of synthesis of studied compound (I) can consider an estimation of acid force of components of synthesis pKa (II) and pKa (III) and a required product pKa (I) and an establishment of dependence between them.

The Purpose of the present work was quantum-chemical calculation of components of synthesis of acetyl chloride (III) and 4-ethyl-2,6,7-trioxa-1-phosphabicyclo[2.2.2]octane (II) and a product 5-acetyloxymethyl-2-chloro-5-ethyl-1,2,3-dioxaphosphorynane (I) by method AB INITIO in basis 6-311G **, a theoretical estimation of their acid force and an establishment of dependence between acid force of components of synthesis pKa (III) and pKa (II) and pKa (I).

## METHODICAL PART

Method AB INITIO in basis 6-311G ** with optimization of geometry on all parameters by the standard gradient method which has been built in in PC GAMESS, used for quantum-chemical calculation of components of synthesis of 5-acetyloxymethyl-2-chloro-5-ethyl-1,2,3-dioxaphosphorynane (I) [1]. Calculation was carried out in approach of the isolated molecule a gas phase. The theoretical estimation of acid force of components of synthesis was carried out under the formula [2]:

$$pKa = 49.04-134.61 \cdot q_{MAX}^{H+},$$

where pKa - a universal parameter of acidity, and $q_{MAX}^{H+}$- the maximal charge on atom of hydrogen of a molecule.

For visual representation of models of components of synthesis program MasMolPlt [3] was used.

# Results of Calculations

The Optimized geometrical and electronic structures, the general energy, electronic energy, lengths of bonds and valent corners of components of synthesis of acetyl chloride (III) and 4-ethyl-2,6,7-trioxa-1-phosphabicyclo[2.2.2]octane (II) and a product 5-acetyloxymethyl-2-chloro-5-ethyl-1,2,3-dioxaphosphorynane (I) are received by method AB INITIO in basis 6-311G ** and presented on Figures 1-3 and in Tables 1-4. Values pKa of all components of synthesis are certain by means of the formula:

$$pKa = 49.04-134.61\cdot q_{MAX}^{H+}$$

$q_{MAX}^{H+}(I) = +0.16$ , $q_{MAX}^{H+}(II) = +0.11$ and $q_{MAX}^{H+}(III) = +0.14$. Accordingly, pKa (I) = 27.5, pKa (II) = 34.2 and pKa (III) = 30.1. By us it has been positioned, that between acid force of components of synthesis pKa (II) and pKa (III) and acid force of a received product pKa (I) there is a following dependence:

$$pKa = pKa\ (II) - \sqrt{pKa(III) + 14.7}$$

The general energy ($E_0$), the electronic energy ($E_{эл}$), the maximal charge on atom of hydrogen - $q_{MAX}^{H+}$ and values pKa-a universal parameter of acidity of components of synthesis pKa (II) and pKa (III) and a received product pKa (I) are presented in Tables 4.

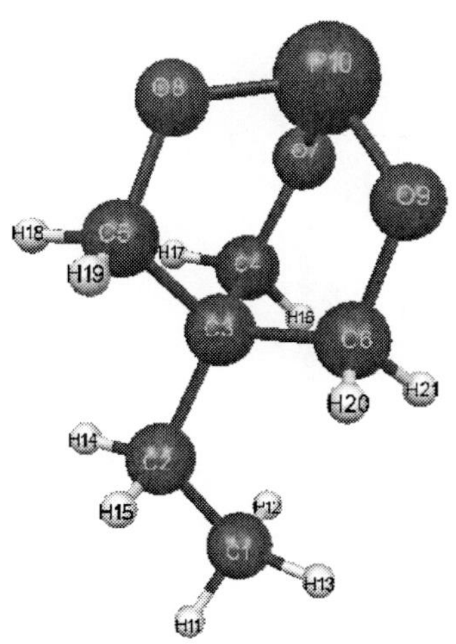


Figure 1. Geometric and electronic structure of a molecule of 4-ethyl-2,6,7-trioxa-1-phosphabicyclo[2.2.2]octane ($E_0$=-2095991 kJ/mol, $E_{el}$=-3994417 kJ/mol).

**Table 1. Optimized lengths of bonds, valency corners and charges of atoms of a molecule 4-ethyl-2,6,7-trioxa-1-phosphabicyclo[2.2.2]octane**

| Bond lengths | R, Å | Valency corners | Grad | Atom | Charge (by Milliken) |
|---|---|---|---|---|---|
| C(1)-C(2) | 1.53 | | | C1 | -0.23 |
| C(2)-C(3) | 1.53 | C(3)C(2)C(1) | 116 | C2 | -0.18 |
| C(3)-C(4) | 1.53 | C(4)C(3)C(2) | 112 | C3 | -0.60 |
| C(4)-C(5) | 1.53 | C(5)C(3)C(2) | 108 | C4 | 0.27 |
| C(5)-C(6) | 1.53 | C(6)C(3)C(2) | 112 | C5 | 0.26 |
| C(4)-O(7) | 1.41 | O(7)C(4)C(3) | 110 | C6 | 0.26 |
| C(5)-O(8) | 1.41 | O(8)C(5)C(3) | 111 | O7 | -0.71 |
| C(6)-O(9) | 1.41 | O(9)C(6)C(3) | 110 | O8 | -0.71 |
| O(8)-P(10) | 1.62 | P(10)O(8)C(5) | 118 | O9 | -0.71 |
| C(1)-H(11) | 1.08 | H(11)C(1)C(2) | 109 | P10 | 1.21 |
| C(1)-H(12) | 1.08 | H(12)C(1)C(2) | 107 | H11 | 0.10 |
| C(1)-H(13) | 1.08 | H(13)C(1)C(2) | 107 | H12 | 0.09 |
| C(2)-H(14) | 1.08 | H(14)C(2)C(3) | 108 | H13 | 0.09 |
| C(2)-H(15) | 1.08 | H(15)C(2)H(14) | 106 | H14 | 0.10 |
| C(4)-H(16) | 1.08 | H(16)C(4)O(7) | 107 | H15 | 0.10 |
| C(4)-H(17) | 1.08 | H(17)C(4)H(16) | 108 | H16 | 0.11 |
| C(5)-H(18) | 1.08 | H(18)C(5)O(8) | 107 | H17 | 0.10 |
| C(5)-H(19) | 1.08 | H(19)C(5)H(18) | 108 | H18 | 0.10 |
| C(6)-H(20) | 1.08 | H(20)C(6)O(9) | 107 | H19 | 0.11 |
| C(6)-H(21) | 1.08 | H(21)C(6)H(20) | 108 | H20 | 0.10 |
| | | | | H21 | 0.11 |

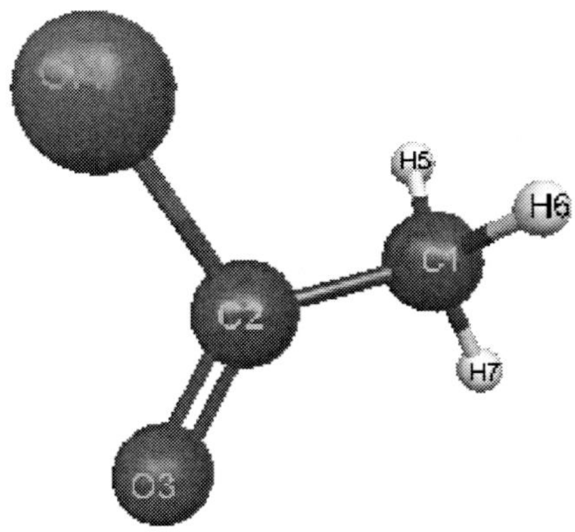


Figure 2. Geometrical and electronic structure of a molecule of acetyl chloride ($E_0$=-1604983 kJ/mol, $E_{el}$=-1998646 kJ/mol).

Quantum-chemical calculation of a molecule 5-acetyloxymethyl-2-chloro-5-ethyl-1,2,3-dioxaphosphorynane (I), and also components of its synthesis – acetyl chloride (III) and 4-ethyl-2,6,7-trioxa-1-phosphabicyclo[2.2.2]octane (II) for the first time is executed by method AB INITIO in basis 6-311G **. The optimized geometrical and electronic structures of these compounds are received. Their acid force is theorized. By us it is shown, that all of them concern to a class weak C-H-acids (pKa> 14). Dependence between acid force of components of synthesis of acetyl chloride pKa (III) and 4-ethyl-2,6,7-trioxa-1-phosphabicyclo[2.2.2]octane (II) pKa (II), and also acid force of a received product 5-acetyloxymethyl-2-chloro-5-ethyl-1,2,3-dioxaphosphorynane (I) is positioned.

**Table 2. Optimized lengths of bonds, valency corners and charges of atoms of a molecule of acetyl chloride**

| Bond lengths | R, Å | Valency corners | Grad | Atom | Charge (by Milliken) |
|---|---|---|---|---|---|
| C(1)-C(2) | 1.50 | | | C1 | -0.24 |
| C(2)-O(3) | 1.15 | O(3)C(2)C(1) | 127 | C2 | 0.30 |
| C(2)-Cl(4) | 1.79 | Cl(4)C(2)C(1) | 112 | O3 | -0.33 |
| C(1)-H(5) | 1.08 | H(5)C(1)C(2) | 109 | Cl4 | -0.14 |
| C(1)-H(6) | 1.08 | H(6)C(1)C(5) | 108 | H5 | 0.14 |
| C(1)-H(7) | 1.08 | H(7)C(1)C(5) | 110 | H6 | 0.14 |
| | | | | H7 | 0.14 |

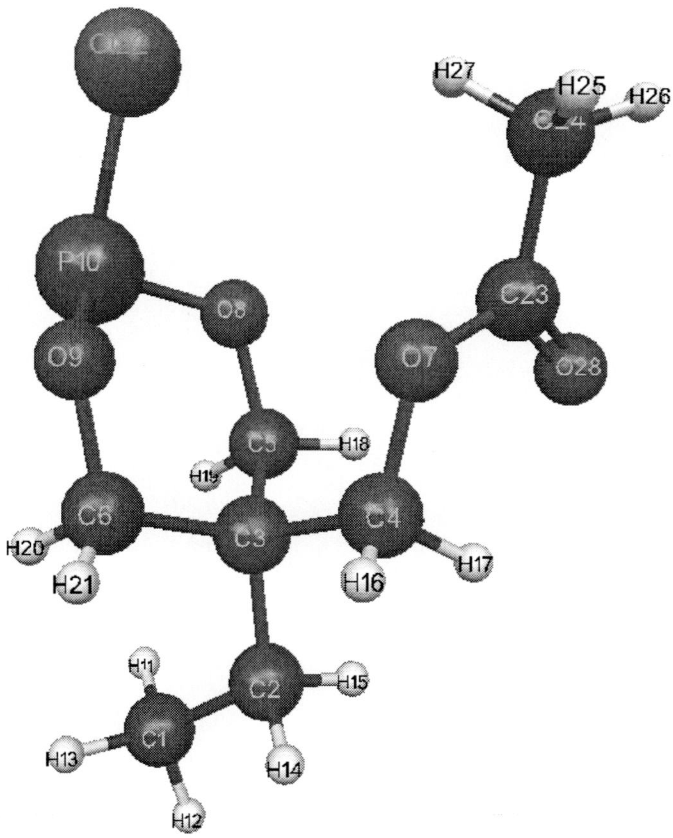


Figure 3. Geometrical and electronic structure of a molecule of 5-acetyloxymethyl-2-chloro-5-ethyl-1,2,3-dioxaphosphorynane ($E_0$ =-3700998 kJ/mol, $E_{el}$ =-7080777 kJ/mol).

**Table 3. Optimized lengths of bonds, valency corners and charges of atoms of a molecule of 5-acetyloxymethyl-2-chloro-5-ethyl-1,2,3-dioxaphosphorynane**

| Bond lengths | R, Å | Valency corners | Degree | Atom | Charge (by Milliken) |
|---|---|---|---|---|---|
| C(1)-C(2) | 1.53 | | | C1 | -0.23 |
| C(2)-C(3) | 1.55 | C(3)C(2)C(1) | 118 | C2 | -0.20 |
| C(3)-C(4) | 1.54 | C(4)C(3)C(2) | 104 | C3 | -0.34 |
| C(4)-C(5) | 1.54 | C(5)C(3)C(2) | 109 | C4 | 0.15 |
| C(5)-C(6) | 1.53 | C(6)C(3)C(2) | 110 | C5 | 0.13 |
| C(4)-O(7) | 1.41 | O(7)C(4)C(3) | 115 | C6 | 0.12 |
| C(5)-O(8) | 1.42 | O(8)C(5)C(3) | 112 | O7 | -0.43 |
| C(6)-O(9) | 1.41 | O(9)C(6)C(3) | 112 | O8 | -0.68 |
| O(8)-P(10) | 1.61 | P(10)O(8)C(5) | 118 | O9 | -0.66 |
| C(1)-H(11) | 1.08 | H(11)C(1)C(2) | 112 | P10 | 1.01 |
| C(1)-H(12) | 1.08 | H(12)C(1)C(2) | 107 | H11 | 0.09 |
| C(1)-H(13) | 1.08 | H(13)C(1)C(2) | 107 | H12 | 0.11 |
| C(2)-H(14) | 1.08 | H(14)C(2)C(3) | 108 | H13 | 0.09 |
| C(2)-H(15) | 1.08 | H(15)C(2)H(14) | 105 | H14 | 0.11 |
| C(4)-H(16) | 1.08 | H(16)C(4)O(7) | 105 | H15 | 0.12 |
| C(4)-H(17) | 1.07 | H(17)C(4)H(16) | 108 | H16 | 0.11 |
| C(5)-H(18) | 1.07 | H(18)C(5)O(8) | 105 | H17 | 0.14 |
| C(5)-H(19) | 1.08 | H(19)C(5)H(18) | 108 | H18 | 0.16 |
| C(6)-H(20) | 1.08 | H(20)C(6)O(9) | 108 | H19 | 0.11 |
| C(6)-H(21) | 1.08 | H(21)C(6)H(20) | 107 | H20 | 0.11 |
| P(10)-Cl(22) | 2.06 | Cl(22)P(10)O(8) | 99 | H21 | 0.12 |
| O(7)-C(23) | 1.32 | C(23)O(7)C(4) | 118 | Cl22 | -0.29 |
| C(23)-C(24) | 1.50 | C(24)C(23)O(7) | 111 | C23 | 0.49 |
| C(24)-H(25) | 1.08 | H(25)C(24)C(23) | 110 | C24 | -0.24 |
| C(24)-H(26) | 1.08 | H(26)C(24)C(25) | 110 | H25 | 0.12 |
| C(24)-H(27) | 1.08 | H(27)C(24)C(25) | 107 | H26 | 0.12 |
| C(23)-O(28) | 1.30 | O(28)C(23)O(7) | 123 | H27 | 0.15 |
| | | | | O28 | -0.48 |

**Table 4. The general energy, energy of bonds, the maximal charge on atom of hydrogen, a universal parameter of acidity of components of synthesis of 5-acetyloxymethyl-2-chloro-5-ethyl-1,2,3-dioxaphosphorynane**

| № | The Component of synthesis | $E_0$, kJ/mol | $E_{el}$, kJ/mol | $q_{MAX}^{H+}$ | pKa |
|---|---|---|---|---|---|
| 1. | 5-Acetyloxymethyl-2-chloro-5-ethyl-1,2,3-dioxaphosphorynane | -3700998 | -7080777 | +0.16 | 27.5 |
| 2. | 4-Ethyl-2,6,7-trioxa-1-phosphabicyclo[2.2.2]octane | -2095991 | -3994417 | +0.11 | 34.2 |
| 3. | Acetyl chloride | -1604983 | -1998646 | +0.14 | 30.1 |

## References

[1] Schmidt M.W., Baldrosge K.K., Elbert J.A. et. al. // *J. Computer Chem.* 1993. Vol. 14. P. 1347-1363.

[2] Babkin V.A., Fedunov R.G., Minsker K.S. et. al. // *Oxidation communication.* 2002. Vol. 25. № 1. P. 21-47.

[3] Bode B.M. and Gordon M.S. // *J. Mol. Graphics Mod.* 1998. Vol. 16. P. 133-138.

In: Chemical and Biochemical Reactions
Editors: G. E. Zaikov, L. Liu et al.
ISBN 978-1-61122-860-1


*Chapter 9*

# STRUCTURE AND PROPERTIES OF POLYMER ADSORPTION LAYERS ON INORGANIC PIGMENTS SURFACE IN AQUEOUS DISPERSION FORMED UNDER ULTRASONIC TREATMENT

***N. Bulychev*[1,2]*, E. Kisterev*[1]*, K. Dirnberger*[2]*,***
***V. Zubov*[3] *and C .D. Eisenbach*[2,4]**

[1]N.S. Kurnakov Institute of General and Inorganic Chemistry of Russian Academy of Sciences, 119991, Leninsky avenue, 31, Moscow, Russia
[2]Institute for Polymer Chemistry, University of Stuttgart, Pfaffenwaldring 55, D-70569 Stuttgart, Germany
[3]Lomonosov Moscow State Academy of Fine Chemical Technology, Pr. Vernadskogo, 86, 117571, Moscow, Russia
[4]Research Institute for Pigments and Coatings, Allmandring 37, D-70569 Stuttgart, Germany

## ABSTRACT

The course and the result of the surface modification of titanium dioxide and ferrous oxide pigments in aqueous dispersion by ethylhydroxyethylcellulose (EHEC) without and with mechanical treatment of the dispersion by ultrasonic power was studied by the electrokinetic sonic amplitude (ESA) method. The evaluation of the ESA data showed that the ultrasonic treatment causes a significant thickness increase of the EHEC layer on the pigment which is primarily attributed to the ultrasonically induced activation of the pigment surface. The thickness of the polymer adsorption layer derived from ESA measurements was confirmed by TEM investigations. The ultrasonic treatment leads to significant changes of the properties of the adsorption layers and is a promising method for the surface modification of pigments with regard to dispersion stability.

**Keywords:** aqueous dispersions, polymers, electrokinetic sonic amplitude method, polymer adsorption layers, ultrasonic treatment

## Aim and Background

Recent developments in research of dispersed systems have shown that mechanical, in particular ultrasonic treatment of aqueous dispersions of pigments in the presence of polymeric stabilizers leads to a significant enhancement of the stability of these dispersions in comparison to dispersions prepared without ultrasonic treatment. However, the effect of the ultrasonic treatment on the pigment-polymer interaction and structure formation of adsorption layers, especially for novel tailor-made amphiphilic copolymers, is still obscure and needs further investigation for elucidation of the phenomena observed. The main aim of this paper is to get more insights on the effects of the ultrasonic treatment on the polymer solid-liquid interface behaviour and the formation of polymer adsorption layers.

## Introduction

In recent years, aqueous colloidal dispersions of pigments have been of increasing interest from both scientific and practical points of view. They are important, ecologically friendly colloidal systems widely used in polygraphic and paint industries. The pigment particles are usually of 200 – 2000 μm diameter and strongly aggregated. In order to achieve a good stabilization in aqueous pigment dispersions, many formulations were proposed [1-8]. Earlier we reported about the role of mechanical, e.g. ultrasonic treatment for obtaining of highly stable dispersions [7,9]. It was shown, that using of polymer surfactants in combination with mechanical action can significantly improve the quality of dispersed systems. However, specific aspects of the pigment-polymer interaction and structure formation of adsorption layers especially under mechanical, i.e., ultrasonic action have not yet been studied in detail.

Electrokinetic sonic amplitude (ESA) measurements have been demonstrated to be a powerful method that could provide the desired information about the process of polymer adsorption [7-9] and was employed in this work. Commercial ethylhydroxyethylcellulose (EHEC) was used as stabilizer of $TiO_2$ and $Fe_2O_3$ aqueous dispersions. The polymer layers created on the particle surfaces in the absence and presence of mechanical (ultrasonic) action were investigated.

## Experimental Part

### Materials

As pigments, the titanium dioxide rutil pigment Kronos 2310 with particle size 0.3 μ, ferrous oxide Bayerfox with particle size 0.3 μ were employed as received. EHEC with molecular weight 60.000 was provided by Berol.

## Techniques

For the preparation of aqueous pigment dispersions, the pigment was added to water or to the aqueous polymer solution and dispersing of the pigment was first achieved by means of a laboratory stirrer (700 rpm for 10 min). When ultrasonification was applied, the system was subsequently treated with ultrasound for 2 min with an ultrasonic generator UZDN-2 or Branson Sonifier B-12 with actual power of 1.5 $W/cm^2$.

The pigment-polymer interaction and the polymer adsorption layer formation were investigated by electrokinetic sonic amplitude (ESA) measurements as described elsewhere [12]. The particle size was measured by both ESA (cf. [12]) and by using Nanosizer (Malvern Instruments). Standard optical equipment was used for refractive index measurements. Transmission electron microscopy (TEM) investigations were carried out according to the standard procedure. The samples of the pigment dispersions were diluted with distilled water by a factor of 100, then dropped on the TEM grid of 3 mm diameter and dried on air for 1 day. The TEM pictures were taken by using a Jeol 2000 FX device under the voltage of 200 kV.

# RESULTS AND DISCUSSION

Data reported in literature [7,9,10], showed that mechanical, in particular, ultrasonic treatment of aqueous dispersions of pigments in the presence of polymeric stabilizers leads to a significant enhancement of the stability of these dispersions as compared to dispersions prepared without ultrasonic treatment. It was proven by IR-analysis that the thickness and stability of the polymer adsorption layers were increased and improved by the ultrasonification [9]. The optimal parameters of the ultrasonic treatment and their influence on the properties of dispersed systems were established [11]. However, the effect of the ultrasonic treatment on the pigment-polymer interaction and polymer adsorption is still obscure and needs further investigation for better elucidation of the phenomena observed.

ESA measurements of aqueous dispersions of $TiO_2$, stabilized by ethylhydroxyethylcellulose (EHEC), give quantitative information about the process of adsorption of polymers as reflected from first the dependence of the dynamic mobility $\mu$ on the relative polymer concentration (cf. [12-14]) as shown in Figure 1 and 2, and second from the comparison of this dependency for systems without and with ultrasonic treatment (Figure 2).

As one can see from the comparison of Figure 1 and 2, ultrasonic treatment has a distinct effect on the behavior of the particulate pigment-polymer suspension. Without ultrasonic action, the formation of a polymeric adsorption layer on the pigment surface seems to be reached at 1 wt.-% EHEC in relation to the pigment concentration (saturation concentration as indicated by the unchanged dynamic mobility with further increase of polymer concentration) followed by polymolecular adsorption between about 2 and 3 %. In the presence of ultrasonic action, first the initial dynamic mobility of the pristine $TiO_2$ is much higher, and the saturation concentration of polymer is only reached at about 15 %. Further addition of polymer does not effect the dynamic mobility.

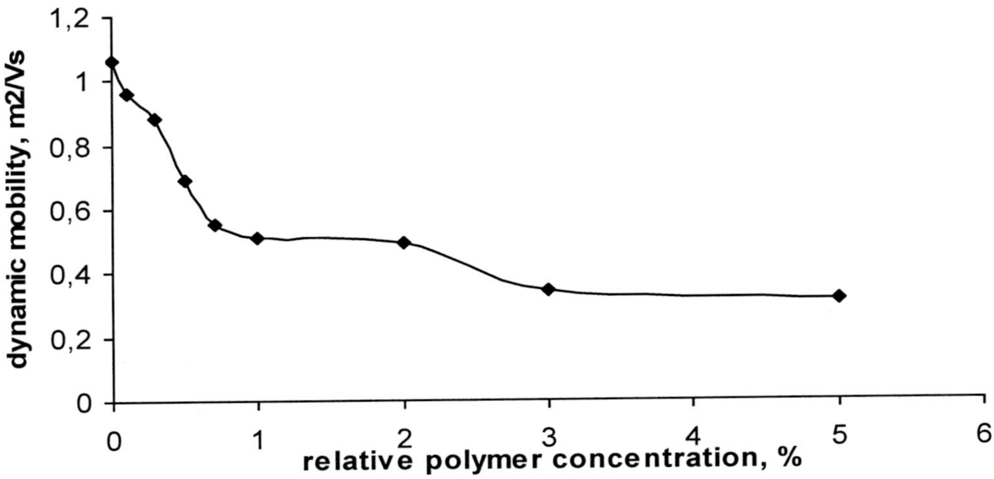


Figure 1. Dependence of the dynamic mobility on the relative concentration of EHEC for 1 wt.-% $TiO_2$ aqueous dispersion without ultrasonic treatment.

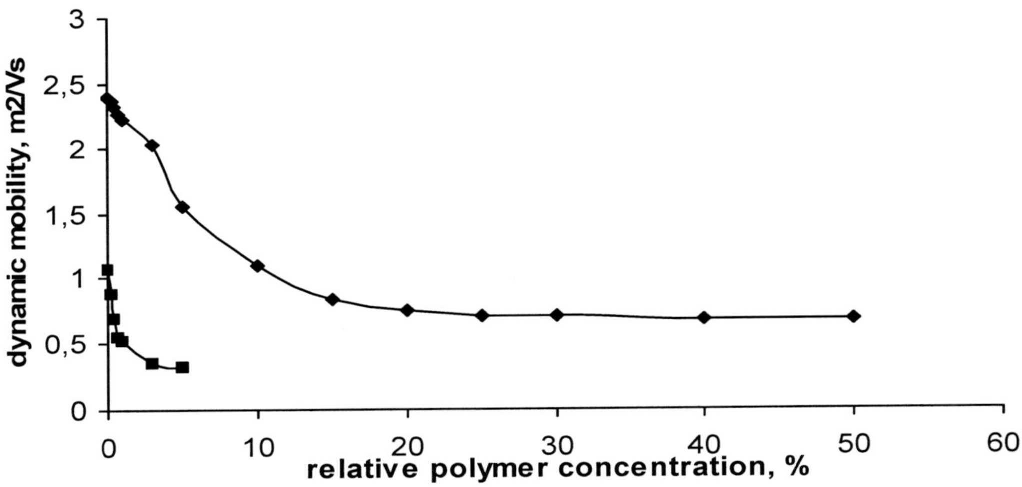


Figure 2. Dependence of the dynamic mobility on the relative concentration of EHEC for 1 wt.-% $TiO_2$ aqueous dispersion with (◆) and without (■, cf. Figure 1) ultrasonic treatment..

However, e.g., for the system with 50 wt.-% polymer relative to the 1 wt.-% pigment dispersion, it was observed that the dynamic mobility was further decreased to 0.35 $m^2/Vs$ after 1 day. This means that even though the ultrasonic power was shut off, polymer deposition still occurred, indicating that polymolecular adsorption in case of treated samples proceeds over a longer period of time. Obviously the ultrasonic action not only increases the pigment surface by creating a more fine dispersion but also activates the pigment surface leading to an ultimately higher polymer adsorption in comparison to the non-treated samples. Similar results were obtained for $Fe_2O_3$ aqueous dispersions as revealed from Figure 3, 4.

Comparing the values of saturation concentration without and after ultrasonic treatment, one can infer that the amount of polymer adsorbed on the particle surface significantly increases when ultrasonification is applied. In this context it has to be considered that the ultrasonic treatment leads to a finer dispersion by decreasing the particle size, and, consequently leading to an increase of the surface area prone to polymer deposition. Thus both the increase in the total surface area of the dispersed particles and a possible ultrasonically induced activation of the particle surface must be considered as being responsible for the observed effects.

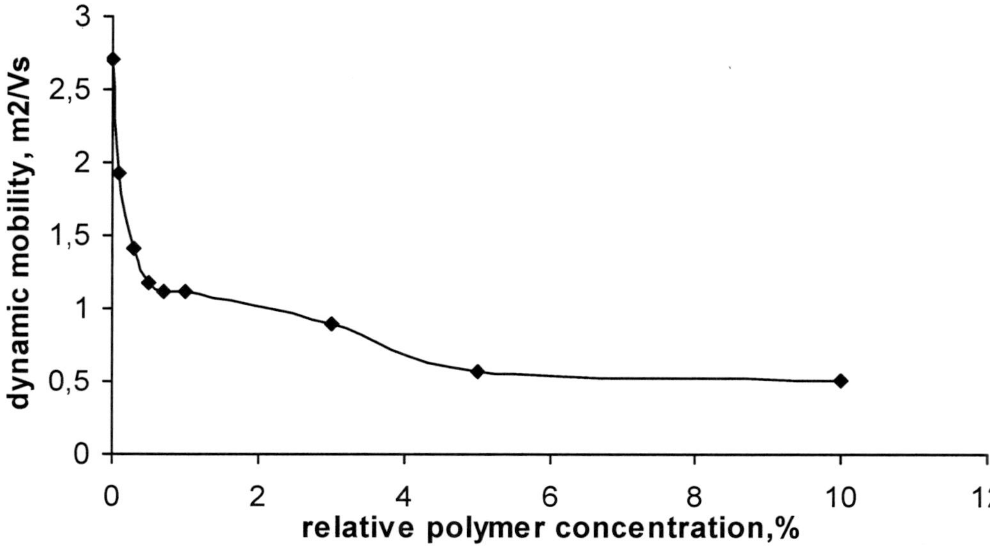


Figure 3. Dependence of the dynamic mobility on the relative concentration of EHEC for 1 wt.-% $Fe_2O_3$ aqueous dispersion without ultrasonic treatment.

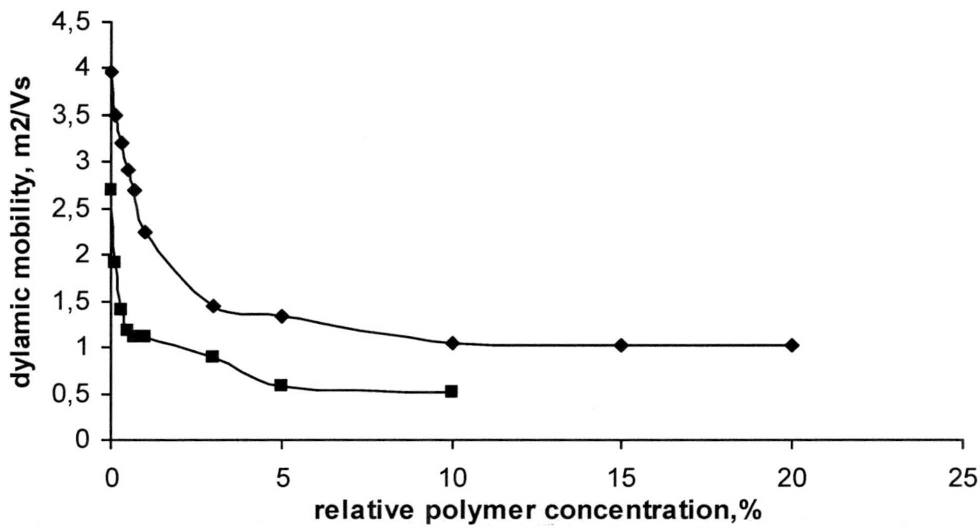


Figure 4. Dependence of dynamic mobility on relative concentration of EHEC for 1 wt.-% $Fe_2O_3$ aqueous dispersion with (◆) and without (■, cf. Figure 1) ultrasonic treatment.

To get some quantitative information about the particulars of the polymer adsorption and to explain the increased amount of polymer adsorbed when ultrasonic treatment was applied, the surface area of the uncoated pigment and of the polymer coated pigment were calculated; these data were compared with information obtained from TEM pictures of the particulate system and further related to data derived from ESA measurements.

First, the total surface area of the particles as related to the particle radius was calculated: The total surface area $\mathbf{S_{tot}}$ of all particles in the dispersion can be expressed by:

$$S_{tot} = S_{part} * n_{part} \tag{1}$$

Here $\mathbf{S_{part}}$ is the surface area of a single particle, and $\mathbf{n_{part}}$ is the number of particles.

Assuming the particles to be spherical, the surface area of one particle can be calculated from the particle radius $\mathbf{r_{part}}$:

$$S_{part} = 4 * \pi * r^2{}_{part} \tag{2}$$

The number of particles in the dispersion as related to the particle radius is given by:

$$n_{part} = \frac{m_{total}}{m_{part}} = \frac{m_{total}}{V_{part} * \rho_{TiO_2}} = \frac{3 * m_{total}}{4 * \pi * r^3{}_{part} * \rho_{TiO_2}} \tag{3}$$

Here $\mathbf{m_{total}}$ is the total mass of the pigment in the dispersion, $\mathbf{m_{part}}$ is the mass of a single particle, $\mathbf{V_{part}}$ is the single particle volume, and $\boldsymbol{\rho_{TiO2}}$ is the density of titanium dioxide.

Thus, the formula for the total surface area of all particles in the dispersion is expressed by:

$$S_{tot} = \frac{4 * \pi * r^2{}_{part} * 3 * m_{total}}{4 * \pi * r^3{}_{part} * \rho_{TiO_2}} = \frac{3 * m_{total}}{r_{part} * \rho_{TiO_2}} \tag{4}$$

The ratio of the total particle surface area of the ultrasonically treated (index 2) and non-treated (index 1) dispersions is given by:

$$\frac{S_{tot_2}}{S_{tot_1}} = \frac{3 * m_{total} * r_{part_1} * \rho_{TiO_2}}{3 * m_{total} * r_{part_2} * \rho_{TiO_2}} = \frac{r_{part_1}}{r_{part_2}} = \frac{d_{part_1}}{d_{part_2}} \tag{5}$$

Thus, the ratio between the values of the particle surface area $S_{tot\,2}/S_{tot\,1}$ can be expressed as the reverse ratio of the average particle diameters, i.e., $\left(d_{part\,2}/d_{part\,1}\right)^{-1}$.

Using this equation and by inserting the values of the particle size at the saturation concentration of ultrasonically treated and non-treated samples of $TiO_2$ and $Fe_2O_3$ dispersion aqueous in EHEC-solution, it is possible to calculate the increasing of particles surface area during the ultrasonic treatment. These data are given in Table 1.

From Table 1 it is seen that for both systems the particle surface area increases by a factor of 3.4 – 4.7 upon ultrasonic treatment of the dispersion. However, the saturation concentrations for these systems increase by a factor of 7 – 15. This indicates that the amount of polymer adsorbed per unit of the particle surface after ultrasonic treatment is higher in comparison with the non-treated samples; this also means an increased thickness of the adsorption layer in the case of ultrasonically treated dispersions.

**Table 1. Effect of ultrasonic treatment on particle diameter ($d_{part}$) and increase of particle surface area ($S_{tot}$) in comparison to the saturation concentration (SC) of polymer adsorption; indices 1, 2: non-treated and ultrasonically treated sample**

| System | Averaged particle diameter, μm | | Ratio between particle surface area with and without ultrasonic treatment $S_{tot\,2}/S_{tot\,1}$ | Ratio between saturation concentration with and without ultrasonic treatment $SC_2/SC_1$ |
|---|---|---|---|---|
| | Without ultrasonic treatment | After ultrasonic treatment | | |
| $TiO_2$ + EHEC | 0.85 | 0.25 | 3.4 | 15 |
| $Fe_2O_3$ + EHEC | 1.4 | 0.3 | 4.7 | 7 |

**Table 2. Ratios between saturation concentration (SC) and corresponding change of dynamic mobility Δμ for pigment dispersion without (index 1) and with (index 2) ultrasonic treatment**

| System | Ratio between saturation concentration after and without ultrasonic treatment $SC_2/SC_1$ | Difference between dynamic mobility of pure pigment and that at the saturation concentration without ultrasonic treatment $\Delta\mu_1$ | Difference between dynamic mobility of pure pigment and that at the saturation concentration after ultrasonic treatment $\Delta\mu_2$ | $\Delta\mu_2/\Delta\mu_1$ |
|---|---|---|---|---|
| $TiO_2$ + EHEC | 15 | 0.5 | 1.7 | 3.4 |
| $Fe_2O_3$ + EHEC | 7 | 1.7 | 2.5 | 1.47 |

The difference of the dynamic mobility $\Delta\mu$ between the pure and EHEC coated pigments (at the saturation concentration) is higher for the ultrasonically treated ($\Delta\mu_2$) than the non-treated ($\Delta\mu_1$) samples (see Table 2). The proportionality between the $\Delta\mu_2/\Delta\mu_1$ ratio (last column Table 2) of the two pigment systems is similar as the ratio $SC_2/SC_1$ of the saturation concentration (second column Table 2).

Thus the correlation of the ESA measurement data with regard to the established saturation concentration and change in the dynamic mobility gives a conclusive picture of the effects of ultrasonic treatment on the pigment/EHEC interaction which is in accordance with the calculations of the particle surface/particle diameter relation (Eq. 5).

As it was proven by refraction index measurements of the supernatant liquid of sedimented pigment dispersions, up to the saturation concentration of the polymer, all polymer added to the system is adsorbed on the particle surface; only when this concentration was exceeded, a change in the index of refraction due to dissolved polymer was observed. Thus, since the amount of polymer adsorbed and the surface area of the particles is known, the thickness of adsorption layers is given by Eq. (6) where the total area of the particles $\mathbf{S_{tot}}$ is calculated by the equation (4):

$$d = \frac{V_{tot}}{S_{tot}} \tag{6}$$

The volume of the adsorption layer $\mathbf{V_{tot}}$ is given by the ratio of the mass of polymer adsorbed $\mathbf{m_{tot\,pol}}$ to the density $\boldsymbol{\rho_{pol}}$ of the polymer (Eq. 7):

$$V_{tot} = \frac{m_{tot\,pol}}{\rho_{pol}} \tag{7}$$

The relationships of Eq. (6) and (7) allow to calculate the thickness of the adsorption layers on the pigment surface for both the ultrasonically treated and untreated pigment dispersions. The data are presented in Table 3. One can see that the thickness of the adsorption layer increases by the ultrasonic treatment for both systems tested. This confirms that during the ultrasonic treatment an activation of the pigment surface occurs.

**Table 3. Thicknesses (d; cf. Eq. (6)) of the EHEC polymer adsorption layer on the $TiO_2$ and $Fe_2O_3$ for treated and non-treated dispersions as calculated on the basis of the saturation concentration obtained from the ESA measurements**

| System | Thickness of adsorption layer without ultrasonic treatment, nm. | Thickness of adsorption layer after ultrasonic treatment, nm. | Ratio between thicknesses of treated and non-treated samples |
|---|---|---|---|
| $TiO_2$ + EHEC | 5 | 17 | 3.4 |
| $Fe_2O_3$ + EHEC | 11 | 16 | 1.5 |

**Table 4. Thickness of inner (Δ) and outer (δ) EHEC polymer adsorption layer on $TiO_2$ and $Fe_2O_3$ pigment for dispersions without and with ultrasonic treatment as calculated from Eq. (8) by using experimental ESA data (see text)**

| System | Thickness (Δ) of the inner adsorption layer in nm | | Thickness (δ) of the outer adsorption layer in nm | |
|---|---|---|---|---|
| | without ultrasonic treatment | with ultrasonic treatment | without ultrasonic treatment | with ultrasonic treatment |
| $TiO_2$ + EHEC | 1.2 | 8.7 | 4.7 | 6.5 |
| $Fe_2O_3$ + EHEC | 2.4 | 3.8 | 7.3 | 7.7 |

Aqueous dispersions of $TiO_2$ and $Fe_2O_3$ stabilized by EHEC were also investigated by transmission electron microscopy (TEM) in order to visualize the change of the dispersed system upon ultrasonic treatment, and also to confirm the above conclusions about the thicknesses of the EHEC polymer adsorption layer.

From the TEM picture Figure 5 it is seen that the dispersion of the $TiO_2$ without added polymer stabilizer and without ultrasonic treatment consists of aggregates of about 5 – 10 particles; the size of such aggregates of the non-treated system is about 1 μm and is in agreement with Nanosizer and ESA data. However, the dispersion of the $TiO_2$, again without EHEC, after ultrasonic treatment consists of extended aggregates (Figure 6) of sizes up to 4 – 5 μm; this size increase is due to the activation of the pigment particle surface by the ultrasonic power. As a result of this coagulation, fast sedimentation of the dispersion occurs. From Figure 5 and 6 it is also evident that the primary particles are spherical in shape; this confirms the assumption made above in the calculation of the particle surface area.

When the dispersion of the $TiO_2$ or $Fe_2O_3$ pigments is carried out with added EHEC, the polymer acts as a stabilizer and primary particle dispersions are obtained. As it is seen from the comparison of the TEM pictures of the $TiO_2$ or $Fe_2O_3$ aqueous dispersions stabilized by EHEC (polymer concentration chosen corresponding to the saturation concentration as obtained from ESA measurements, Figure 1 – 4) which were obtained without (Figure 7 and 9) and with ultrasonic treatment (Figure 8 and 10), the ultrasonification leads to an increased thickness of the EHEC coating layer around the primary particles.

These findings are in agreement with the much higher polymer saturation concentration required in the ultrasonically treated dispersions (see Figure 2 and 4) as compared to the non-treated systems (see Figure 1 and 3) which has already been discussed above (Table 1 and Table 2). The thickness of the adsorption layers strongly increases from 4 – 6 μm of the not ultrasonically treated dispersions to about 15 – 20 μm when ultrasonic treatment was applied. This is in rough agreement with the values calculated above from ESA data (see Tab. 3).

Thus, the TEM investigations allow to conclude that mechanical, i.e., ultrasonic treatment has a pronounced effect on the pigment-polymer interaction in the dispersed system: the amount of polymer adsorbed and the thickness of the adsorption layer is significantly increased due to the activation of the particle surface.

In order to further elucidate the origin of this polymer adsorption layer growth, ESA measurements under variation of the frequency of the applied alternating electrical field have been carried out. These data can be used to get quantitative information about the structure of the adsorption layers according to the polymer gel layer theory (cf. [15]). In this theory it is assumed that the polymer adsorption layer consists of an inner layer of polymers adsorbed on the particle surface and an outer layer formed by the interpolymer interactions; the inner layer has a higher density and its formation is a direct result of pigment-polymer interactions.

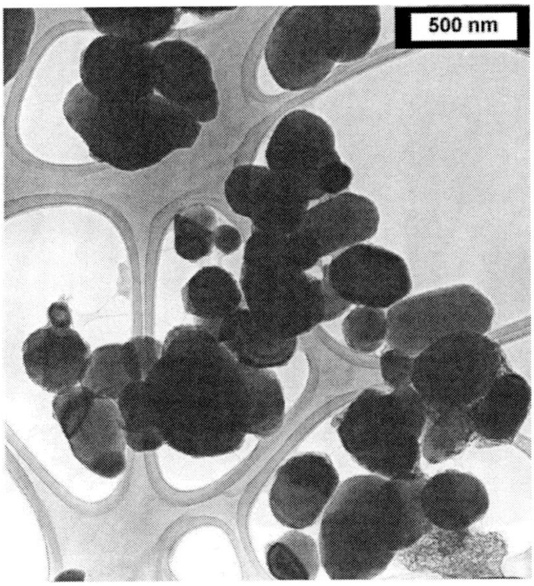


Figure 5. TEM picture of $TiO_2$ aqueous dispersion without polymer stabilizer and without mechanical (ultrasonic) treatment..

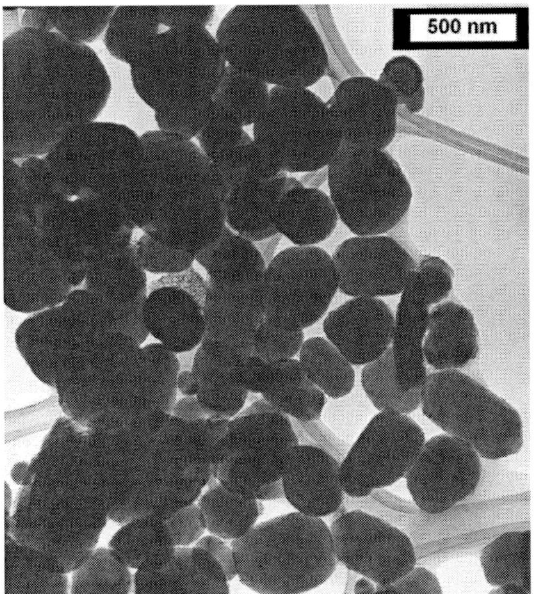


Figure 6. TEM picture of $TiO_2$ aqueous dispersion without polymer stabilizer after ultrasonic treatment.

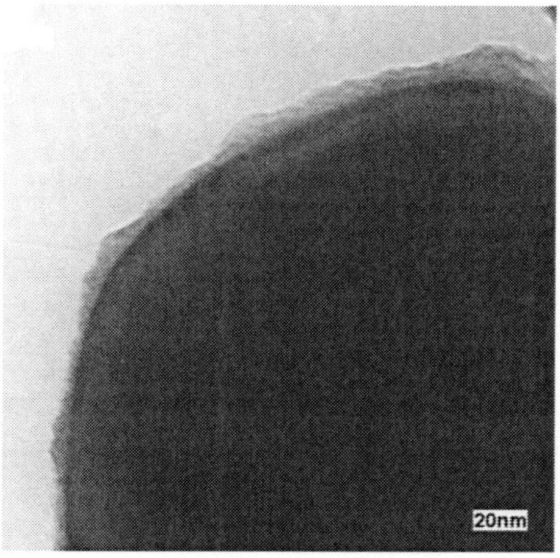


Figure 7. TEM picture of $TiO_2$ primary particle aqueous dispersion stabilized by EHEC (1 wt.-% relative pigment) without mechanical (ultrasonic) treatment.

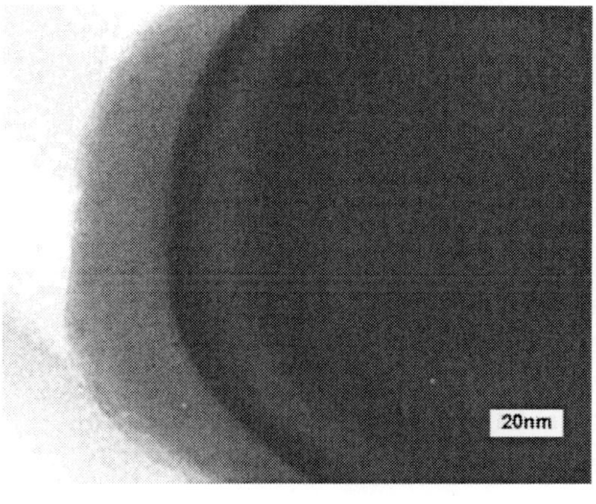


Figure 8. TEM picture of $TiO_2$ primary particle aqueous dispersion stabilized by EHEC (15 wt.-% relative pigment) after ultrasonic treatment.

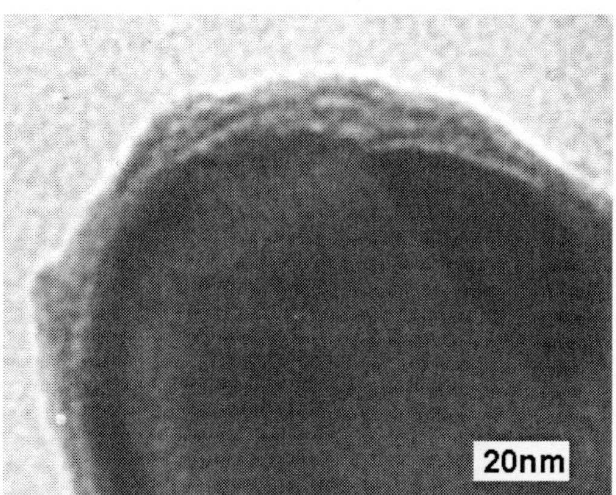


Figure 9. TEM picture of $Fe_2O_3$ primary particle aqueous dispersion stabilized by EHEC (1 wt.-% relative pigment) without mechanical (ultrasonic) treatment.

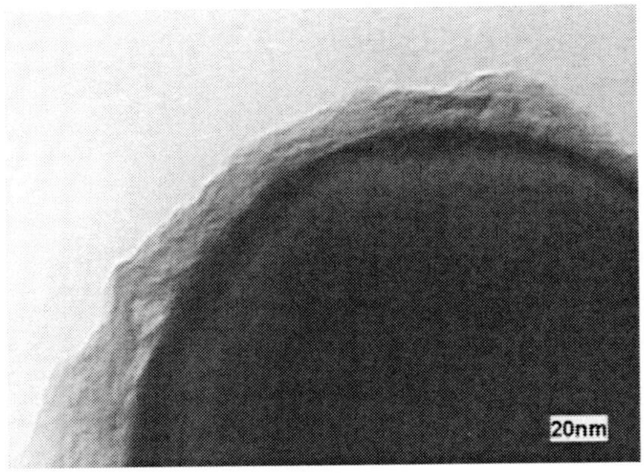


Figure 10. Picture of $Fe_2O_3$ aqueous dispersion stabilized by EHEC (7 wt.-% relative to pigment) after ultrasonic treatment.

The procedure given in literature [16] for the calculation of the thicknesses of the inner ($\Delta$) and outer ($\delta$) polymer adsorption layers after the polymer gel layer theory was applied for non-treated and ultrasonically treated dispersions of $TiO_2$ and $Fe_2O_3$, stabilized by EHEC. The formula for calculating the relative dynamic mobility $\mu/\mu_0$ (ratio of the dynamic mobility $\mu$ measured for polymer containing pigment dispersion to the dynamic mobility $\mu_0$ of the polymer free dispersion) is given in Eq. (8):

$$\frac{\mu}{\mu_0} = \ell^{-\kappa\Delta}\left\{-\frac{i\omega\alpha/\gamma - k^2}{k^2-\kappa^2} * \left[\ell^{-\kappa\delta}\left(1+\frac{\kappa}{k}\tanh k\delta\right) - \frac{\kappa^2}{k^2\cosh k\delta}\right] - \frac{i\omega\alpha}{\gamma k^2}\right\} \qquad (8)$$

Here the quantity k is given by:

$$k^2 = \frac{\alpha}{\eta}\left(1 + \frac{i\omega\eta}{\gamma}\right) \qquad (9)$$

and the quantity $\gamma$ by:

$$\gamma = \omega_0 \eta \qquad (10)$$

with $\Delta$ – thickness of inner adsorption layer, $\delta$ – thickness of outer adsorption layer, $\omega$ – frequency, $\eta$ – dynamic viscosity, $\kappa$ – Debye-Huckel parameter, $\alpha$ – drag coefficient, $\omega_0$ – relaxation frequency.

In our case one has to take into account the differences in the particle size and particle inertia factor in the dispersion which was proven not to be the same for both the ultrasonically treated and non-treated samples without and with the addition of EHEC polymer. For ultrasonically not-treated samples, the difference is due to the disjoining forces (Deryagin's effect), for treated samples this deflocculation results from both the ultrasonic treatment as well as the stronger polymer adsorption which decreases of particle interaction strength (Rebinder's effect). The inertia factor G(r) can be calculated by the formula 11:

$$G(r) = \frac{1+(1+i)\sqrt{r/2}}{1+(1+i)\sqrt{r/2}+i(r/9)(3+2(\Delta\rho/\rho_m))} \quad (11)$$

where **r** is particle radius, $\boldsymbol{\rho_m}$ is the density of the solvent, $\boldsymbol{\Delta\rho}$ is the difference between densities of the particles and the solvent

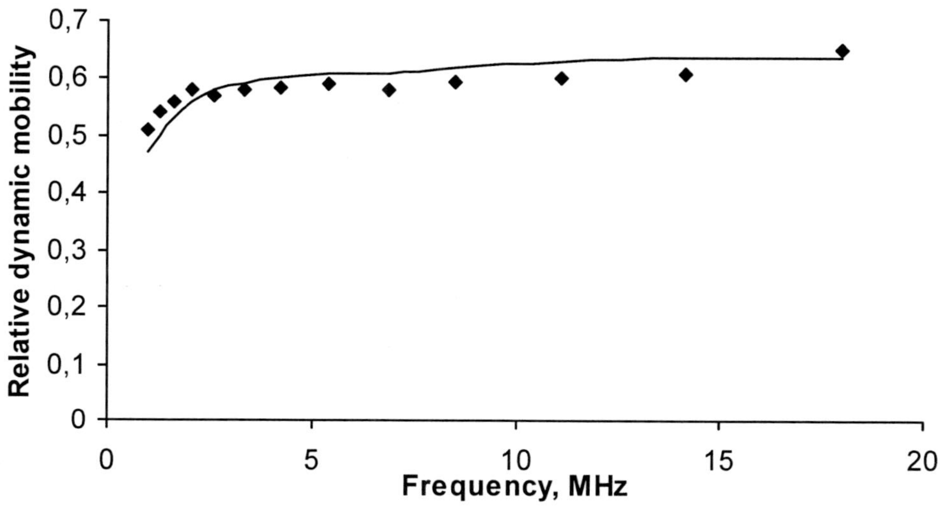


Figure 11. Dependence of the relative dynamic mobility on frequency for aqueous $TiO_2$ dispersions stabilized by EHEC (1 wt.-% relative to pigment) without ultrasonic treatment. ◆: experimental data, solid line: calculated theoretical curve.

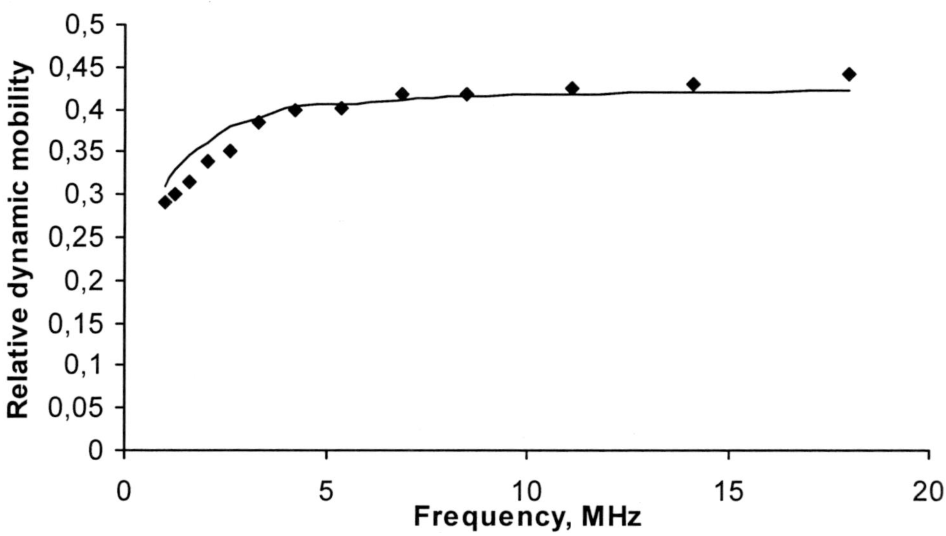


Figure 12. Dependence of the relative dynamic mobility on frequency for aqueous $TiO_2$ dispersions stabilized by EHEC (15 wt.- % relative to pigment) without ultrasonic treatment. ◆: experimental data, solid line: calculated theoretical curve.

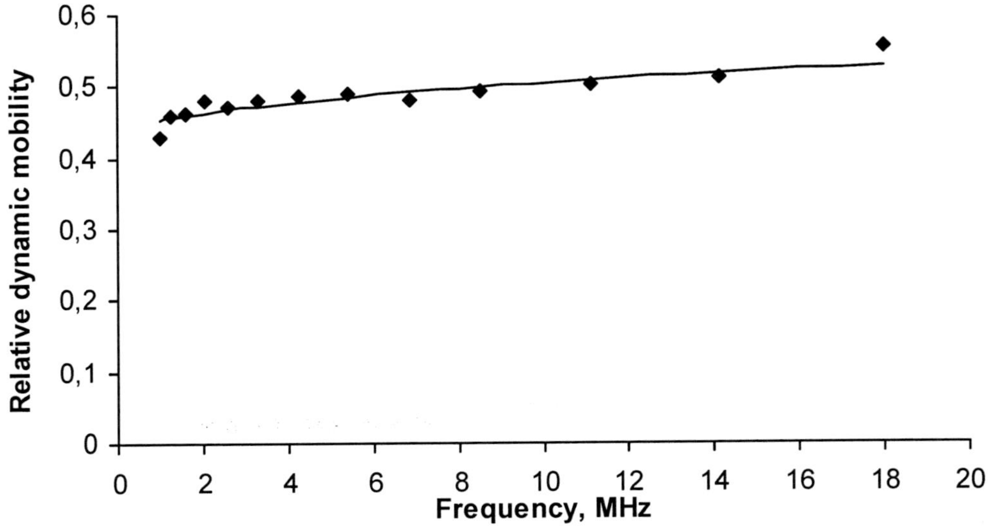


Figure 13. Dependence of the relative dynamic mobility on frequency for aqueous $Fe_2O_3$ dispersions stabilized by EHEC (1 wt.-% relative to pigment) without ultrasonic treatment.◆: experimental data, solid line: calculated theoretical curve.

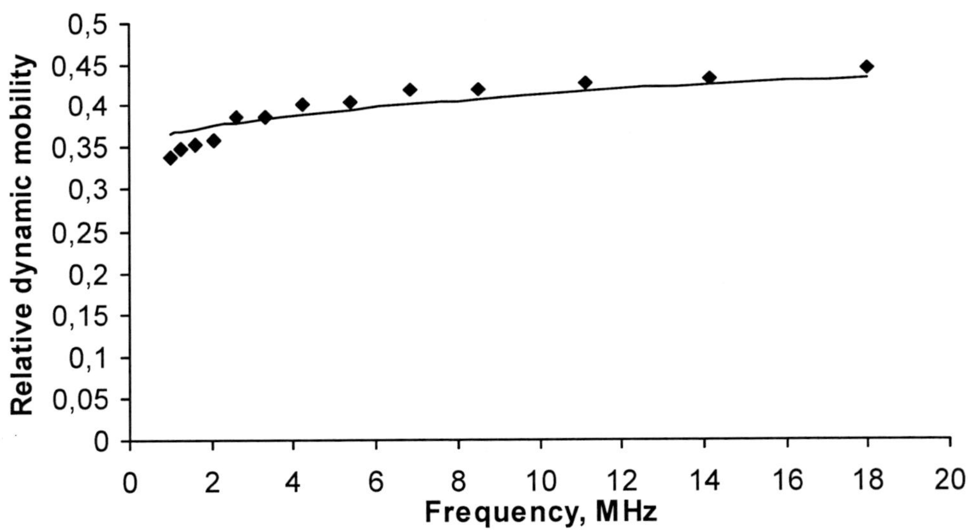


Figure 14. Dependence of the relative dynamic mobility on frequency for aqueous $Fe_2O_3$ dispersions stabilized by EHEC (7 wt.-% relative to pigment) without ultrasonic treatment. ◆: experimental data, solid line: calculated theoretical curve.

First, the relative dynamic mobility, i.e., the ratios between experimentally obtained values of the dynamic mobility of pigment particles covered by polymer and those of non-covered particles were established for the whole frequency range (data points in Figure 11 – 14). Then, using the equation (8), the thicknesses of inner (Δ) and outer (δ) adsorption layers were calculated. These data are compiled in Table 4. Inserting averaged values into the formula (8) one gets the theoretical dependency of the relative dynamic mobility (ratio between dynamic mobilities of polymer-coated and pure particles) from the frequency of the applied electrical field (lines in Figure 11 – 14). The comparison of the experimental and theoretical data allows to check the correctness of the calculations; it is evident that for all cases the theoretical curves were in good agreement with the experimental data.

The total thickness of the EHEC polymer adsorption layer as given by the sum of the inner and outer adsorption layer as calculated by applying the gel layer theory (see data Table 4) is in fair agreement with the values based on the polymer saturation concentration established by the concentration dependency of the ESA signals (see Table 3). The comparison the values of thicknesses of the inner and outer adsorption layers of the dispersion without and with ultrasonic treatment reveals that for both systems investigated the ultrasonic treatment predominantly effects the thickness of the inner layer which is especially prominent for the $TiO_2$ pigment. This reflects the activation of the pigment surface by the ultrasonic power leading to an increased adsorption of polymer; on the other hand, the outer layer which results from interpolymer interactions is less effected and maintains relatively unchanged in thickness.

## CONCLUSIONS

ESA and TEM measurements have shown that the improved colloidal stability of ultrasonically treated dispersed pigment/polymer stabilizer systems as known from sedimentation studies is due to special ultrasonic treatment related pigment-polymer interactions. For aqueous $TiO_2$ and $Fe_2O_3$ dispersions stabilized by EHEC, the saturation concentrations of polymer surfactants, i.e., the maximum amount of adsorbed polymer were established, and the increasing of the thickness of the adsorption layers upon ultrasonic treatment was proven by ESA measurements and confirmed by TEM investigations of the particulate systems.

By means of the ESA technique insights into the ultrasonically induced change of the thickness and structure of the polymer adsorption layers were obtained by applying the gel layer theory: In particular, the data allow to conclude that ultrasonic treatment of aqueous inorganic pigment dispersions is a powerful method for pigment surface modification leading first to an activation of the pigment surface; as a consequence, improvement of the pigment-polymer interaction is achieved which results in the creation of polymer adsorption layers of high thickness.

Thus ultrasonic treatment enables to obtain stable and high-dispersed pigment suspensions with modified pigment surface; this method opens new perspectives to modify the pigment surface more efficiently.

## REFERENCES

[1] "*Surfactants and Polymers in Aqueous Solution*", 2nd edition, K. Holmberg, B. Jönsson, B. Kronberg, B Lindman, J. Wiley & Sons, Chichester, UK 2003.

[2] R.R. Netz, D. Andelman. Polymer adsorption in aqueous media, *Physics Reports,* 380 (2003) 1-95.

[3] P. Somasundaran, S. Krishnakumar. Adsorption of surfactants and polymers at the solid-liquid interface, Colloids and Surfaces, A: *Physicochemical and Engineering Aspects,* 123(1997)491-513.

[4] Theo G.M. Van de Ven. Surface modification of pigments with temperature-responsive polymer grafted by plasma-induced polymerization, *Advances in Colloid and Interface Science*, 48 (1994) 121-140.

[5] P. Somasundaran, Xiang Yu. Surface modification strategies for multicomponent polymer systems, Advances in Colloid and Interface Science, 53 (1994) 33-49.

[6] V. P. Zubov, I. F. Kuzkina, I. I. Ivankova, O. J. Schmitz, Effect of Ultrasonic Treatment on Stability of $TiO_2$ Aqueous Dispersions, *European Coating Journal,* 1998, 12, 954.

[7] V. P. Zubov, I. F. Kuzkina, I. I. Ivankova, O. J. Schmitz, Aqueous Dispersions of Copper Phthalocyanine, *European Coating Journal,* 1998, 11, 856.

[8] M. Antonietti, M. C. Weissenberger, Influences of monomer and oligomer structures in their adsorption on metal oxide surfaces, *Macromol. Rapid. Com.* 1997, 18, 295.

[9] V. P. Zubov, N. V. Serebryakova, I. A. Arutyunov, I. F. Kuzkina, N. A. Bulychev and Yu. A. Khrustalev, Effect of mechanical treatment of inorganic pigments surfaces on stability of their aqueous dispersions in the presence of ethylhydroxyethylcellulose, *Colloid Journal (Translated from Kolloidny Zhurnal)*, 2004, 66, 302.

[10] N. A. Bulychev, I. A. Arutunov, V. P. Zubov, B. Verdonck, T. Zhang, E. J. Goethals, F. E. Du Prez, Block Copolymers of Vinyl Ethers as Thermo-Responsive Colloidal Stabilizers of Organic Pigments in Aqueous Media, *Macromolecular Chemistry and Physics,* 2004, 205 (18), 2457-2463.

[11] N. A. Bulychev, V.N. Fomin, I. A. Arutunov, V. P. Zubov, O.V. Abramov, E.V. Kisterev, Effect of mechanical action on surface modification in aqueous pigment dispersions, *Materialovedenie*, 2005, №4, 20.

[12] C. D. Eisenbach, Ch. Schaller, T. Schauer, K. Dirnberger, Surface modification of pigments by adsorption of amphipolar copolyelectrolytes, ACS Symposium Series 881 "Particle Sizing and Characterization", American Chemical Society, 2004, 215.

[13] C. Schaller, A. Schoger, K. Dirnberger, T. Schauer, C. D. Eisenbach, Stabilizing pigments in full-mixing-systems, *Macromol. Symp.* 2002, 179, 173.

[14] C. Schaller, T. Schauer, K. Dirnberger, C. D. Eisenbach, Synthesis and stabilizing properties of amphipolar polyelectrolytes, *Eur. Phys. J.* 2001, E 6, 365.

[15] R.W. O'Brien, Determination of Particle Size and Zeta-Potential, *J. Fluid Mech.*, 1988, 190, 71.

[16] M.L. Carasso, W.N. Rowlands, R.W. O'Brien, The Effect of Neutral Polymer and Nonionic Surfactant Adsorption on the Electroacoustic Signals of Colloidal Silica, *J. Colloid and Interface Science,* 1997, 193, 200-214.

In: Chemical and Biochemical Reactions
Editors: G. E. Zaikov, L. Liu et al.
ISBN 978-1-61122-860-1


*Chapter 10*

# PIGMENT SURFACE MODIFICATION IN PIGMENT DISPERSIONS WITH THERMO-RESPONSIVE POLY(METHYLVINYLETHER) COPOLYMERS IN COMBINATION WITH ULTRASONIC TREATMENT

***N. Bulychev*[1,2]*, E. Kisterev*[1]*, O. Confortini*[3]*, F. E. Du Prez*[3]*, V. Zubov*[4] *and C. D. Eisenbach*[2,5]*.***

[1]N.S. Kurnakov Institute of General and Inorganic Chemistry of Russian Academy of Sciences, 119991, Leninsky avenue, 31, Moscow, Russia,
[2]Institute for Polymer Chemistry, University of Stuttgart, Pfaffenwaldring 55, D-70569 Stuttgart, Germany,
[3]Department of Organic Chemistry, Polymer Chemistry Research Group, Ghent University, Krijgslaan 281 S4, B-9000 Ghent, Belgium,
[4]Lomonosov Moscow State Academy of Fine Chemical Technology, Pr. Vernadskogo, 86, 117571, Moscow, Russia,
[5]Research Institute for Pigments and Coatings, Allmandring 37, D-70569 Stuttgart, Germany.

## ABSTRACT

The process of surface modification of hydrophobic organic pigments (copper phthalocyanine (CuPc) and carbon black) as well as a hydrophilic inorganic pigment (titanium dioxide) in aqueous dispersions by employing tailor-made thermo-responsive copolymers, and the colloidal stability have been studied. The pigment surface modification is achieved by conventional adsorption and by (thermo)precipitation of amphiphilic methyl vinyl ether (MVE) containing polyvinylether block and PMVE graft copolymers with poly(ethylene oxide) side chains exhibiting a lower critical solution temperature (LCST). The effect of mechanical treatment of the pigment dispersion by ultrasonic power alone or in combination with the LCST property was investigated. The course of the pigment surface coating process was followed by the Electrokinetic Sonic

Amplitude (ESA) method. The temperature-controlled sorption of PMVE-g-PEO graft copolymers on both inorganic and organic pigment surface was investigated. It was found that ultrasonic treatment together with LCST thermoprecipitation is a promising method for the surface modification of pigments with regard to dispersion stability.

**Keywords:** thermo-responsive copolymers, ultrasonic treatment, aqueous dispersions, inorganic pigments, organic pigments, lower critical solution temperature

## AIM AND BACKGROUND

In numerous investigations, a variety of block copolymers based on poly(methyl vinyl ether) (PMVE) has been already used for the stabilization of CuPc and carbon black aqueous dispersions. PMVE belongs to the class of thermo-responsive polymers and exhibits a lower critical solution temperature (LCST) at 37 °C. The introduction of such a thermo-responsive segment in an otherwise temperature-insensitive polyvinylether provided an efficient strategy for the design of a new class of "intelligent" polymer surfactants with stabilization activity which is responsive to environmental conditions. The optimal surfactants showed stability properties that are even better than those of the commercial Pluronics. In this paper, the process of the adsorption of P(MVE-b-IBVE) block copolymers and poly(ethylene oxide) (PEO) side chain containing P(MVE-g-EO) graft copolymers on the surface of hydrophobic copper phthalocyanine (CuPc) or carbon black and hydrophilic $TiO_2$ particles in aqueous dispersion will be studied by advanced methods at various polymer concentrations.

## INTRODUCTION

In recent years, polymer surfactants have been of increasing interest as stabilizers for particulate systems, especially for emulsions and suspensions. Aqueous colloidal dispersions of pigments are important, ecologically friendly colloidal systems widely used in polygraphic and paint industries. The pigment particles are usually of 20 – 200 nm diameter and may be quite hydrophobic. In order to achieve a good stabilization in aqueous pigment dispersions, many formulations were proposed [1 - 8]. Earlier we reported about the role of mechanical, e.g. ultrasonic treatment to obtain highly stable dispersions [7, 9]. It was shown that using of polymer surfactants in combination with ultrasonic action can significantly improve the quality of dispersed systems. However, some aspects concerning pigment-polymer interaction and formation of adsorption layers under mechanical action needed additional elucidation.

The colloid stabilization of aqueous dispersions with polymer surfactants is believed to be a consequence of adsorption of the amphiphilic macromolecules on the particle surface resulting in mono- or multi-layers of certain structure and thickness which provide certain sterical and/or electrostatic stabilization effects. Polymer adsorption from aqueous solution on a particle surface is a result of specific interactions of various active sites on the particle surface with corresponding sites (groups) of the macromolecule. Therefore the chemical structures of the stabilizers are believed to be adjusted to the nature (e.g. hydrophilicity, charge etc.) of each type of the particles.

Thermo-responsive "smart" polymers exhibiting lower critical solution temperature (LCST) in aqueous solutions change the hydrophilic-hydrophobic balance when passing through LCST (they become more hydrophobic above LCST). One can expect that this property may result in temperature-controlled sorption and structure of the adsorption layer, and thus allowing colloid stabilization of aqueous dispersions. In other words, temperature-controlled changes in the hydrophobic-hydrophilic balance of those "smart" polymers can be used to adjust surfactant behavior of the polymer in order to achieve colloidal stabilization effects for various particles without changes of chemical structure or composition of the stabilizer.

Detailed information about the interaction of copolymers with the pigment surface, e.g., adsorption/desorption phenomena as reflected from the dynamic mobility $\mu_D$ of the dispersed particle, can be obtained by applying the Electrokinetic Sonic Amplitude (ESA) technique [10 - 17]. With this method, a potential is measured from pressure waves generated by the movement of charged particles in an oscillating electric field. The variation of this potential with frequency can be used for determining $\mu_D$ of particles reflecting the formation of thin polymer layers on their surface and for calculating the zeta potential as well as the particle size. For fixed low frequencies as applied in this work, $\mu_D$ is affected by the surface charge of the particle only [16, 18 - 20]. The study of the change of the dynamic mobility $\mu_D$ of aqueous $TiO_2$ dispersion upon addition of amphipolar polyelectrolytes has shown that ESA is a powerful method to reveal the course of polymer-pigment interactions and how this is related to the molecular architecture of the employed polyelectrolytes [19, 20].

# EXPERIMENTAL PART

## Materials

As pigments, β-copper phthalocyanine (β-CuPc) with primary particle size 0.1 μm and carbon black (channel type) FW 285 with primary particle size 11 nm were chosen. The PMVE-copolymers of various well-defined structures, molecular weights and narrow polydispersities, were synthesized by "living" cationic polymerization techniques as described elsewhere [22].

## Techniques

Colloidal stabilization of the aqueous dispersions was monitored by sedimentation measurements of 1 % dispersions of CuPc and carbon black. Block copolymers used for stabilization of CuPc are insoluble in water. Therefore, for the preparation of the aqueous dispersions, the polymers were predissolved in ethyl acetate (5 %) and then introduced to the aqueous phase together with the pigment (ethyl acetate:water volume phase ratio was 1:10) [22]. After mixing with a laboratory stirrer (700 rpm for 10 min), the system was treated with ultrasound for 2 min with an ultrasonic generator UZDN-2 or Branson Sonifier B-12 with actual power of 1.5 W/cm$^2$. The pigment-polymer interaction was investigated by electrokinetic sonic amplitude (ESA) method. ESA measurements were carried out as described earlier [18]. Transmission electron microscopy (TEM) investigations were carried out according to the standard procedure. The samples of the pigment dispersions were diluted with distilled water by a factor of 100, then dropped on the TEM grid of 3 mm diameter and dried on air for 1 day. The TEM pictures were taken by using a Jeol 2000 FX device under the voltage of 200 kV.

# RESULTS AND DISCUSSION

## Pigment Surface Modification with Graft Copolymers

### *Sedimentation Studies*

First, P(MVE-g-EO) graft copolymers with a PMVE backbone and PEO side chains have been investigated. These polymers are hydrophilic below the LCST of PMVE and become amphiphilic above the LCST of PMVE. Hydrophilic pigments (e.g. $TiO_2$) are known to need hydrophilic surfactants for stabilization in aqueous dispersions, whereas hydrophobic pigments (CuPc and carbon black) need amphiphilic polymers to achieve stabilization of their aqueous dispersions. This opens a perspective to use these PMVE graft copolymers as universal smart surfactants for surface modification of both polar inorganic and non-polar organic pigments by taking advantage of the temperature-controlled sorption.

The stabilizing activity of the series of PMVE-g-PEO polymers in $TiO_2$ and CuPc aqueous dispersions were investigated at different temperatures. The results are given in Table 1. The half-time of the sedimentation, i.e. the period of time which is necessary for the border between colored and transparent zones in sedimentation measurements to reach 50% of the test-tube height was chosen as a measure of the stabilization activity.

Below the LCST of PMVE all copolymers are hydrophilic and thus, as one can see from Table 1, with exception of G1 act as efficient stabilizers of $TiO_2$ aqueous dispersions.

Regarding the relative difference of the hydrophilicity of the PEO grafts and the PMVE backbone it is assumed that the more hydrophilic PEO primarily interacts with the hydrophilic $TiO_2$ surface; i.e., the PEO grafts act as anchors. As far as the interaction of the graft copolymer with the hydrophobic CuPc is considered, it is possible to assume a preferred interaction with the less hydrophilic PMVE backbone.

**Table 1. Sedimentation stability of $TiO_2$ and CuPc aqueous dispersions with PMVE-g-PEO graft copolymers as a function of temperature**

| Code | Number of PEO grafts/Mn | Half-time of sedimentation | | | |
|---|---|---|---|---|---|
| | | 20 °C | | 50 °C | |
| | | $TiO_2$, days | CuPc, min. | $TiO_2$, min. | CuPc, days |
| G1 | 9/2000 | 5 | – | – | |
| G2 | 3/5000 | 30 | ↑ | ↑ | 18 |
| G3 | 3/5000 | 30 | 10 – 15 | 30 – 40 | 20 |
| G4 | 7/5000 | 25 | ↓ | ↓ | 23 |
| G5 | 13/5000 | 25 | – | – | 20 |

However, after increasing the temperature above the LCST of the PMVE block, fast coagulation occurs and the size of precipitating particles becomes about 7 - 8 μm. This must be attributed to changes in the conformation of the PMVE backbone upon exceeding of the LCST (collapse of coil) and to interaction phenomena, i.e., the PMVE cannot act as a sterical stabilizing element anymore. As a consequence, the more hydrophilic PEO grafts which act as anchors at the $TiO_2$ surface are immobilized and buried. This means a relative hydrophobization of the polymer layer around the $TiO_2$ particle and particle coagulation causing sedimentation might result. On the other hand, a subsequent cooling below the LCST in combination with ultrasonic action allows to redisperse the system. The stability of the obtained dispersions is comparable to that of non-heated samples. This means that the agglomerates were broken by the action of ultrasonic power and that the PMVE backbone returned to the previous swollen state.

The poor colloidal stabilization with G1 polymer which contains only low molecular PEO segments is probably due to its weak adsorption on the $TiO_2$ particles and may indicate that adsorption is mainly due to interaction of high molecular PEO segments with $TiO_2$ surface. So one can conclude that PEO side chains of the copolymer are responsible for pigment-polymer interaction causing adsorption of PEO grafts on the pigment particles surface whereas the PMVE backbone allows for steric stabilization.

In the case of CuPc aqueous dispersions, reverse stabilization behavior is observed: above the LCST, all polymers act as stabilizers of CuPc aqueous dispersions, while fast coagulation is observed after temperature decrease to temperature below LCST.

This is attributed to the function of the collapsed PMVE backbone above the LCST as anchor to the CuPc surface. Comparing the hydrophilicity of PEO grafts and PMVE backbone below and above LCST, it is meaningful to assume that the collapsed PMVE above LCST is less hydrophilic than the expanded PMVE coil below LCST. This means that the exceeding of the LCST results in the formation of an amphiphilic copolymer with hydrophilic PEO coil and comparatively hydrophobic collapsed PMVE. As a consequence, above LCST the CuPc dispersion is stabilized by the interaction of "hydrophobic" collapsed PMVE with the CuPc surface.

When subsequently lowering the temperature below LCST, the adsorption of the graft copolymer on the CuPc surface is significantly weekend because of the dissolution (expanding) of the PMVE part of the graft copolymer which acts as anchor in the collapsed state only.

For both dispersion types, the best results are obtained for the graft copolymers with the longest PEO grafts (5000 vs. 2000). On the other hand, the number of grafts has not a significant influence on the sedimentation stability within the investigated range (between 3 and 13).

***Electrokinetic Sonic Amplitude (ESA) Measurements***

The adsorption of the PMVE-g-PEO copolymers on the $TiO_2$ and CuPc surface was investigated by the ESA method. For these measurements, the copolymer G3 (acting as the best stabilizer) was chosen (Figures 1 – 2). Other copolymers showed similar results.

From Figure 1 it is evident that the pigment-polymer suspensions show quite different behavior as a function of their preparation method. Without ultrasonic action, saturation of the surface adsorption layer seems to be reached at 1 % of PMVE-g-PEO in relation to the pigment $TiO_2$. After ultrasonic action, initial dynamic mobility of the $TiO_2$ is much higher, and saturation concentration of polymer is only reached at about 5 %. The differences in the initial values of the dynamic mobility and following differences in its changes due to the sorption of the polymer can be attributed to mechanical activation of the surface with ultrasonic treatment. Further addition of polymer did not effect the dynamic mobility. Thus, comparing the saturation concentrations for mechanically treated and non-treated samples, one can conclude that the ultrasonic action not only increases the pigment surface due to the obtained finer dispersion, but also leads to a higher thickness of the adsorption layer in comparison to the non-treated samples.

As already described above, fast coagulation of the $TiO_2$ dispersion was observed after exceeding the LCST of PMVE (Table 1). ESA examination of these coagulated dispersions enabled to prove that at the saturation concentrations, the dynamic mobility and ξ-potential of the particle surface did not change noticeably in comparison to those at room temperature; this indicates the absence of polymer desorption. This is also further prove of the earlier statement that pigment-polymer interaction at room temperature is apparently due to the adsorption of the PEO chains, while the PMVE backbones are preferably exposed into the aqueous media.

Preliminary experiments have shown that hydrophilic polymers don't interact with hydrophobic CuPc surface in aqueous media. In case of PMVE-g-PEO copolymers, ESA measurements proved the absence of adsorption below LCST (Figure 2, curve 1). However as it is shown by the curve 2, above LCST of PMVE, the copolymer PMVE-g-PEO is able to adsorb on the CuPc surface in aqueous dispersions that results in high stabilization of CuPc dispersion. Saturation of the adsorption layer is reached at about 3 %. After decreasing of the temperature, the values of dynamic mobility and ξ-potential are similar to those for pure CuPc, which indicates the existence of desorption processes. This observation is in agreement with the fast coagulation process that was mentioned earlier.

## Pigment Surface Modification of CuPc with Block Copolymers

### *ESA Studies*

AB and ABA block copolymers of methylvinyl and isobutylvinyl ethers (PMVE–PIBVE and PIBVE–PMVE–PIBVE) exhibit a quite different thermo-responsive behavior as compared to the graft copolymers. They are amphiphilic below the LCST and become quite hydrophobic above the LCST. This gives a possibility to study their stabilizing properties in aqueous dispersions of organic pigments below and above LCST and to gain some insight into the pigment-polymer interaction and structure of the polymer layers in these systems at various temperatures.

First, aqueous dispersions of CuPc stabilized by $PMVE_{65}$ – $PIBVE_{10}$ and $PIBVE_{22}$ – $PMVE_{75}$ – $PIBVE_{22}$ (as the best colloidal stabilizers at room temperature [21]) were studied by ESA measurements (Figures 3 and 4).

Figures 3 and 4 represent the changing of the dynamic mobility of surface modified CuPc dispersion with increasing amount of polymer added to the system in the surface modification. As one can see, the pigment-polymer suspensions, stabilized by di- and triblock PMVE containing copolymers below the LCST (curves 1) show quite similar behavior. The saturation of surface layer seems to be reached at 5 % of $PMVE_{55}$ – $PIBVE_{10}$ and 3 % of $PIBVE_{22}$ – $PMVE_{75}$ – $PIBVE_{22}$ in relation to the pigment. Further addition of block copolymer did not noticeably increase the dynamic mobility. That means that formation of thick polymolecular layers hardly occurs. The carbon black dispersions showed similar behavior.

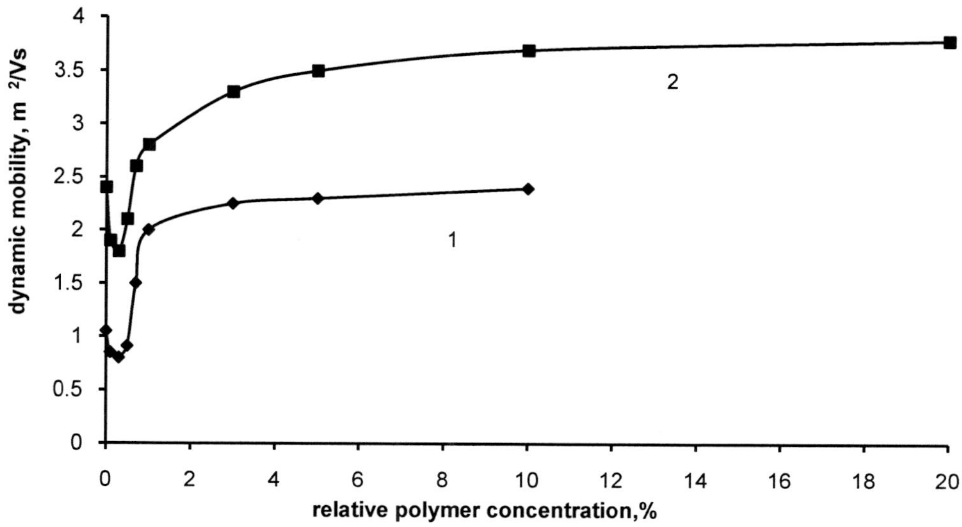


Figure 1. Dependence of the dynamic mobility on the relative concentration of PMVE-g-PEO (G3) for $TiO_2$ aqueous dispersion without and after ultrasonic treatment (temperature 20 °C).

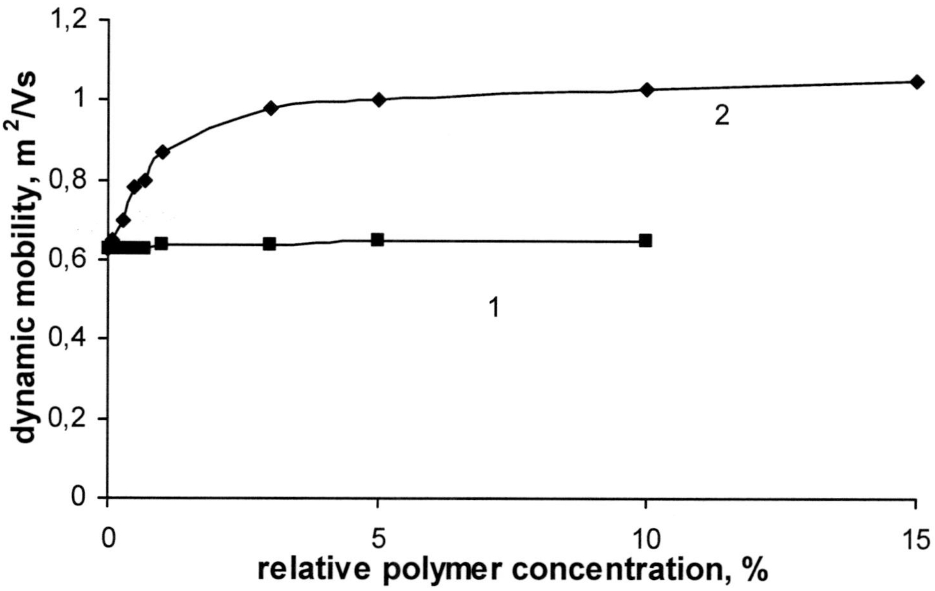


Figure 2. Dependence of the dynamic mobility on the relative concentration of PMVE-g-PEO (G3) for CuPc aqueous dispersion after ultrasonic treatment (temperature 50 °C).

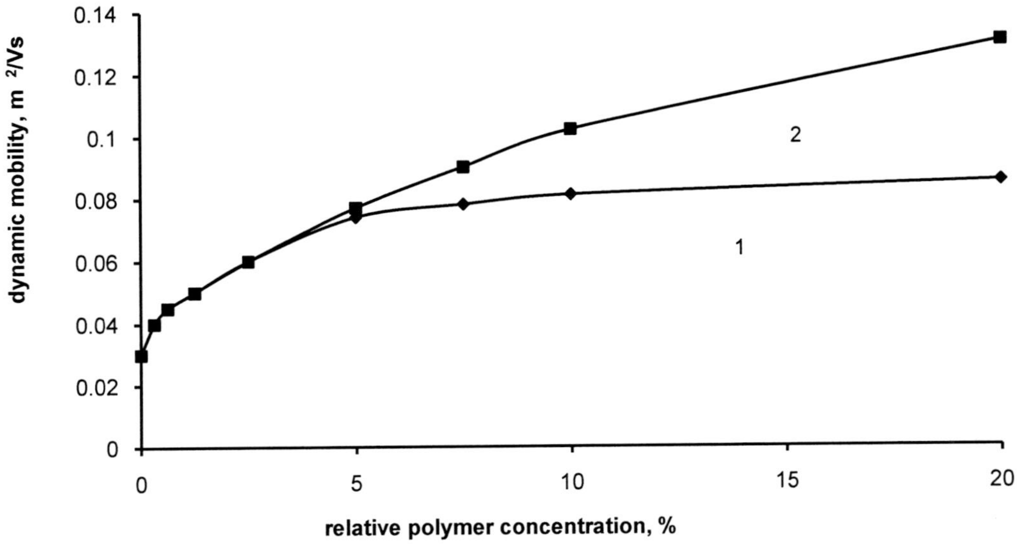


Figure 3. Dependence of dynamic mobility of surface treated CuPc on the relative concentration of $PMVE_{55} - PIBVE_{10}$ in aqueous dispersion in combination with ultrasonic treatment for pigment surface modification. 1 – conventional procedure, 2 – LCST thermoprecipitation.

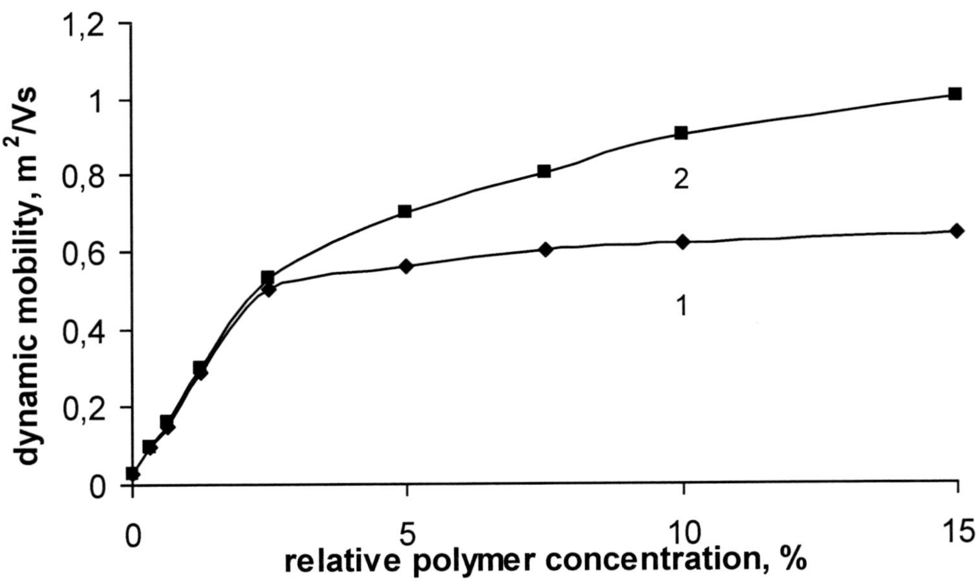


Figure 4. Dependence of dynamic mobility of surface treated CuPc on the relative concentration of $PIBVE_{22} - PMVE_{75} - PIBVE_{22}$ in aqueous dispersion in combination with ultrasonic treatment for pigment surface modification. 1 – conventional procedure, 2 – LCST thermoprecipitation.

As PMVE in aqueous solutions exhibits a lower critical solution temperature (LCST = 37 °C), it was of interest to study the creation of a surface layer by the LCST technique in combination with ultrasonic treatment, which may increase the thickness of the layers and stabilization of the dispersion in comparison to the polymer deposition below LCST. The LCST technique alone has been shown to be quite efficient in pigment surface modification [23].

In this work, the procedure of dispersion preparation was slightly changed. First, in constant temperature experiments (below LCST), the isothermal condition during the ultrasonic treatment is maintained by water cooling. In comparative experiments carried out without cooling, the heat generated by the ultrasonic treatment caused the temperature of the system to rise up to 45 – 50 °C. As this temperature is above the LCST of PMVE, the block copolymers of PMVE became hydrophobic and were completely precipitated on the particle surface. The thickness of the deposited layers indeed increased (Figures 3 and 4, curves 2) and saturation concentration could not be established due to the thermoprecipitation (curves 2), which is ascribed to the fact that all polymer added to the solution precipitates on the particle surface. Thus, the more polymer is in the system, the thicker is the deposited layer, as is reflected by the higher values of dynamic mobility. This procedure gives the possibility to create surface polymer layers with controlled thickness.

## Transmission Electron Microscopy (TEM) Studies

In order to visualize the surface modification of the pigments in aqueous dispersion by treatment with ultrasound in the presence of LCST-polymers, the aqueous dispersions were investigated by transmission electron microscopy (TEM).

The pictures of CuPc particles stabilized by di- and triblock PMVE copolymers after mechanical (ultrasonic) treatment only are depicted in Figures 5 and 6. One can identify the polymer layers (grey zones) created by di- and triblock PMVE copolymers. The thickness of these layers is limited by the amount of polymer, which can be deposited on the particle surface up to saturation of polymer layer. The excess of the polymer will apparently create micelles and will not precipitate on the particles. As also seen from Figures 5 and 6, the thickness of the polymer layer created by triblock copolymer is higher as compared to the layer in the presence of the diblock copolymer. This is in agreement with the difference between the values of dynamic mobility obtained by ESA measurements (Figures 3 and 4).

Figure 5. Picture of CuPc particle in aqueous dispersion stabilized by $PMVE_{55} – PIBVE_{10}$ after ultrasonic treatment. Scale 20 nm.

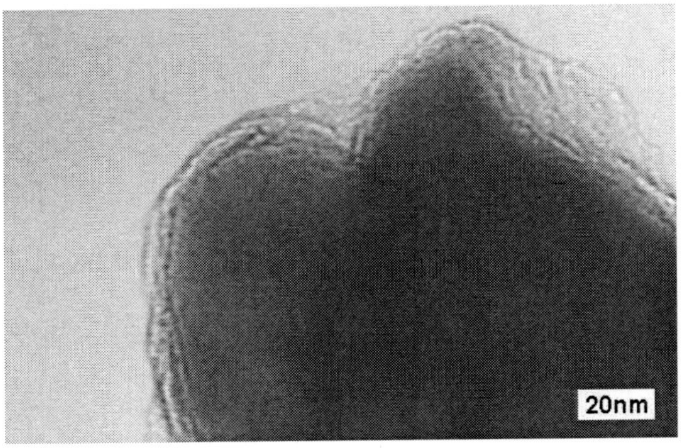


Figure 6. Picture of CuPc particle in aqueous dispersion stabilized by $PIBVE_{22} - PMVE_{75} - PIBVE_{22}$ after ultrasonic treatment. Scale 20 nm.

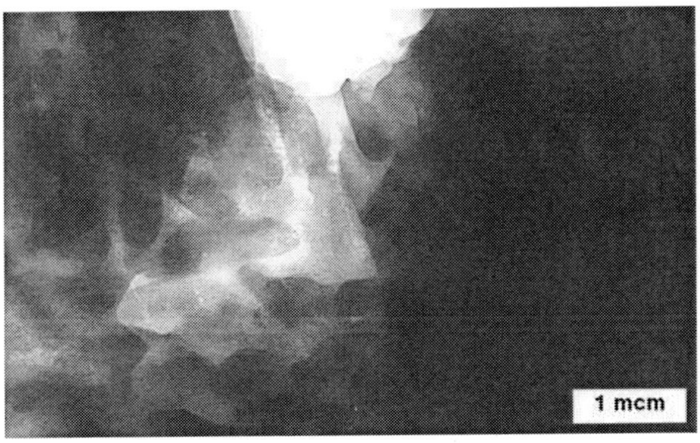


Figure 7. Picture of CuPc aqueous dispersion stabilized by $PMVE_{55} - PIBVE_{10}$ after ultrasonic treatment according to LCST-thermoprecipitation technique. Scale 1 μm.

Figure 8. Picture of CuPc particle in aqueous dispersion stabilized by $PMVE_{55} - PIBVE_{10}$ after ultrasonic treatment according to LCST-thermoprecipitation technique. Scale 20 nm.

In Figures 7 and 8, the TEM pictures of CuPc particles as covered by $PMVE_{55} - PIBVE_{10}$ according to the thermoprecipitation technique are given. From Figure 7 it is inferred that thermoprecipitation of the polymer, which becomes more hydrophobic above the LCST, leads to fast coagulation of the dispersion. The particles are strongly aggregated, with aggregate size of 5 – 10 μm. The polymer layer is much thicker in comparison to the layer thickness resulting from conventionally prepared systems (Figure 8). After decreasing of the temperature below LCST, the PMVE copolymer becomes amphiphilic and partial desorption of the polymer is observed; thus the system can be redispersed. In order to retain the thickness of the adsorption layer, one can consider a system in which the polymer can be crosslinked to prevent the desorption process (cf. [23]).

## CONCLUSIONS

The temperature-controlled sorption of the PMVE-g-PEO copolymers on both inorganic and organic pigment surfaces was investigated. It was shown that depending on the temperature those polymers can adsorb on both hydrophilic and hydrophobic pigment surface. This gives a possibility to use these PMVE graft copolymers as universal smart surfactants for surface modification of both polar inorganic and non-polar organic pigments and for temperature-controlled interaction.

Block copolymers containing PMVE and hydrophobic blocks were shown to act as temperature-controlled stabilizers of CuPc, carbon black and $TiO_2$ aqueous dispersions. Mechanical treatment of dispersed pigment systems proven not only enhances the dispersion stability, but also influences the pigment-polymer interaction. It was further observed that the precipitation of polymers or polymer segments on the particle surface by exceeding the LCST is much more efficient than isothermal deposition below LCST for creation of thick polymolecular polymer layers on the particles surface.

## REFERENCES

[1] *"Surfactants and Polymers in Aqueous Solution"*, 2nd edition, K. Holmberg, B. Jönsson, B. Kronberg, B Lindman, J. Wiley & Sons, Chichester, UK 2003.

[2] R.R. Netz, D. Andelman. Polymer adsorption in aqueous media, *Physics Reports,* 380 (2003) 1-95

[3] P. Somasundaran, S. Krishnakumar. Adsorption of surfactants and polymers at the solid-liquid interface, Colloids and Surfaces, A: P*hysicochemical and Engineering Aspects,* 123(1997)491-513

[4] Theo G.M. Van de Ven. Surface modification of pigments with temperature-responsive polymer grafted by plasma-induced polymerization, *Advances in Colloid and Interface Science*, 48 (1994) 121-140

[5] P. Somasundaran, Xiang Yu. Surface modification strategies for multicomponent polymer systems, *Advances in Colloid and Interface Science,* 53 (1994) 33-49

[6] V. P. Zubov, I. F. Kuzkina, I. I. Ivankova, O. J. Schmitz, Effect of Ultrasonic Treatment on Stability of $TiO_2$ Aqueous Dispersions, *European Coating Journal,* 1998, 12, 954.

[7] V. P. Zubov, I. F. Kuzkina, I. I. Ivankova, O. J. Schmitz, Aqueous Dispersions of Copper Phthalocyanine, *European Coating Journal,* 1998, 11, 856.

[8] M. Antonietti, M. C. Weissenberger, Influences of monomer and oligomer structures in their adsorption on metal oxide surfaces, *Macromol. Rapid. Com.* 1997, 18, 295.

[9] V. P. Zubov, N. V. Serebryakova, I. A. Arutyunov, I. F. Kuzkina, N. A. Bulychev and Yu. A. Khrustalev, Effect of mechanical treatment of inorganic pigments surfaces on stability of their aqueous dispersions in the presence of ethylhydroxyethylcellulose, *Colloid Journal (Translated from Kolloidny Zhurnal)*, 2004, 66, 302.

[10] R.W. O'Brien, Determination of Particle Size and Zeta-Potential, *J. Fluid Mech.*, 1988, 190, 71.

[11] R.W. O'Brien, B.R. Midmore, A. Lamb, R.J. Hunter, Interface characteristics in dispersed pigment systems. Faraday Discuss, *Faraday Discuss. Chem. Soc.,* 1990, 90, 301.

[12] M. Loewenberg, R.W. O'Brien, Electroacoustic Signals of Colloidal Silica, *J. Colloid Interface Sci.,* 1992, 150, 158.

[13] P.F. Rider, R.W. O'Brien, Dynamic Mobility of a Colloidal Particle, *J. Fluid. Mech.,* 1993, 257, 607.

[14] H. Maier, J.A. Baker, J.C. Berg, Elucidating pigment interactions with ESA, *J. Colloid Interface. Sci.,* 1987, 119, 512.

[15] N.P. Miller, J.C. Berg, Novel methods to determine particle sizes and interface characteristics in dispersed pigment systems, *Colloids and Surfaces*, 1991, 59, 119.

[16] M.L. Carasso, W.N. Rowlands, R.W. O'Brien, The Effect of Neutral Polymer and Nonionic Surfactant Adsorption on the Electroacoustic Signals of Colloidal Silica, *J. Colloid and Interface Science,* 1997, 193, 200-214.

[17] R.W. O'Brien, The Effect of a Adsorbed Polyelectrolyte Layer on the Dynamic Mobility of a Colloidal Particle, *Part. Syst. Charact.,* 2002, 19, 1.

[18] C. D. Eisenbach, Ch. Schaller, T. Schauer, K. Dirnberger, Surface modification of pigments by adsorption of amphipolar copolyelectrolytes, ACS Symposium Series 881 "Particle Sizing and Characterization", American Chemical Society, 2004, 215.

[19] C. Schaller, A. Schoger, K. Dirnberger, T. Schauer, C. D. Eisenbach, Stabilizing pigments in full-mixing-systems, *Macromol. Symp.* 2002, 179, 173.

[20] C. Schaller, T. Schauer, K. Dirnberger, C. D. Eisenbach, Synthesis and stabilizing properties of amphipolar polyelectrolytes, *Eur. Phys. J.* 2001, E 6, 365.

[21] N. A. Bulychev, I. A. Arutunov, V. P. Zubov, B. Verdonck, E. J. Goethals, F. E. Du Prez, Block Copolymers of Vinyl Ethers as Thermo-Responsive Colloidal Stabilizers of Organic Pigments in Aqueous Media, *Macromol. Chem. Phys.,* 2004, 205 (18), 2457.

[22] E.J. Goethals, W. Reyntjens, X. Zhang, B. Verdonck, T. Loontjens, Thermoadjustable colloid stabilizers, *Macromol. Symp.* 2000, 157, 93.

[23] T. Schauer, C. D. Eisenbach, Organic polymer treatment - the way to modern pigments, *ECJ*, 2003, 3, 114.

In: Chemical and Biochemical Reactions
Editors: G. E. Zaikov, L. Liu et al.
ISBN 978-1-61122-860-1


***Chapter 11***

# STATIONARY KINETICS OF PHOTOINITIATED COPOLYMERIZATION OF BIFUNCTIONAL (METH)ACRYLATES TILL HIGH CONVERSIONS. KINETIC MODEL OF THE PROCESS

***G. I. Khovanets'*, Yu. G. Medvedevskikh, I. Yu. Yevchuk and G. E. Zaikov†***

Department of Physico-Chemistry of Combustible Minerals L. M. Lytvynenko Institute of Physico-organic Chemistry and Coal Chemistry NAS of Ukraine, 3a, Naukova St., 79053, Lviv, Ukraine

*N. M. Emanuel Instituteof Biochemical Physics RAS, 4, Kosygin St., 119991, Moscow, Russia

## ABSTRACT

This paper represents the results of the investigations of bifunctional (meth)acrylates copolymerization kinetics by the method of lazer interferometry. The kinetics of the stationary photoinitiated copolymerization has been studied in the system: 1,6-hexanediol diacrylate – triethyleneglycol dimethacrylate (HDDA – TGM-3) till high conversions depending on the system composition, concentration of photoinitiator and UV – irradiation intensity. The obtained experimental data have been interpreted in accordance with the conception of microheterogeneity of the polymerizing system. In the concrete variant of this conception which takes into account a presence of at least two different reaction zones, where the process of copolymerization occurs, the general kinetic model of photoinitiated copolymerization of bifunctional (meth)acrylates till high conversions has been offered.

**Keywords:** photoinitiated copolymerization, bifunctional (meth)acrylates, kinetic model, high conversions

---

* E-mail: hop_vfh@ukr.net
† E-mail: chembio@sky.chph.ras.ru

# INTRODUCTION

Numerous publications on radical photoinitiated copolymerization of mono- and polyfunctional monomers indicate the actuality of investigations of kinetic regularities of acrylic and methacrylic monomers copolymeriozation till high conversions and derivation of proper kinetic models. The wide choice of acrylic and methacrylic acids esters forms the basis for the obtaining of polymeric products with various physical and chemical properties (see, for example, the E. H. Riddle's review [1]). Such well known properties of homopolymers of acrylic and methacrylic esters, as optical transparency, brilliance and durability are characteristic for copolymers as well [2]. By the choice of the proper components and adjusting of their ratio it is possible to obtain copolymers, which meet different practical requirements. Nowadays synthetic copolymers are produced in an industrial scale and are widely used in different branches of national economy. The interest in investigation of kinetic conformities of copolymerization is still great, since properties and possible application of copolymers are caused to a considerable degree by conditions of their synthesis.

At present time chemistry and technology of polymers must solve the problem of making of different wares during the process of polymer synthesis. Polymerization by UV – irradiation is widely used lately for obtaining of polymeric materials which have application in electronics, optics, medicine and other industries [3-7]. The properties of polymers must meet rather more high requirements.

Three-dimensional radical copolymerization of polyfunctional monomers is important method of obtaining of polymer nets which meet the requirements of modern technique. It is known that the nature of net-forming comonomer substantially influences on the structurally–physical evolution of reaction environment and as a result – on the copolymers properties. Copolymerization of such comonomers allows to obtain copolymers with additional properties, whereas valuable characteristics peculiar to homopolymer are retained.

However, experimental works on kinetics of photoinitiated copolymerization of polyfunctional (meth)acrylates are episodic and are devoted mainly to the study of physico-mechanical properties and to the practical application of the obtained products. Theoretical works which should describe a process till high conversions are absent.

Therefore the aim of this work was to get statistically significant experimental material on kinetics of photoinitiated copolymerization of bifunctional (meth)acrylates till high conversions and to derivate the kinetic model of the process on the basis of experimental data in accordance with the conception of microheterogeneity of polymerizing system.

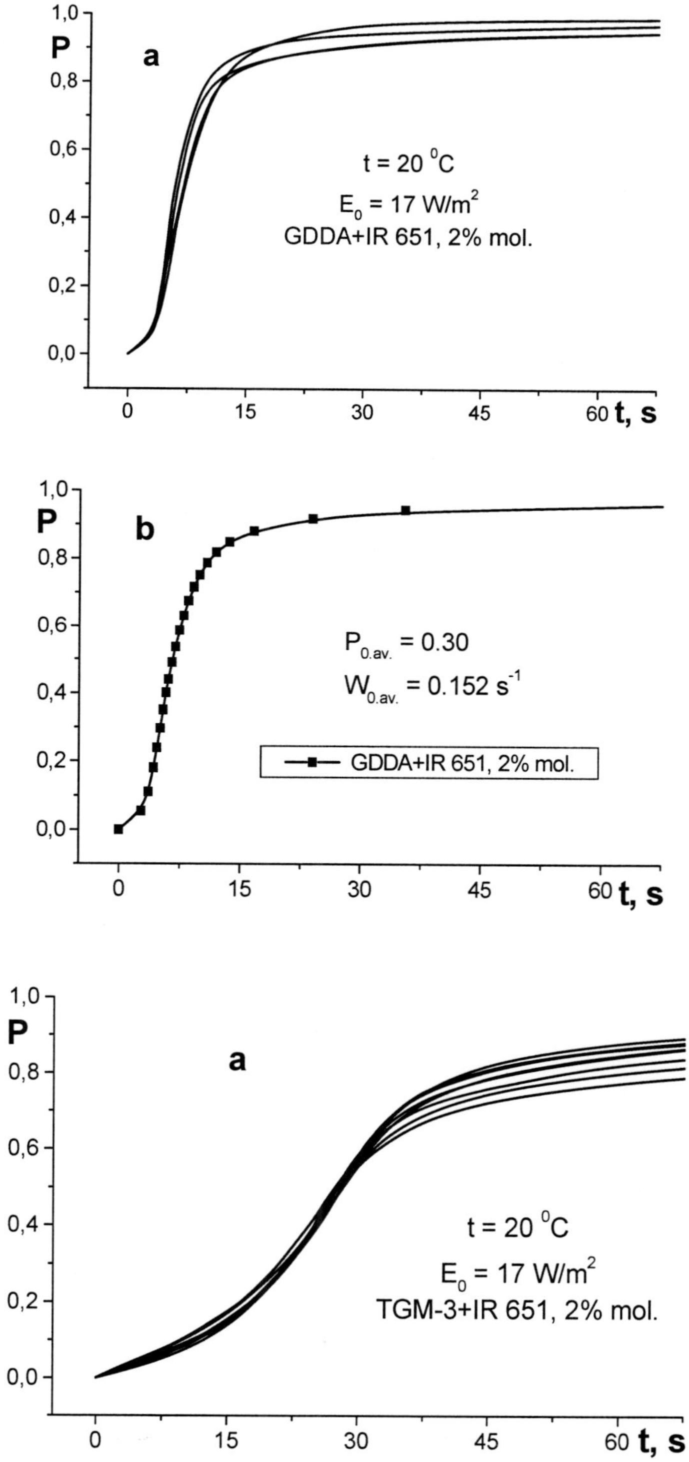


Figure 1. (Continued)

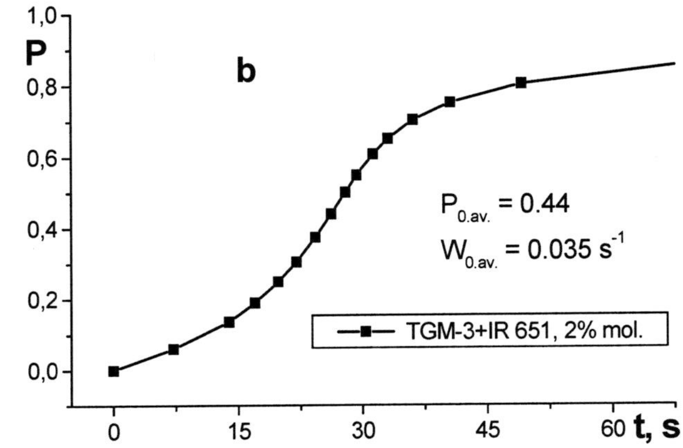


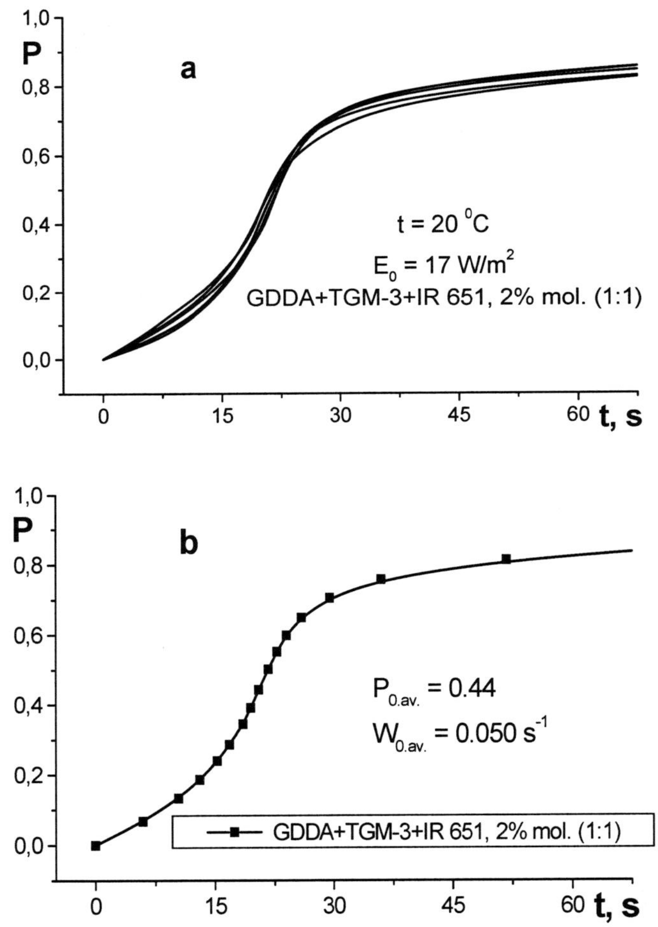


Figure 1. The initial kinetic curves of copolymerization of the system HDDA – TGM-3 (a) and the averaged results (b).

## 2. Experimental Regularities of Photoinitiated Copolymerization of Di(meth)acrylates till High Conversions

We have studied kinetics of photoinitiated copolymerization of bifunctional (meth)acrylates in the system: 1,6-hexanediol diacrylate – triethyleneglycol dimethacrylate (HDDA – TGM-3) at molar ratio of the components HDDA : TGM-3 4:1, 2:1, 1:1, 1:2 and 1:4 in thin layers till high conversions depending on concentration of photoinitiator 2,2–dimethoxy–1,2–diphenylethane–1–on (IRGACURE 651) (1 and 2 % mol.) and UV-irradiation intensity (7, 17 and 48 $W/m^2$).

Monomers were purified according to the method described in [8] by mixing them with the previously activated powder $Al_2O_3$ and subsequent centrifugation. Purification was performed with the purpose to remove inhibitors from monomers. The absence of the latent period on the obtained kinetic curves indicates that we have achieved this aim.

Kinetics of copolymerization was studied by laser interferometry (modified method of dilatometry) [9] in thin layers, covered by glass to prevent the access of oxygen, which insignificantly distorts UV-irradiation spectrum of the lamp DRT-400 [10]. The relative integral conversion P was estimated by the ratio between the current contraction of composition layer and the maximal contraction (at $t \to \infty$). The results of the experiments were represented in the form of the integral kinetic curves conversion vs time, as it is shown in Figure 1.

One can notice the fluctuation sensitiveness of the copolymerization process of di(meth)acrylates), that decreases the reproduction of the kinetic curves obtained under the same conditions of the process (see Figure 1). Therefore we obtained several (most often 7 – 12) kinetic curves for every condition of copolymerization, which were then averaged, as it is shown for some examples represented in Figure 1.

Some of the results of experimental study of photoinitiated copolymerization of the system HDDA – TGM-3 depending on its composition, photoinitiator concentration $c_0$ and UV-irradiation intensity $E_0$ are represented in Figure 2 in the form of averaged kinetic curves.

Comparison of the kinetic curves of copolymerization of di(meth)acrylates HDDA and TGM-3 at the given $E_0$ and $c_0$ depending on the composition of copolymerizing system is shown in Figures 3 – 6.

The integral kinetic curves of the process of photoinitiated copolymerization of 1,6-hexanediol diacrylate and triethyleneglycol dimethacrylate, represented in Figures 3 – 6 in the form of the relative conversion P vs time t, regardless of the polymerizing system composition are typically S-shaped and consist only of two parts: autoacceleration and autodeceleration.

As one can see from Figures 3 – 6, if the ratio HDDA : TGM-3 is changed in behalf of dimethacrylate TGM-3, the shape of integral experimental dependencies conversion vs time changes from characteristic for diacrylate HDDA to characteristic for dimethacrylate TGM-3. The main role in the change of kinetic curves shape belongs to dimethacrylate TGM-3: already at the ratio HDDA : TGM-3 = 2 : 1 the kinetic curve of copolymerization becomes like the curve for polymerization of dimethacrylate,

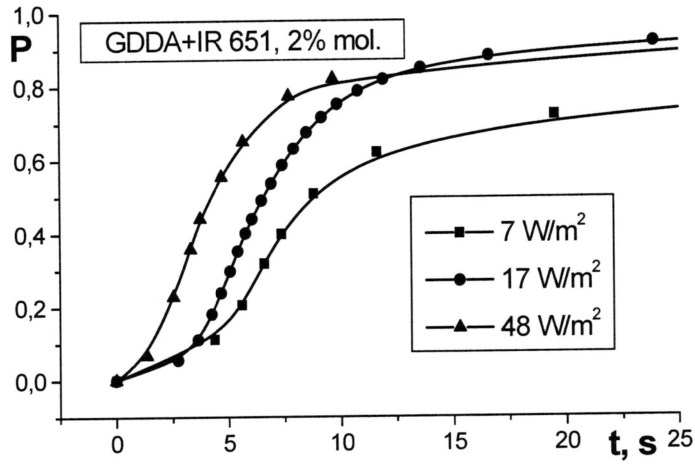


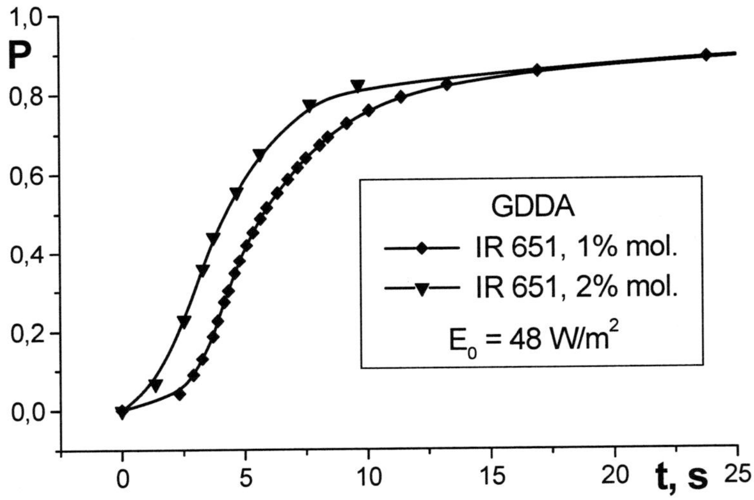


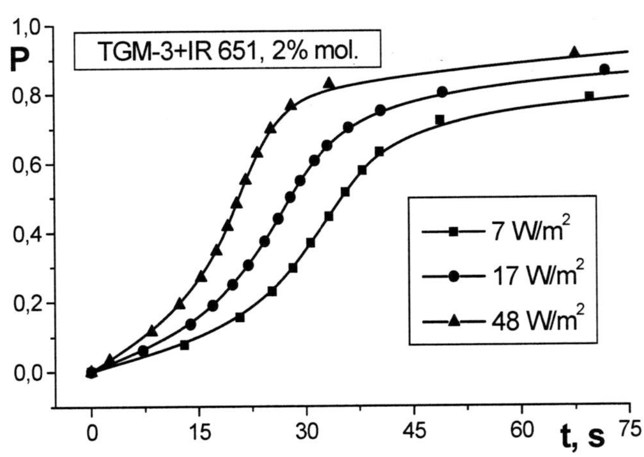


Figure 2 (Continued)

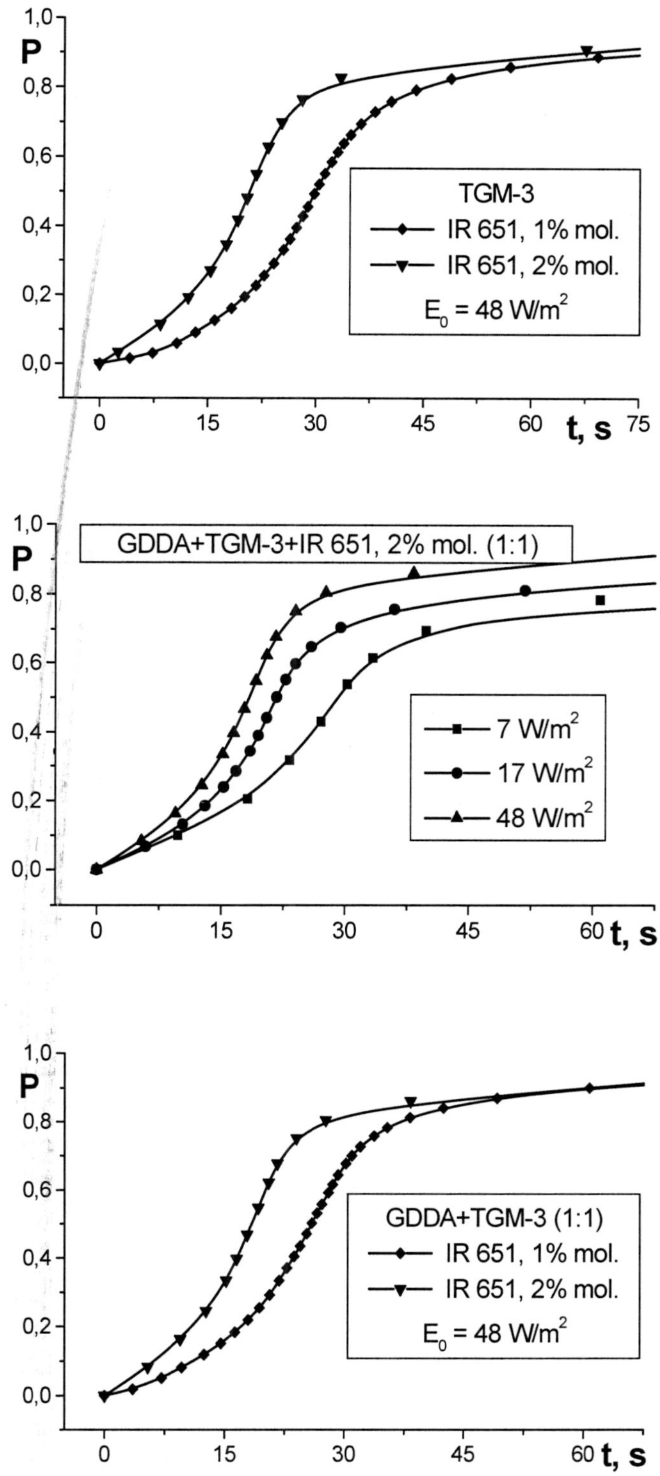


Figure 2. The averaged kinetic curves of photoinitiated copolymerization of HDDA and TGM-3 at different UV-irradiation intensity and photoinitiator concentrations.

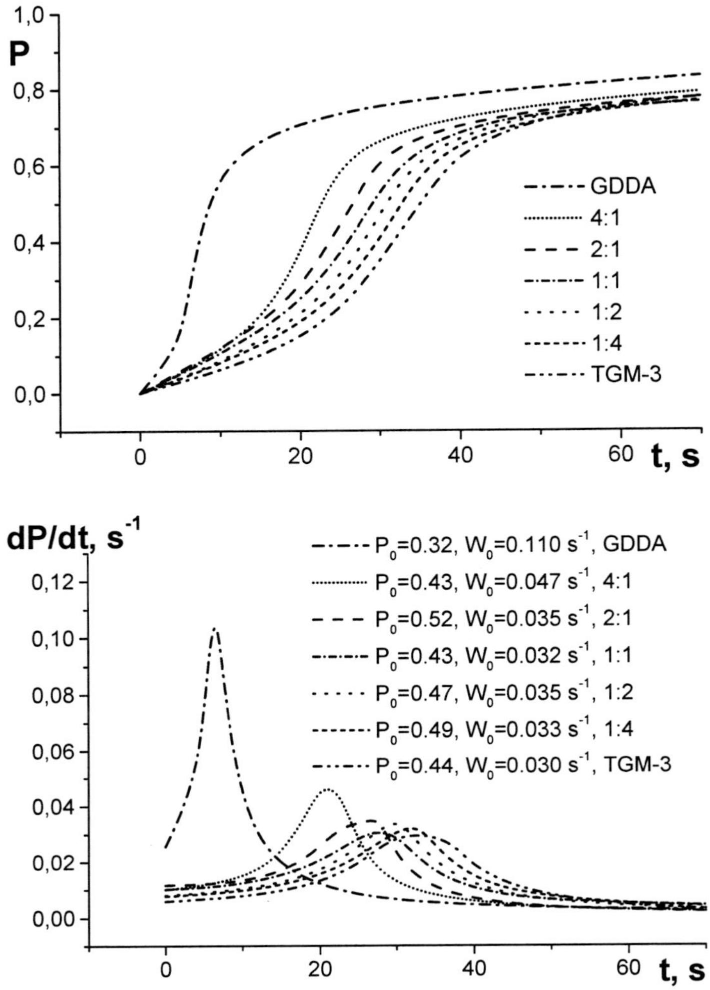


Figure 3. The integral kinetic curves and their differential anamorphoses of photoinitiated copolymerization of the system HDDA – TGM-3 depending on its composition ([IR 651] = 2.0 % mol., T = 293 K, $E_0 = 7$ W/m$^2$).

The feature of copolymerization of the system HDDA : TGM-3 is also the fact that the process rate more significantly depends on slower component, that is TGM-3; thus with the growth of TGM-3 fraction in the initial composition conversion $P_0$ (with which the maximal copolymerization rate $W_0$ corresponds) sharply increases at simultaneous decreasing of $W_0$. It evidences the decline of the contribution of the process at interface into the total one. That may be caused by two reasons. The first reason consists in the fact that the low rate constant of the chain propagation $k_{pii}$ during TGM-3 homopolymerization leads to the low values of the crosslinking rate constants $k_{pij}$ of chain propagation during copolymerization. The second one can be explained by different structure of the layer at the interface: monomer – polymer solution with HDDA and TGM-3 micrograins.

The represented experimental kinetic curves of copolymerization are similar to those of homopolymerization, for which kinetic models have been derived, that satisfactorily describe them [11].

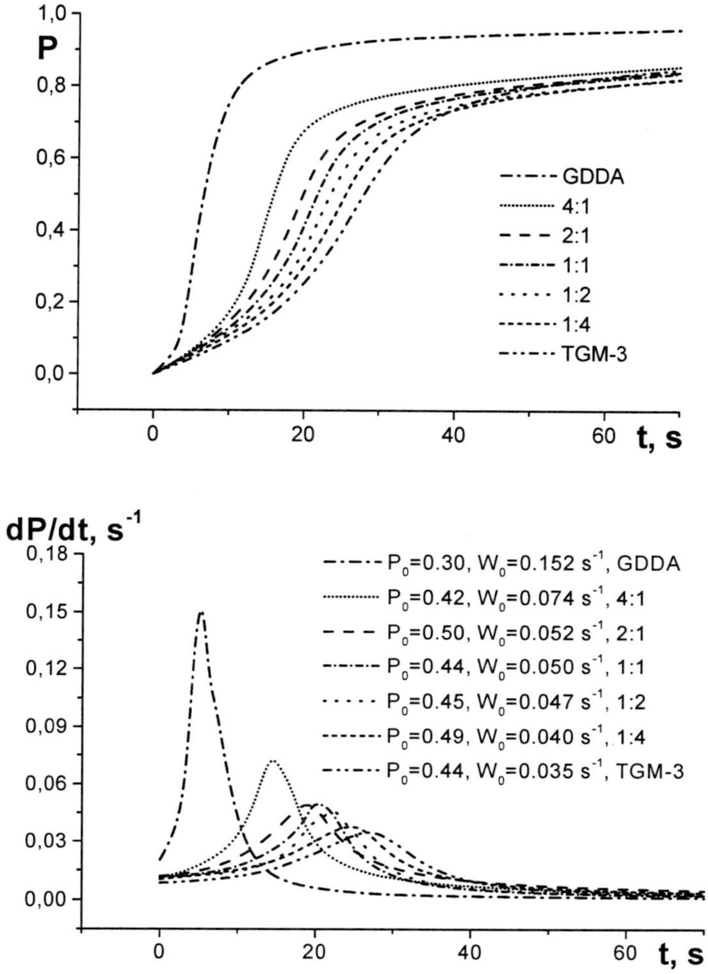


Figure 4. The integral kinetic curves and their differential anamorphoses of photoinitiated copolymerization of the system HDDA – TGM-3 depending on its composition ([IR 651] = 2.0 % mol., T = 293 K, $E_0 = 17$ W/m$^2$).

## 3. Kinetic Model of Photoinitiated Copolymerization of Bifunctional (Meth)acrylates till High Conversions

During copolymerization of bifunctional monomers solid phase appears to be networked polymer with low solubility in liquid monomer phase ($\leq 1$ %). Therefore precipitation of micrograins of networked polymer, that is, polymer phase, begins at very low conversions ~ 1 %. A new reaction zone is forming – the interphase layer at the interface: liquid monomer phase (LP) - solid polymer phase (PP).

Monomer solubility in networked polymer is low, therefore the process of polymerization in polymer phase may be neglected. Consequently, the process of polymerization of polyfunctional monomers is realized in two reaction zones: the volume of liquid monomer phase (MP), where dissolved polymer is practically absent, and in the interphase layer at the interface: MP - solid polymer phase (PP), where monomer solubility is insignificant.

Assuming, that the specific rate $\omega_i$ of i-monomer expense in every reaction zone is another, the general rate of expense of every monomer per the total system volume may be written in the form:

$$W_i = -\frac{d[A_i]}{dt} = \omega_{vi}(1-\varphi_s) + \omega_{vsi} F_{vs}(1-\varphi_s)\varphi_s . \tag{1}$$

Here $\omega_{vi}$ and $\omega_{vsi}$ are specific partial rates of copolymerization of i-monomer in MP and PP volumes respectively, $\varphi_s$ is the current volume fraction of PP micrograins; $F_{vs}$ is coefficient, which takes into account fractal properties of micrograin surface. In liquid monomer phase volume copolymerization occurs according to the classic kinetic scheme with the square chain termination:

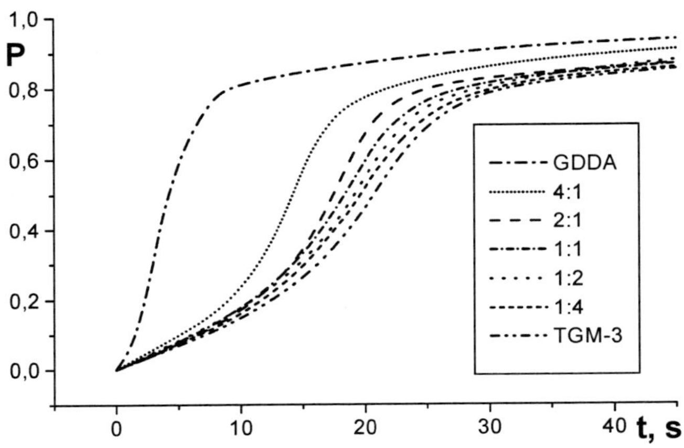


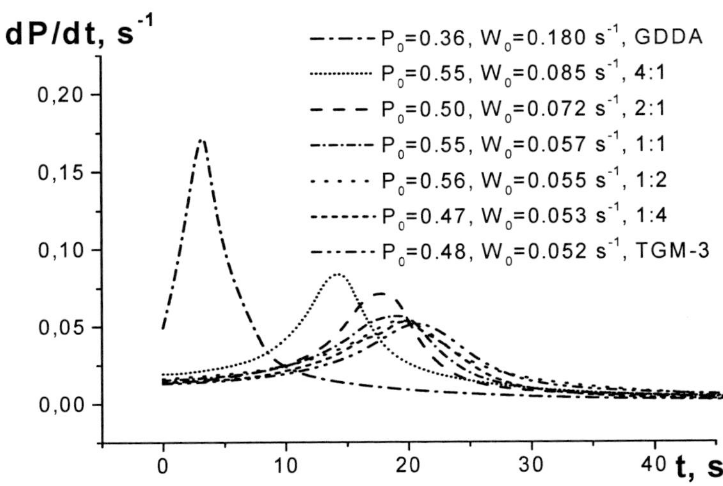


Figure 5. The integral kinetic curves and their differential anamorphoses of photoinitiated copolymerization of the system HDDA – TGM-3 depending on its composition ([IR 651] = 2.0 % mol., T = 293 K, $E_0$ = 48 W/m$^2$).

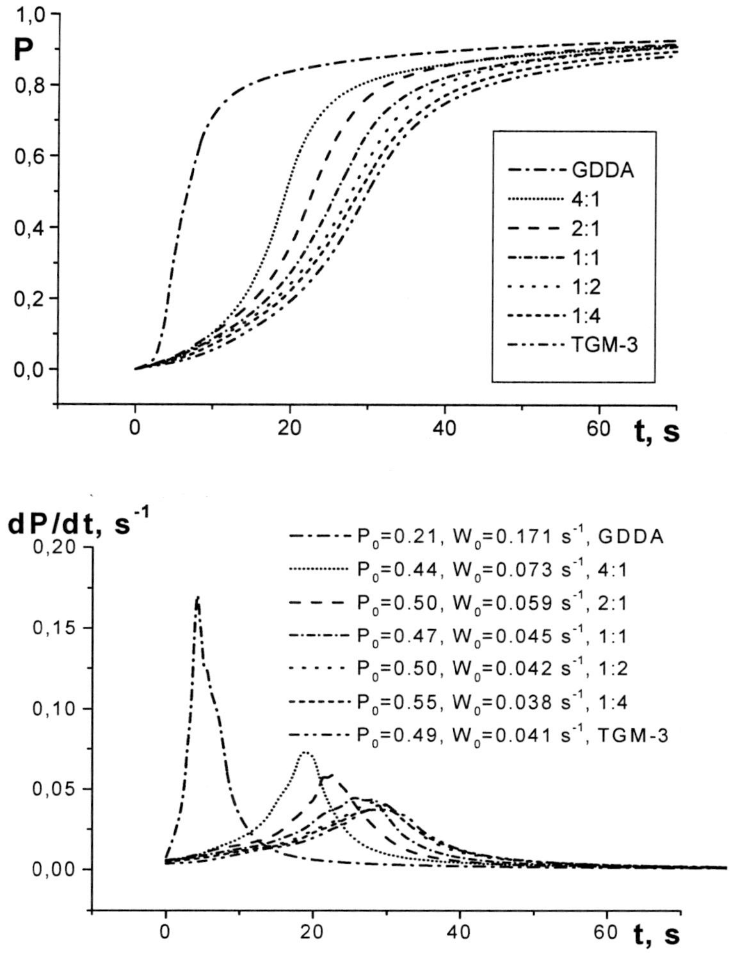


Figure 6. The integral kinetic curves and their differential anamorphoses of photoinitiated copolymerization of the system HDDA – TGM-3 depending on its composition ([IR 651] = 1.0 % mol., T = 293 K, $E_0$ = 48 W/m$^2$).

chain propagation

$$A_i + R_j \xrightarrow{k_{pij}} R_i,\ i = 1,2, \tag{2}$$

chain termination

$$R_i + R_j \xrightarrow{k_{tij}} \text{product},\ i = 1,2\ . \tag{3}$$

Therfore specific rates $\omega_{vi}$ of copolymerization by i-monomer in MP can be described by the known Majo-Walling equation

$$\omega_{v_i} = \frac{(k_{pii}k_{pij}[A_{vi}]^2 + k_{pij}k_{pji}[A_{vi}][A_{vj}])v_{in}^{1/2}}{\{k_{tii}(k_{pij}[A_{vi}])^2 + 2k_{tij}k_{pij}k_{pji}[A_{vi}][A_{vj}] + k_{tjj}(k_{pji}[A_{vj}])^2\}^{1/2}}, \tag{4}$$

where $k_{pii}$, $k_{tii}$ are rate constants of polymer chain propagation and termination respectively; $[A_{vi}]$ is i-monomer concentration in MP; $V_{in}$ is an initiation rate. Solid polymer phase at the interface with liquid monomer phase forms solid structure of interface layer. That is why in the interphase layer at the interface MP – PP segmental and translation mobilities of macroradicals diminish sharply, increasing the role of diffusion braking of the square chain termination. At the same time the probability of trapping or self-burial of growing macroradical in polymeric matrix increases. Hence, in accordance with the already known notions [11-14], we assume, that in the interphase layer a chain termination is linear and appears as an act of macroradical propagation leading to a trap, what can be represented by following kinetic scheme:

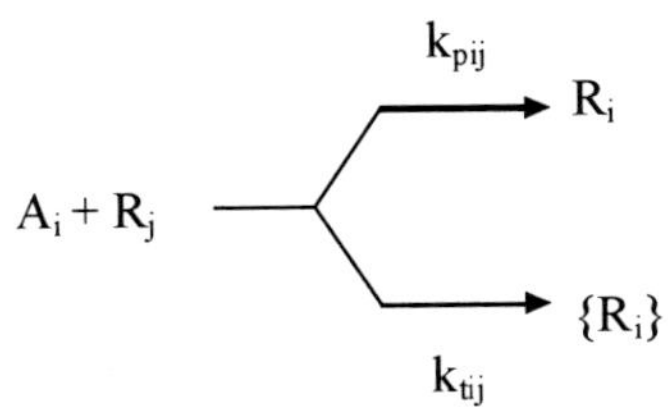


(5)

According to this scheme, specific rate $\omega_{vs}$ of copolymerization by i-monomer in the volume of reaction zone of the interphase layer can be described by equation [15, 16]:

$$\omega_{vsi} = \frac{(k_{pii}k_{pij}[A_{vsi}]^2 + k_{pij}k_{pji}[A_{vsi}][A_{vsj}])v_{in}}{k_{tii}k_{pij}[A_{vsi}]^2 + (k_{tij}k_{pij} + k_{tji}k_{pij})[A_{vsi}][A_{vsj}] + k_{tjj}k_{pji}[A_{vsj}]^2}. \quad (6)$$

We assume, that monomer concentrations in the interphase layer are equal to their concentrations in the volume of monomer phase, that is, $[A_{vsi}] = [A_{vi}]$ and $[A_{vsj}] = [A_{vj}]$.

To describe kinetics of copolymerization by eq. (1) taking into account expressions (4) and (6) for specific rates, it is necessary to enter also the rate of precipitation of solid polymer phase $d\varphi_s / dt$. The expression for $d\varphi_s / dt$ can be obtained from balance equations taking into account that monomer concentrations in PP are close to zero: $[A_{si}] \approx 0$, $[A_{sj}] \approx 0$:

$$\frac{d\varphi_s}{dt} = \frac{W_i + W_j}{[A_{vi}] + [A_{vj}]}. \quad (7)$$

The equations (1), (4), (6) and (7) together completely describe kinetics of copolymerization of two bifunctional monomers till high conversions by the model of two reaction zones.

## 4. Conclusions

Experimental kinetic curves of photoinitiated copolymerization of bifunctional monomers till high conversions, except for features, are similar to those of homopolymerization, hence, their simulation was performed in terms of conception of different reaction zoned taking into account kinetic equation of copolymerization at linear chain termination.

Kinetic curves of copolymerization in the system of bi-bifunctional monomers at the change of system composition retain the shape peculiar to homopolymerization of monomers and consist of only two parts – stages of autoacceleration and autobraking.

In the system HDDA – TGM-3 additions of slower component (TGM-3) have more significant effect on copolymerization rate diminishing it. It points out that decreasing of copolymerization rate takes place both due to the small value of the rate constant of chain propagation $k_{pii}$ and due to the smaller values of crosslinking rate constants of chain propagation $k_{pij}$ as compared with the rate constant of chain propagation of more rapid component.

The represented features of copolymerization of bifunctional monomers have been quantitatively expressed in kinetic model of photoinitiated copolymerization of bifunctional (meth)acrylates till high conversions offered in this work.

## References


[1] Riddle E. H. *Monomeric Acrylic Esters / E. H. Riddle.* – New York: Reinhold, 1954, p. 221

[2] The Polymeric Materials Encyclopedia: *Synthesis, Properties and Applications* / [ed. by Salamone J. C.]. – New York: CRC Press, Inc. Boca Raton, Vol. 7, 1996, p.p. 5532–5544.

[3] Semenyuk I. V. *Rozrobka tekhnologiyi oderzhannya akrylatnykh monomeriv dlya vyhotovlennya optychnykh dyskiv* // Thesis for a candidate's degree in technical sciences, speciality 05.17.04, State university "Lviv polytekhnika", Lviv, 1995. – p. 17 (in Ukrainian).

[4] Lachinov M. B., Guliashvili T. T., Chkheidze N. R. et al. *Kineticheskiye zakonomernosti radikalnoy polimerizatsiyi monomerov ftoralkilmetakrilatnogo ryada v masse* // Vysokomol. Soed., Vol. 40 (A), No 2, 1998, p. 215 – 220 (in Russian).

[5] Lovell L. G., Newman S. M., Bowman C. N. The Effects of Light Intensity, Temperature, and Comonomer Composition on the Polymerization Behavior of Dimethacrylate Dental Resins // *J. Dent. Res.,* Vol. 78, No 8, 1999, p.p. 1469-1476.

[6] Berlin A. A., Korolev G. V., Kefeli T. Ya., Sivergin Yu. M. *Akrilovyje oligomery i materialy na ikh osnove* // Moscow, Khimiya, 1983, p. 232 (in Russian).

[7] Shybanov V. V., Kostenko T. A. *Fotopolimerizuyushchiyesya materialy dlya izgotovleniya pechatnykh form* // Moscow, Kniga, 1991, p. 40 – (Nauch.-issled. tsentr "Informpechat", Poligraf. prom. Obzornaya informatsiya) (in Russian).

[8] Zagladko E. A. *Kinetychna model tryvymirnoyi radykalnoyi polimeryzatsiyi v blotsi // Dissertation for a candidate's degree in chemical sciences,* speciality 02.00.06, Lviv, 1998, p. 104 (in Ukrainian).

[9] Grishchenko V. K., Maslyuk A. F., Gudzera S. S. *Zhydkiye fotopolimerizuyushchiyesya kompozitsiyi* // Kiyev, Naukova dumka, 1985, p. 208 (in Russian).

[10] Lecamp L., Lebaudy P., Youssef B. et al. Influence of UV radiation wavelenght on conversion and temperature distribution profiles within dimethacrylate thick material during photopolymerization // *Polymer, Vol.* 42, No 21, 2001, p.p. 8541–8547.

[11] Medvedevskikh Yu. G., Kytsya A. R., Bazylyak L. I. et al. Stationary and nonstationary kinetics of the photoinitiated polymerization // Utrecht, Boston: Brill Academic Publishers, *(New Concepts in Polymer Science),* 2004, p. 313.

[12] Dyachkov A. I., Efimov A. L., Efimov L. I. O mekhanizme obryva tsepey pri radikalnoy polimerizatsiyi metilmetakrilata pri glubokikh konversiyakh // *Vysokomol. soed.*, Vol. 25(A), No 10, 1983, p.p. 2176–2181 (in Russian).

[13] Kurdikar D. L., Pepas N. A. A kinetic study of diacrylate photopolymerization // *Polymer,* Vol. 35, No 5, 1994, p.p. 1004–1011.

[14] Brun E. B., Ivanov V. A., Kaminskiy V. A. Kinetika postpolimerizatsiyi pri glubokikh stepenyakh prevrashcheniya // *Proceedings of AS of USSR.* Vol. 291, No 3, 1986, p.p. 618–622 (in Russian).

[15] Medvedevskikh Yu. G., Khovanets' G. I., Yevchuk I. Yu., Zaikov G. I. Regularities of photoinitiated copolymerization of (meth) acrylates till high conversions. Kinetic model of the process at linear chain termination // *Polymer Research Journal,* Vol. 1, Iss. 3, 2008, p.p. 213–223.

[16] Medvedevskikh Yu., Khovanets' G., Yevchuk I. Kinetic model of photoinitiated copolymerization of monofunctional monomers till high conversions // *Chemistry & Chemical Technology,* Vol. 3, No 1, 2009, p.p. 1–6.

In: Chemical and Biochemical Reactions
Editors: G. E. Zaikov, L. Liu et al.
ISBN 978-1-61122-860-1


*Chapter 12*

# INVESTIGATION OF THE ANTIOXIDANT ACTIVITY DYNAMIC OF THE MINT WATER EXTRACT BY ELECTROCHEMICAL METHODS

***N. N. Sazhina*[1], *V. M. Misin*[1,*] *and E. I. Korotkova*[2,†]**
[1]Emanuel Institute of Biochemical Physics, Russian Academy of Sciences
4 Kosygin Street. 119334 Moscow, Russia
[2]Tomsk Politechnik University. 30 Lenin Street. Tomsk, Russia

## ABSTRACT

In the present work the total content of antioxidants and their activity relatively oxygen and its radicals are measured in water extract of mint peppery (*Mentha piperita L.*) by two electrochemical methods: ammetric and voltammetric. It is established that these parameters, and also voltammogramms, defining mechanism of oxygen reduction, are essentially changed during the first hours after extraction. The explanation of the received results from the point of view of mint structure is offered.

**Keywords:** antioxidant, ammetric and voltammetric methods; total phenolic content; polyphenols, oxygen radicals

## AIM AND SCOPE

Definition of the total antioxidant content and antioxidant activity relatively oxygen and its radicals in water extract of mint peppery (*Mentha piperita L.*) by two electrochemical methods.

* E-mail: misin@sky.chph.ras.ru
† E-mail: eikor@mail.ru

## INTRODUCTION

Now the great attention is given to research of the quantity and activity of antioxidants (AO) in herbs, foodstuff, drinks and other objects. It is known that the increase in activity of oxidation processes in an organism leads to infringement of structure and properties of lipid membranes. There is a direct dependent between the superfluous quantity of free radicals in organism and occurrence of dangerous diseases [1].

Antioxidants are biologically active substances, which remove excessive free radicals, decreasing the lipid oxidation. Therefore a detailed research of total antioxidant activity (TAA) and mechanisms of interaction with oxygen and its radicals are highly prospective. [2].

Mint peppery (*Mentha piperita L.*) is one of the most widespread types of herbs. It is widely used in medicine, cookery, household, pharmaceutical and cosmetic industry. Mint is applied as an outside and inside medical pharmaceutical in the form of infusions and teas for treatment of gastro enteric diseases, as a demulcent at palpitation, depression and sleeplessness, as sedative means, as anesthesia at burns and stings of insects, removes stress and a headache. Antioxidant properties of a mint allow to prevent occurrence of a cataract and other illnesses connected with ageing of an organism. Mint peppery contains essence (2-3 %) in which basic component is the menthol defining taste and anesthezing properties of mint, and also other substances – a ethers, phellandrene, pinene, jasmole, piperitone, menthofuran etc. There are also tannic and resinous substances, carotin (0.01 %), ascorbic acid (0.01 %), routines (0.015 %) and other polyphenolic combinations [3].

In the literature there are publications devoted to measurement of the AO quantity and activity in extracts of mint [4-7]. However it, basically, difficult chemical methods, and the results received by these methods are badly comparable among themselves, including the absence of uniform measure units. Use of two operative electrochemical methods has allowed studying the antioxidant activity dynamic of mint after extraction.

## EXPERIMENTAL PART

The purpose of the present work was to measure the antioxidant activity (AOA) of a water mint extract by a voltammetric method and to study the mechanism of influence of mint components on process of electrochemical oxygen reduction (ER $O_2$). In parallel in mint extract measurements of the total phenol type antioxidant content have been spent by an ammetric method.

Object of research was the water extract of the mint peppery (*Mentha piperita L.*), collected and dried in 2008 year in the Tver region. For extract reception the dry grass was crushed in a mortar till the particles size 1-2 mm. Further this grass (0.5 g) was filled by the distilled water with T=95$^0$C and volume 50 ml and maintained within 10 minutes without thermo stating. Then an extract was carefully filtered through the paper filter «a dark blue tape» and if necessary diluted before measurements [7]. Acidity of an extract (pH) has monitored as 6.6.

**Table. Groups of biologically active substances divided according mechanisms of influence on oxygen and its radicals**

| Group number | 1-group | 2-group | 3-group | 4-group |
|---|---|---|---|---|
| Substance names | Catalasa, phtalocyanines of metals, humin acids. | Phenol nature substances, vitamins A, E, C, B, flavanoids, ubichinon, glucose. | N, S, Se-containing substances, amines, amino acids, active aldehydes | Superocsidismutase (SOD), porphyry metals, cytochrome C |
| Influence on ER $O_2$ process | Increase in current of ER $O_2$, potential shift in negative area | Decrease of current of ER $O_2$, potential shift in positive area. | Decrease of current of ER $O_2$, potential shift in negative area. | Increase in current of ER $O_2$, potential shift in positive area |
| The prospective electrode mechanism. | EC* mechanism with the following reaction of hydrogen peroxide disproportion and partial regeneration of molecular oxygen. | EC mechanism with the following chemical reaction of interaction of AO with active oxygen radicals | CEC mechanism with the precede and following chemical reactions of interaction of AO with oxygen and its active radicals | EC mechanism with catalytic oxygen restoration through formation of intermediate complexes. |

* The note: E – an electrode stage of process, C – chemical reaction.

In the present work two electrochemical methods were used. The ammetric method [8] allows defining the total phenol type antioxidant content in investigated samples. Measurements were spent on the device «Zwet-Jauza-01-AA» [9]. The essence of the given method consists in measurement of the electric current arising at oxidation of investigated substance (or mixes of substances) on a surface of a working electrode at constant potential +1,3V. At this potential there are oxidation only OH-groups natural phenolic type antioxidants [7-9]. The received signal from the tested sample is compared to a signal of an individual antioxidant – gallic acid, measured under the same conditions. The measurement error is no more than 10%, measurement time - 10-15 minutes [9].

The voltammetric method uses an electroreduction of oxygen (ER $O_2$) as modeling reaction. This process is similar to oxygen reduction in tissues and in stuffs of plants. It proceeds at the working electrode in several stages with formation of the reactive oxygen species (ROS), such as $O_2^{\cdot-}$ and $HO_2^{\cdot-}$ [10]:

$$O_2 + e^- \rightleftarrows O_2^{\cdot-} \quad E^o=+0.012 \text{ V} \tag{1}$$

$$O_2^{\cdot-} + H^+ \rightleftarrows HO_2^{\cdot-} \tag{2}$$

$$HO_2^{\cdot -} + H^+ + e^- \rightleftarrows H_2O_2 \; E^o{=}{+}0.682 \text{ V} \qquad (3)$$

$$H_2O_2 + 2H^+ + 2e^- \rightleftarrows 2H_2O \; E^o{=}{+}1.770 \qquad (4)$$

For determination of total antioxidant activity in work [10] it is offered to use the first wave of ER $O_2$ corresponding to stages (1) – (3) when on a surface of a mercury film electrode active oxygen radicals and hydrogen peroxide as an end product are formed. It takes attention that antioxidants of varied nature were divided on 4 groups according their interaction with oxygen and its radicals (table) [11].

As antioxidant activity criterion of investigated substances the kinetic criterion $K$ is used which reflects quantity of oxygen and the active oxygen radicals which have reacted with AO (or their mixes) in a minute of time: $K = \frac{C_0}{t}(1 - \frac{I}{I_0})$ , mkmol/l·min, where $I$, $I_0$ — currents of electroreduction of oxygen, accordingly, at presence and at absence of AO in a solution of supporting electrolyte, $C_0$ — initial concentration of oxygen in a solution (mkmol/l), i.e. solubility of oxygen in supporting electrolyte under normal conditions, $t$ — time of an exposition of a working electrode at constant potential under the limiting current of the oxygen, characterizing course of reaction of interaction of AO with active oxygen radicals.

For voltammetric study of total antioxidant activity of samples automated voltammetric analyzer "Analyzer of TAA" ("SPE Polyant" Tomsk, Russia) was used. As supporting electrolyte the phosphate buffer solution in volume of 10 ml with known initial concentration of molecular oxygen was used, where was added dosed out ($\delta$=10 – 500 mkl) quantity of the investigated sample [10]. The electrochemical cell (V=10 ml) was connected to the analyzer and consisted of a working mercury film electrode (MFE), a silver- silver chloride reference electrode with KCl saturated ($Ag|AgCl|KCl_{sat}$) and a silver- silver chloride auxiliary electrode. An open type cell can be used in this investigation. The reference electrode and the working electrode were held in the electrochemical cell. The working electrode potential was initially set at 0 V for about 120s. During this step the solution was stirred by a magnetic stirrer. After stirring the potential was scanned negatively, causing oxygen reduction, which gives a current first wave at E = -0.1 ÷ -0.3V. Its value is proportional to the amount of oxygen in the solution bulk. Potential rate scan was 0.1 V/s, potential range was E = 0 to –0.9 V. The voltammetric method of research has good sensitivity. It is simple and cheap. However, as in any similar electrochemical method, the disorder of instrument readings at identical measurements has appeared high enough (to 1,5-2,0 times). Therefore for each sample 3-4 repetitions have been spent, and results were averaged. The root-mean-square deviation of kinetic criterion K from average value for all investigated samples has made no more than 30 %.

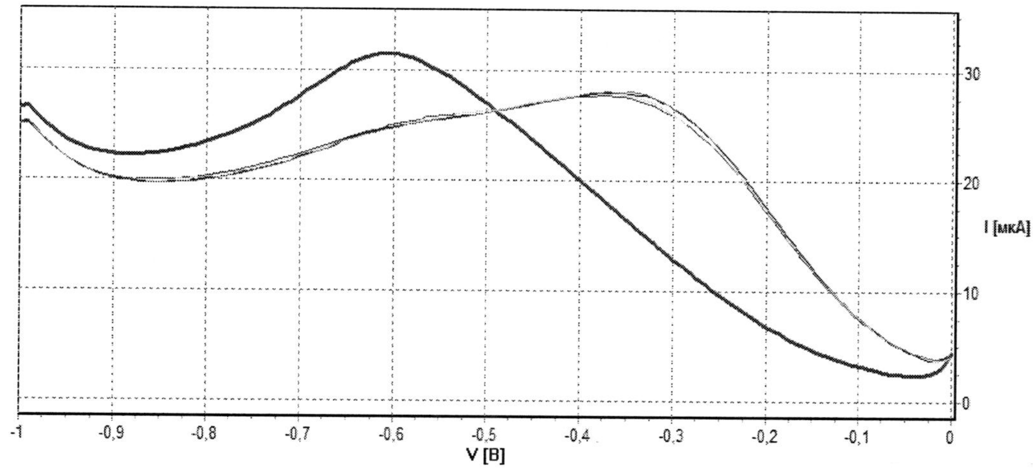


Figure 1. Voltammogramms of current of oxygen electroreduction at absence (left curve) and at presence(right curve) of mint extract in cell at t=5 min after mint extraction. Extract dose δ=100 mkl in 10 ml of a buffer solution.

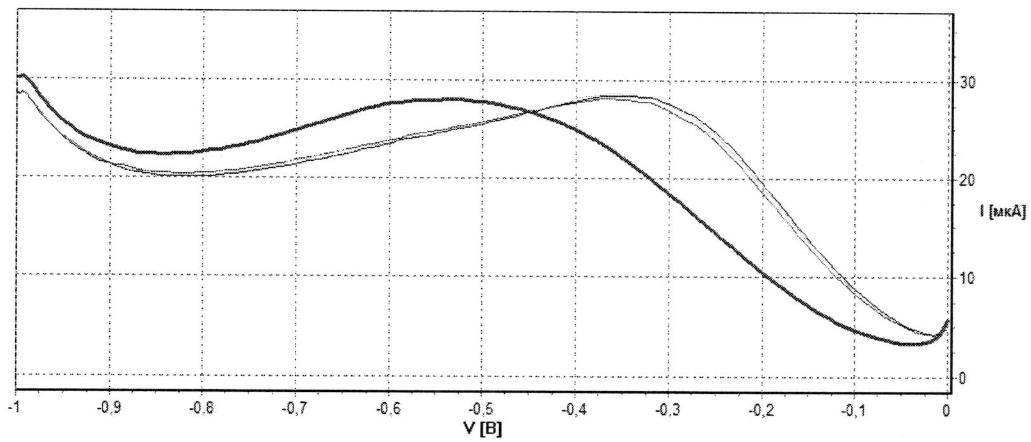


Figure 2. Voltammogramms of current of oxygen electroreduction at absence (left curve) and at presence(right curve) of mint extract in cell at t=30 min after mint extraction. Extract dose δ=100 mkl in 10 ml of a buffer solution.

## RESULTS AND DISCUSSIONS

On Figures 1-3 voltammogramm of the samples, received at various times after mint extraction are resulted: after extraction t=5 min (Figure 1), t=30 min (Figure 2) and t=60 min (Figure 3). It is visible that the fresh mint extract (Figure 1) "works" on the mechanism of phenol nature AO, reducing a current and shifting potential in positive area (2nd group in the table). However, approximately in an hour after extraction (Figure 3), character of interaction of components of mint with oxygen and its radicals varies, following the mechanism, characteristic for substances of 4th group in the table. Transition from one mechanism to another occurs approximately through 30 min (Figure 2) and the kinetic criterion *K* becomes thus close to zero. Shift of potential of a current half wave of oxygen restoration (Δ) remains approximately identical (0.20 – 0.25 V). At the further storage of an extract in vitro voltammogramm character essentially does not vary, and the module of kinetic criterion $|K|$ grows to values 2.0±0.5 mkmol/l·min in 3 hours after extraction.

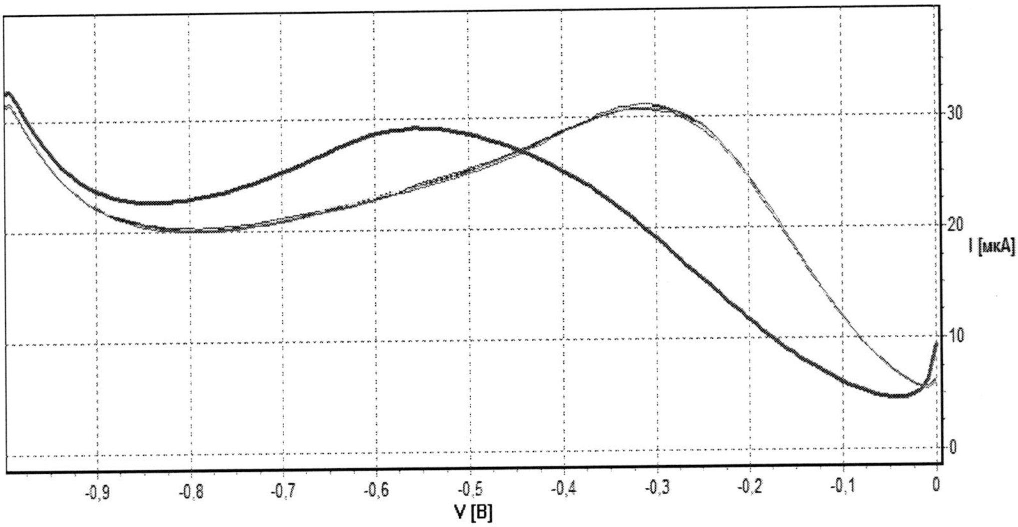


Figure 3. Voltammogramms of current of oxygen electroreduction at absence (left curve) and at presence(right curve) of mint extract in cell at t=60 min after mint extraction. Extract dose δ=100 mkl in 10 ml of a buffer solution.

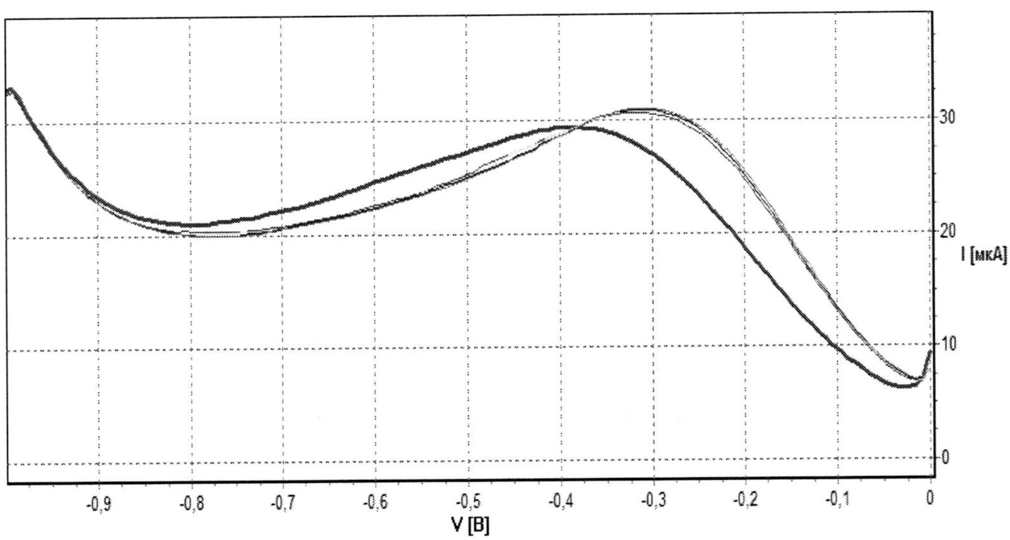


Figure 4. Voltammogramms of current of oxygen electroreduction at absence (left curve) and at presence (right curve) of mint extract in cell at t>60 min after mint extraction. Extract dose δ=25 mkl in 10 ml of a buffer solution.

On Figure 3-5 the example of dependence of kinetic criterion $K$ and potential shift $\Delta$ on concentration of a mint extract in the supporting electrolyte is shown at storage times of an extract over 1 hour. During experiment the influence of extract dose $\delta$ in supporting electrolyte on the $|K|$ and $\Delta$ was investigated. As results $|K|$ and $\Delta$ decrease from $|K|=1.5\pm0.5$ mkmol/l·min, $\Delta=0.25$ V at $\delta=100$ mkl (Figure 3), to $|K|=1.3\pm0.3$ mkmol/l·min and $\Delta=0.10$ V for $\delta=25$ mkl (Figure 4) and $|K|=0.8\pm0.2$ mkmol/l·min, $\Delta=0.05$ V for $\delta=10$ mkl (Figure 5). At increase $\delta$ above 100 mkl the voltammogramm character small varies, and at $\delta>300$ mkl decrease $|K|$ is observed.

In parallel with voltammogramm registration for the same mint extract the total phenolic antioxidant content C (in units of gallic acid) has been measured from an extract storage time t by ammetric method (Figure 6). Dependence C on t testifies to notable enough falling C after extraction and filtering of a mint extract (approximately 20 % for 2 hours of storage). It, possibly, is connected with destruction of the unstable phenol substances containing in an extract. Therefore, apparently, voltammogramm character and values of kinetic criterion (Figures 1-3) during the first moment after extraction could establish as influence of classical phenolic antioxidant on the process of electroreduction of oxygen and its radicals. (Figure 6).

As it has been mentioned in the introduction, one of the basic components of mint essence is menthol $C_{10}H_{20}O$ [12], voltammogramm character therefore has been studied at introduction in a buffer solution in cell a water menthol solution. Menthol (60 mg) has been dissolved in 100 ml of hot water and doses of this solution 200 – 500 mkl were added in 10 ml of a buffer solution. Voltammogramms have been registrated under the same conditions, as a mint extract. On Figure 7 the typical example of such voltammogramm is presented from which it is visible that menthol "works" as classical AO, decreasing the current of oxygen electroreduction. For a dose of a menthol solution $\delta=500$ mkl the kinetic criterion $K$ has made $1.5\pm0.4$ mkmol/l·min, slowly decreasing at reduction of menthol concentration. Such behavior has also a cyclohexanol (similar menthol substance) which activity at the same concentration is included into the same interval of values $K$, as for menthol in a similar way operates. For such components of mint as routines and ascorbic acid, their voltammogramms are a classical example of 2nd groups of antioxidants from table [13]. It is possible to conclude that the total mechanism of interaction of mint components with oxygen and its radicals is defined by the substances not listed above, and other components, for example some substances of the phenolic nature giving shift of potential in a positive side and dominating in the chemical mechanism for a fresh extract. At the further storage of an extract, these unstable substances destruct and the general character of oxygen reduction is defined already by the substances containing in mint and entering into 4th group of the table. Probably, they are any metal complexes which operate as catalysts of passing chemical reactions.

## CONCLUSIONS

The total content of antioxidants and their activity relatively oxygen and its radicals are measured in water extract of mint peppery (*Mentha piperita L.*) by two electrochemical methods: ammetric and voltammetric. It is established that these parameters, and also voltammogramms, defining mechanism of oxygen reduction, are essentially changed during the first hours after extraction. The explanation of the received results from the point of view of mint structure is offered.

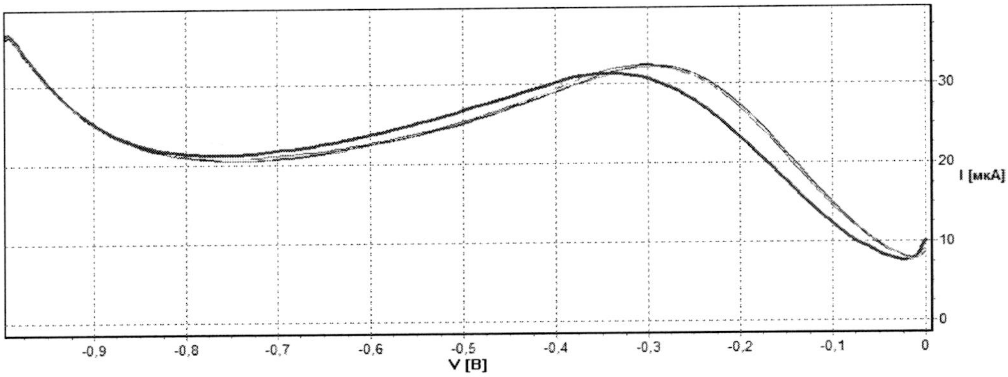


Figure 5. Voltammogramms of current of oxygen electroreduction at absence (left curve) and at presence (right curve) of mint extract in cell at t>60 min after mint extraction. Extract dose δ=10 mkl in 10 ml of a buffer solution.

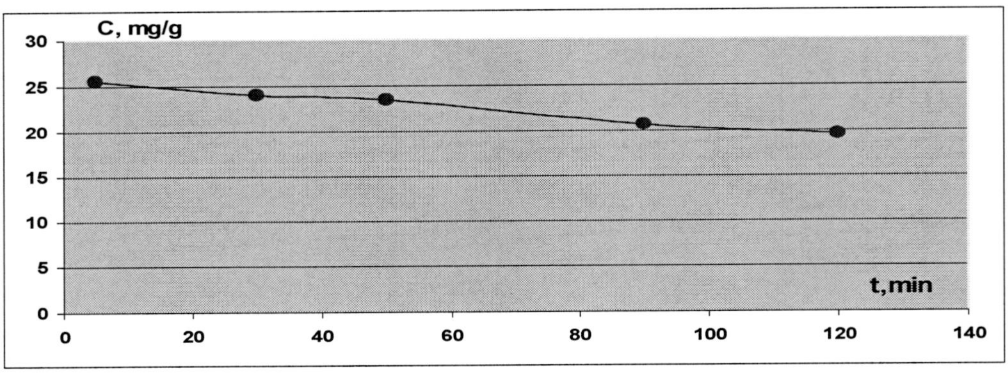


Figure 6. The total phenol antioxidant content C in a mint extract depending on extract storage time t.

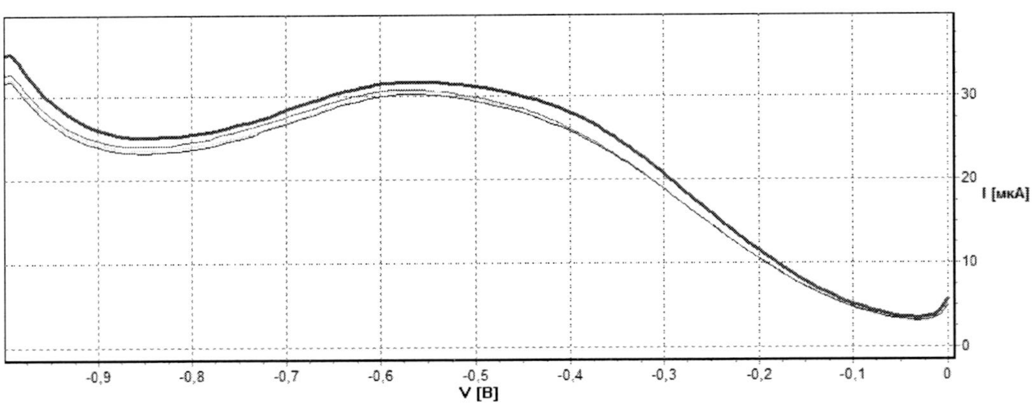


Figure 7. Voltammogramms of current of oxygen electroreduction at absence (top curve) and at presence (bottom curve) of water menthol solution. Menthol solution dose δ=500 mkl in 10 ml of a buffer supporting electrolyte.

## REFERENCES


[1] Burlakova E.B. In *Bioantioxidant: yesterday, today, tomorrow. Biological Kinetics. Reviews*. M. Chemistry. 2005. V. 2. P. 10-45.

[2] Dillard C.J., German J.B. Phytochemicals: nutraceuticals and human health // *J. Sci. Food Agric*. 2000. V. 80. P. 1744-1756.

[3] Dudchenko L.G., Kozjakov A.S., Krivenko V.V. Gingery-aromatic and gingery-flavour plants //K: Naukova dumka. 1989. P. 304.

[4] Proestos C., Chorianopoulos N., Nychas G.-J. E., Komaitis M. RP-HPLC analysis of the phenolic compounds of plant extracts. Investigation of their antioxidant capacity and antimicrobial activity // *J. Agric. Food Chem.* 2005. V. 53. N 4. P. 1190-1195.

[5] Autoui Ali K., Mansouri A., Boskou G., Refalas P. Tea and herbal infusions: Their antioxidant activity and phenolic profile // *Food Chem.* 2005. V. 89. N 1. P. 27-36.

[6] Katalinic V., Vilos M., Kulisic T., Jukic M. Screening of 70 medicinal plant extracts for antioxidant capacity and total phenols // *Food Chem.* 2006. V. 94. N 4. P. 550-557.

[7] Misin V. M, Sazhina N.N., Zavjalov A.J., Yashin JA.I. Measurement of the phenol content in extracts of medicinal grasses and their mixes by ammetric method. // *Chemistry of vegetative raw materials*. 2009. №4. P. 127-132.

[8] Jashin A.J., Jashin J.I., Chernousova N.I., Pahomov V.P. Express electrochemical method of antioxidant activity definition in foodstuff. // *Beer and waters*. 2004. №6. P. 44-46.

[9] Yashin A.Ya. Inject-flowing system with ammetric detector for selective definition of antioxidants in foodstuff and drinks. // *Russian chemical magazine.* 2008. V. LII. №2. P. 130-135.

[10] Korotkova E.I., Karbainov Y.A., Avramchik O.A. Investigation of antioxidant and catalytic properties of some biological-active substances by voltammetry. // *Anal. and Bioanal. Chem.* 2003. V. 375. №. 1-3. P. 465-468.

[11] Korotkova E.I., Lukina A.N., etc. In *Estimation methods of antioxidant activity of biologically active substances in medical and preventive appointment.* The collection of reports. Moscow. 2004. P. 182-194.

[12] Handbook of Chemistry and Physics. 71st edition. // CRC Press. Ann Arbon. Michigan. 1990.

[13] Korotkova E.I. The voltammetric method for definition of total antioxidant activity in objects of an artificial and natural origin. // Dissertation of the Doctor of Chemistry. Tomsk. 2009.

In: Chemical and Biochemical Reactions
Editors: G. E. Zaikov, L. Liu et al.
ISBN 978-1-61122-860-1


*Chapter 13*

# THE STRUCTURAL MODEL OF IMPACT TOUGHNESS OF PARTICULATE-FILLED POLYMER NANOCOMPOSITES

***G. V. Kozlov*[1], *Z. Kh. Aphashagova*[1], *A. Kh. Malamatov*[1] *and G. V. Zaikov*[2]**
[1]Kabardino-Balkarian State University, Nal'chik-360004, Chernishevsky st., 173, Russian Federation
[2]N.M. Emanuel Institute of Biochemical Physics of Russian Academy of Sciences, Moscow-119334, Kosygin st., 4, Russian Federation

## ABSTRACT

There was offered the quantitative structural model of particulate-filled polymer composites impact toughness, based on the fractal analysis ideas. The indicated model demonstrated good correspondence to the experimental data. It has been shown, that the action of nanofiller as nucleator, resulting to crystallinity degree and amorphous phase structure change, exercises the main influence on impact toughness value.

**Keywords:** Nanocomposite, polyethylene, calcium carbonate, structure, fractal analysis, impact toughness

## INTRODUCTION

The authors of paper [1, 2] found out that the introduction of particulate nanofiller (calcium carbonate, $CaCO_3$) in high density polyethylene (HDPE) results to nanocomposite HDPE/$CaCO_3$ impact toughness $A_p$ increase in comparison with neat polymer on about 20 %. The authors [1, 2] fulfilled the detailed fractographic analysis of this effect and explained the observed $A_p$ increase by nanocomposite HDPE/$CaCO_3$ plastic deformation mechanism change in comparison with neat HDPE. Not going into details of the mentioned analysis, one should note some reasons for doubts about its correctness. In Figure 1 the schematic diagrams load-time (*P-t*) are adduced for two cases of polymeric materials samples fracture in impact testing: by both instable (a) and stable cracks (b). As it is known [3], $A_p$ value is characterized by the area under *P-t* diagram, which gives mechanical energy consumed with samples fracture. The polymeric materials macroscopic fracture process, defined by the magistral crack propagation, begins at the greatest load *P*. From schematic diagrams *P-t* it follows that fracture process proper does not influence practically on the value $A_p$ in case of crack instable propagation and influences only partly – in case of stable crack. Although the authors [1, 2] made impact testing on the instrumented apparatus, allowing to obtain diagrams *P-t*, these diagrams were not adduced. Besides, the structural aspect of fracture process in papers [1, 2] is being considered with secondary structures (crazes, shearing zones, etc.) using. Their interconnection with neat nondeformed material structure is purely speculative. It is obvious, that at such analysis method it is impossible to obtain quantitative relationships structure-properties (that is the main task of polymers physics [4]). Therefore the purpose of the present paper is quantitative structural analysis of HDPE and nanocomposite HDPE/$CaCO_3$ impact testing results within the frameworks of fractal models.

## EXPERIMENTAL

The experimental data were accepted according to paper [2]. High density polyethylene with the molecular number and weight average to $1.46\times10^4$ and to $5.50\times10^4$, respectively, was used as binding. The powder-like $CaCO_3$, having particles size 50-60 nm, was used as nanofiller. $CaCO_3$ content makes 5 mass. %.

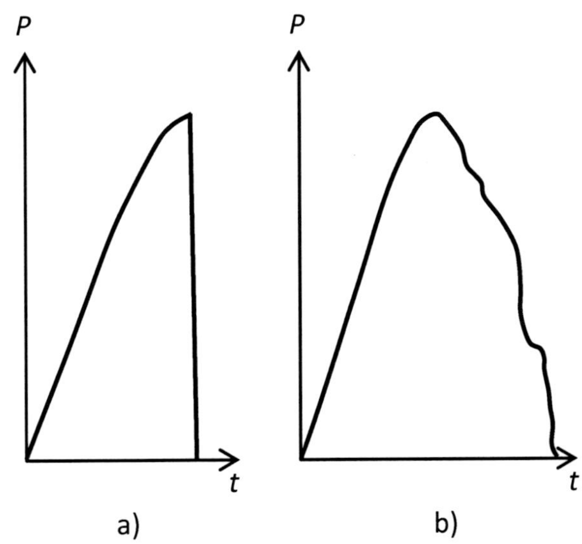


Figure 1. Schematic diagrams load-time (*P-t*) in instrumented impact tests. The fracture by instable crack (a) and stable one (b).

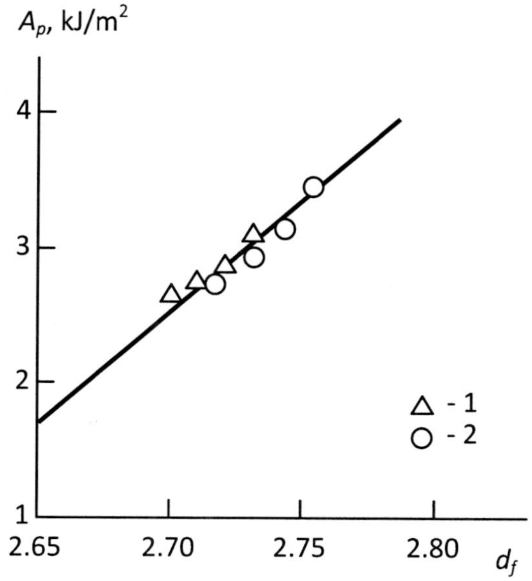


Figure 2. The dependence of impact toughness $A_p$ on structure fractal dimension $d_f$ for HDPE (1) and nanocomposite HDPE/$CaCO_3$ (2).

The impact tests were carried out by Izod method on the samples with sizes 63×12.7×3.0 mm. The samples have a notch with length of 2.5 mm and tip radius of 0.1 mm. The tests were carried out on impact tester Tinius Olsen (Model 899) with impact velocity 1 m/s in the testing temperature range of 233-293 K.

The samples crystallinity degree $K$ was determined with the aid of differential scanning calorimetry (DSC).

## RESULTS AND DISCUSSION

As it is known [5], the fractal dimension $d_f$ is the most general informant of an object structure (in our case – polymeric material) and the true structural characteristic, describing structure elements distribution in space. The value $d_f$ can be determined according to the equation [6]:

$$d_f = 3 - 6\left(\frac{\varphi_{cl}}{C_\infty S}\right)^{1/2}, \quad (1)$$

where $\varphi_{cl}$ is a relative fraction of local order domains (clusters) in polymeric material structure, $C_\infty$ is characteristic ratio, which is equal to 7 for polyethylenes [7], $S$ is macromolecule cross-sectional area, which is equal to 14.3 Å$^2$ for HDPE [8].

In its turn, the value $\varphi_{cl}$ is determined according to the following percolation relationship [9]:

$$\varphi_{cl} = 0.03(1-K)(T_m - T)^{0.55}, \quad (2)$$

where $K$ is cristallinity degree, equal to 0.48 and 0.55 for neat HDPE and nanocomposite HDPE/$CaCO_3$, respectively [2], $T_m$ is melting temperature, equal to ~ 406 and 405 K for the mentioned materials, respectively [2], $T$ is testing temperature.

Let us note, that $d_f$ calculation according to the equation (1) gives values, corresponding to other methods of this parameter estimation. So the value $d_f$ can be calculated alternatively according to the following equation [10]:

$$d_f = (d-1)(1+\nu), \quad (3)$$

where $d$ is dimension of Euclidean space, in which fractal is considered (it is obvious, that in our case $d$=3), $\nu$ is Poisson's ratio, estimated with the aid of the relationship [11]:

$$\frac{\sigma_Y}{E} = \frac{1-2\nu}{6(1+\nu)}, \quad (4)$$

where $\sigma_Y$ is yield stress, $E$ is elasticity modulus.

The estimations according to the equations (1) and (4) have given the following values $d_f$ at testing temperature 293 K: for HDPE 2.73 and 2.68, for nanocomposite HDPE/$CaCO_3$ – 2.75 and 2.73, respectively. As one can see, the good enough correspondence is obtained – the discrepancy by $d_f$ fractional part, which has the main information amount about structure, does not exceed 7 %.

In Figure 2 the dependence $A_p(d_f)$ for the studied polymeric materials is adduced, which has turned out to be linear, common for the near HDPE and nanocomposite HDPE/$CaCO_3$ and is described by the following empirical correlation:

$$A_p = 13.5(d_f - 2.5), \text{ kJ/m}^2. \quad (5)$$

From the equation (5) it follows, that at $d_f$=2.5 the value $A_p$=0. The mentioned fractal dimension corresponds to the ideally brittle fracture condition [10], that defines the condition $A_p$=0. For real solids the greatest fractal dimension of their structure is equal to 2.95 [10], that allows to determine the greatest value $A_p$ according to the equation (5), which is equal to ~ 6.1 kJ/m$^2$.

As Kausch has shown [12], energy dissipation at an impact grows at polymeric materials molecular mobility level increase. Within the frameworks of fractal analysis this level can be characterized with the aid of the fractal dimension $D_{ch}$ of a polymer chain part between its fixation points (chemical cross-links, physical entanglements nodes, clusters, etc.) [6]. Such analysis method was applied successfully for the value $A_p$ description it case of particulate-filled nanocomposites phenylone/β-sialone [13]. The value $D_{ch}$ can be determined with the aid of the following equation [6]:

$$D_{ch} = \frac{\ln N_{cl}}{\ln(4-d_f)-\ln(3-d_f)}, \tag{6}$$

where $N_{cl}$ is a statistical segments number per chain part between clusters, which is determined as follows. Firstly the density of physical entanglements cluster network $\nu_{cl}$ is determined [6]:

$$\nu_{cl} = \frac{\varphi_{cl}}{C_\infty l_0 S}, \tag{7}$$

where $l_0$ is the main chain skeletal bond length, which is equal to 1.54 Å for polyethylenes [7].

Then the estimation of polymer chains total length per polymer volume unit $L$ was carried out as follows [6]:

$$L = S^{-1}. \tag{8}$$

The chain part length between clusters $L_{cl}$ is determined according to the equation [6]:

$$L_{cl} = \frac{2L}{\nu_{cl}}. \tag{9}$$

The statistical segment length $l_{st}$ is determined as follows [14]:

$$l_{st} = l_0 C_\infty. \tag{10}$$

And at last the value $N_{cl}$ can be determined as ratio [6]:

$$N_{cl} = \frac{L_{cl}}{l_{st}}. \tag{11}$$

In Figure 3 the dependence of impact toughness $A_p$ on fractal dimension $D_{ch}$ for the studied materials is adduced. As it should be expected, $A_p$ growth at $D_{ch}$ increase is observed, analytically described by the following relationship:

$$A_p = 6.75(D_{ch}-1),\ \text{kJ/m}^2. \tag{12}$$

The equation (12) allows to determine the greatest value $A_p$ for the studied materials at the condition $D_{ch}$=2.0: this value is equal to 6.75 kJ/m$^2$, that is close to the cited above estimation according to the equation (5) – the average discrepancy makes less than 10 %.

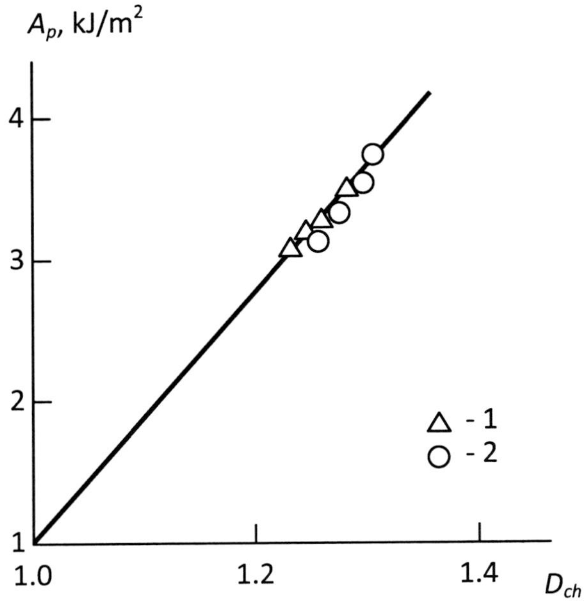


Figure 3. The dependence of impact toughness $A_p$ on fractal dimension $D_{ch}$ of chain part between clusters for HDPE (1) and nanocomposite HDPE/$CaCO_3$ (2).

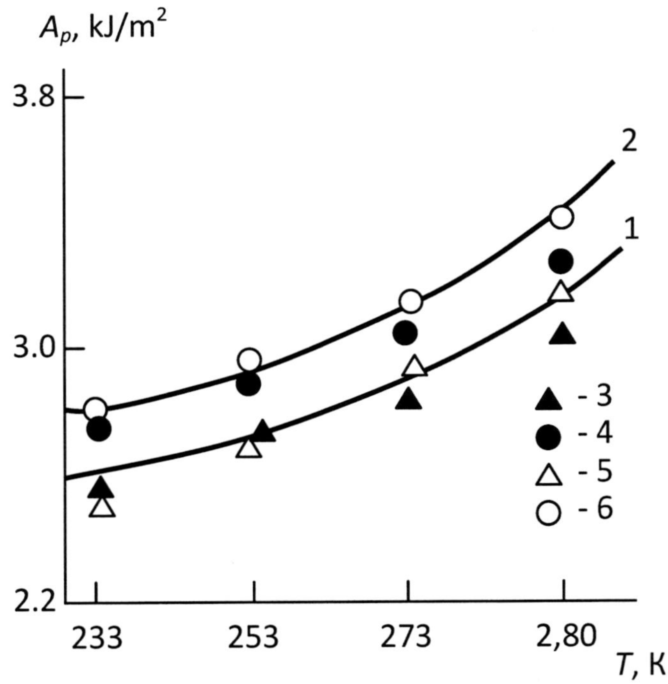


Figure 4. The comparison of experimental (1, 2) and calculated according to the equations (5) (3, 4) and (12) (5, 6) temperature dependences of impact toughness $A_p$ for HDPE (1, 3, 5) and nanocomposite HDPE/$CaCO_3$ (2, 4, 6).

Let us consider the condition of zero impact toughness reached at $d_f$=2.50, but not at $d_f$=2.0 (2.0≤$d_f$≤3.0 [10]). As it is known [15], at $\varphi_{cl}$ growth irrespective of its causes the polymers structure in general and HDPE in particular reaches its quasiequilibrium state, when $\varphi_{cl}$ growth is balanced by entropic tightness of polymer chains and then ceases. As it was shown in paper [15], for HDPE dimension of the structure quasiequilibrium state was equal to 2.50. The calculation according to the adduced above methodics (the equations (2) and (6)-(11)) shows that in this case $D_{ch}\approx 1.0$ and, consequently, polymeric material becomes ideally brittle.

The equations (5) and (12) usage allows to estimate the theoretical values $A_p^T$ and compare them with the experimental magnitudes $A_p$ of this parameter. Such comparison is presented in Figure 4, from which good correspondence of theory and experiment follows (for calculation according to the equation (5) the average discrepancy of $A_p^T$ and $A_p$ makes 3.5 %, according to the equation (12) – 3.9 % that is essentially lower than the usual error of impact toughness experimental determination, which makes ~ 10 %).

Let us consider the physical grounds of impact toughness increasing effect for nanocomposite HDPE/$CaCO_3$ in comparison with matrix polymer HDPE. $A_p$ increase, as follows from the plots of Figures . 2 and 3, is defined by enhancement of dimensions $d_f$ and $D_{ch}$, respectively. In its turn, as it follows from the equations (1) and (6), the mentioned dimension enhancement is defined by $\varphi_{cl}$ reduction, since molecular characteristics $C_\infty$ and $S$ are accepted constant. $\varphi_{cl}$ reduction, as follows from the equation (2), depends only on crystallinity degree $K$ raising, since the values $T_m$ for HDPE and nanocomposite HDPE/$CaCO_3$ are practically the same. This means, that the increasing $A_p$ effect for nanocomposites HDPE/$CaCO_3$ in comparison with matrix polymer is due to the action of nanofiller $CaCO_3$ as nucleator, promoting to $K$ enhancement [1, 2].

## CONCLUSIONS

Therefore, the quantitative correlation of impact toughness of high density polyethylene and particulate-filled nanocomposite on its basis with structural characteristics of these materials is obtained within the frameworks of fractal analysis. The impact toughness enhancement is due to the action of nanofiller as nucleator and corresponding change of amorphous phase structure. The theoretical calculation showed good correspondence to the experimental data.

## REFERENCES

[1] Tannitu M., Misra R.D.K. // Mater. Sci. Engng., 2005. – V. A405, № 1. – P. 178-193.

[2] Deshmane C., Yuan Q., Misra R.D.K. // *Mater. Sci. Engng.,* 2007. – V. A452-453, № 3. – P. 592-601.

[3] Bucknall C.B. Toughened Plastics. – London: *Applied Science,* 1977. – 318 p.

[4] Bartenev G.M., Frenkel S.Ya. *Physics of Polymers.* – Leningrad: Khimiya, 1990. – 432 p. (in Russia)
[5] Kuseev I.R., Samigullin G.Kh., Kulikov D.V., Zakirnichnaya M.M. *Complex Systems in Nature and Engineering.* – Ufa: Publishers ISNTU, 1997. – 225 p. (in Russia)
[6] Kozlov G.V., Novikov V.U. // *Uspekhi Fizicheskich Nauk,* 2001. – V. 171, № 7. – P. 717-764.
[7] Aharoni S.M. // *Macromolecules*, 1983. – V. 16., № 9. – P. 1722-1728.
[8] Aharoni S.M. // *Macromolecules,* 1985. – V. 18., № 12. – P. 2624-2630.
[9] Kozlov G.V., Gazaev M.A., Novikov V.U., Mikitaev A.K. // *Pis'ma v ZhTF,* 1996. – V. 22., № 16. – P. 31-38.
[10] Balankin A.S. *Synergetics of Deformable Body. Moscow: Publishers of Ministry Defence SSSR,* 1991. – 404 p. (in Russia)
[11] Kozlov G.V., Sanditov D.S. Angarmonic Effects and Physical-Mechanical Properties of Polymers. Novosibirsk: *Nauka,* 1994. – 261 p. (in Russia)
[12] Kausch H.H. *Polymer Fracture.* Berlin, Heidelberg, New York: Springer-Verlag, 1978. – 430 p.
[13] Aphashagova Z.Kh, Kozlov G.V., Burya A.I., Malamatov A.Kh. // *Mater. of Khar'kovskoy Nanotekhnologicheskoy Assamblei – 2007.* – Khar'kov: KhSU, 2007. – P. 204-205.
[14] Wu S. // *J. Polymer Sci.: Part B: Polymer Phys.,* 1989. – V. 27, № 4. – P. 723-741.
[15] Kozlov G.V., Zaikov G.E. // In book: Fractals and Local Order in Polymeric Materials. New York: *Nova Science Publishers,* Inc., 2001. – P. 55-63.

In: Chemical and Biochemical Reactions
Editors: G. E. Zaikov, L. Liu et al.

ISBN 978-1-61122-860-1


*Chapter 14*

# THE BEHAVIOUR FEATURES OF POLYMER NANOCOMPOSITES FILLED WITH CALCIUM CARBONATE

***G. V. Kozlov*[1], *Z. Kh. Aphashagova*[1], *A. Kh. Malamatov*[1] *and G. V. Zaikov*[2]**

[1]Kabardino-Balkarian State University,Nal'chik – 360004, Chernyshevskiy st., 173, Russian Federation

[2]N.M. Emanuel Institute of Biochemical Physics of Russian Academy of Sciences, Moscow-119334, Kosygin st., 4, Russian Federation

## ABSTRACT

The most typical behaviour features of polymer nanocomposites filled with disperse calcium carbonate were considered. The quantitative analysis was carried out within the frameworks of structural models: the cluster model of polymers amorphous state structure and fractal analysis. It has been shown that all the considered nanocomposites properties changes were defined by polymeric matrix structure variations, which are due to nanofiller introduction.

**Keywords:** Polymer, nanocomposite, elasticity modulus, interfacial adhesion, yielding, plasticity, fractal analysis

## INTRODUCTION

Lately a scientists interest to particulate-filled polymer nanocomposites, where the nanofiller is calcium carbonate, has been increased. Nanocomposites on the basis of polyethylene [1-8], polypropylene [9, 10], poly (vinyl chloride) [11, 12], butadiene rubber [13] etc. were studied. These studies of nanocomposites polymer/calcium carbonate allow to reveal a number of their specific features, which do not correspond to the known ones at present for polymer composites. Let us note some of them. The elasticity modulus enhancement at calcium carbonate ($CaCO_3$) introduction in the mentioned matrices is usually small even at using of binding agents [9] and for polypropylene it makes up no more than 50 % at $CaCO_3$ contents 40 mass. %. Let us remind that similar processing of mineral microfillers allows to increase an elasticity modulus of epoxy polymer on 300-400 % at the same filler content [14]. Secondly, $CaCO_3$ contents increasing does not practically increase nanocomposites yield stress and sometimes even reduces it [5, 6]. But the most interesting feature of the considered nanocomposites is their plasticity, characterized by either limiting strain at failure $\varepsilon_f$ [12], or impact toughness $A_p$ [5, 6, 9, 13], at $CaCO_3$ contents growth increasing, which often has extreme character. Let us remind that brittleness of polymers at the introduction of various fillers into them (including nanofillers [11]) is one of the most essential deficiencies of polymer composites in general [14]. And at last, melts of nanocomposites filled with $CaCO_3$ often discover anomalous behaviour [15]. A sufficiently large number of works [1-13, 15], which have mainly qualitative character, taking into consideration characteristics of secondary structures, obtained at specimens fracture, but giving no interrelation with nanocomposites neat structure, were devoted to the nanocomposites mentioned these features. Therefore the purpose of the present paper is quantitative analysis of the indicated effects on the example of polymer nanocomposites filled with calcium carbonate.

## EXPERIMENTAL

As polymeric matrix homopolymer of polypropylene (PP) of make F1 supplied by firm Langang Petrochemical Company, China, was used. The disperse $CaCO_3$ was purchased from Solvay firm, particles size makes up 70 nm, nanofiller contents varies within the limits 0-40 mass. %. $CaCO_3$ particles were treated by 1 mass. % of coupling agentliquid silane of make KH550, γ-aminopropyl triethoxy silane, supplied by firm Hengda, China [9].

Nanocomposites PP/$CaCO_3$ were prepared on a screw extruder with a length/diameter ratio of 32, and a screw diameter of 25 mm (T55J-25/32, China) at temperature 503 K and screw rotation rate of 140 rpm. Then the prepared nanocomposites were palletized and the obtained pellets were molded in Nissei PS40E5ASE casting machine into two types of specimens for mechanical tests: the samples for the tests on three-pointed bending have a shape of restangular bars with length of 80 mm, width of 10 mm and thickness of 4 mm; samples for impact testing have the same shape, analogous width and thickness and length of 60 mm. On the last ones V-shaped notch with length of 2 mm and tip radius of 0.25 mm was marked.

Flexural tests were made on apparatus Shimadzu AG-10TA Universal Testing Machine at temperature 293 K and the crosshead speed of 2 mm/min. Impact tests were made according to Izod method at rate 3.5 m/s. The reported data were average of five individual measurements [9].

## RESULTS AND DISCUSSION

As the data of paper [9] have shown, the contents of $CaCO_3$ $W_f$ increasing within the range 0-40 mass. % results to nanocomposites elasticity modulus $E_n$ enhancement from 1387 up to 2086 MPa, i.e. approximately in 1.5 times. The fractal conception of polymer composites elasticity [16] supposes, that this parameter increasing is due to modification ("disturbance") of polymeric matrix structure at filler introduction. The indicated disturbance degree can be characterized by the excess energy localization regions dimension $D_f$, connected with fractal (Hausdorff) dimension $d_f$ of composites structure according to the equation [17]:

$$D_f = 1 + \frac{1}{3 - d_f}. \tag{1}$$

In its turn, the value $d_f$ is determined according to the following relationship [17]:

$$d_f = (d-1)(1+\nu), \tag{2}$$

where $d$ is dimension of Euclidean space, in which a fractal is considered (obviously, in our case $d$=3), ν is Poisson's ratio, estimated according to the mechanical tests results with the aid of the formula [18]:

$$\frac{\sigma_Y}{E} = \frac{1-2\nu}{6(1+\nu)}, \tag{3}$$

where $\sigma_Y$ is yield stress, $E$ is elasticity modulus.

In Figure 1 the dependences $E_n(D_f)$ for four nanocomposites are adduced: PP/$CaCO_3$, poly (vinyl chloride)/calcium carbonate (PVC/$CaCO_3$) with $CaCO_3$ particles size of 44 nm, high density polyethylene/calcium carbonate (HDPE/$CaCO_3$) with particles sizes of 50 and 1200 nm. As it can be seen, all indicated correlations are described by linear dependences with different slope. To obtain the common correlation for all indicated polymer material is possible by normalization of the adduced in Figure 1 straight lines with the aid of ratio $E_{pol}/E_{PP}$, where $E_{pol}$ and $E_{PP}$ are elasticity modulus of arbitrary matrix polymer and PP, respectively. Then the adduced elasticity modulus $E_n^{red}$ of the nanocomposite is described by the following generalized formula:

$$E_n^{red} = 0.32\left(\frac{E_{pol}}{E_{PP}}\right)(D_f - 1). \tag{4}$$

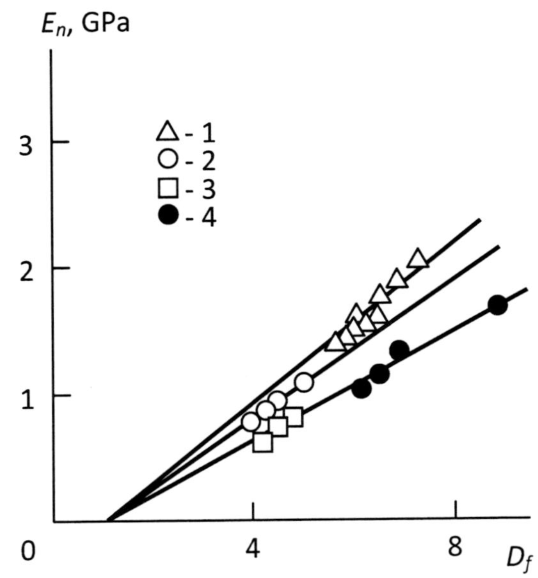


Figure 1. The dependences of elasticity modulus $E_n$ on fractal dimension $D_f$ for nanocomposites on the basis of PP (diameter of $CaCO_3$ nanoparticles $D_p$=70 nm) (1), HDPE ($D_p$=50 nm) (2), HDPE ($D_p$=1200 nm) (3) and PVC ($D_p$=44 nm) (4).

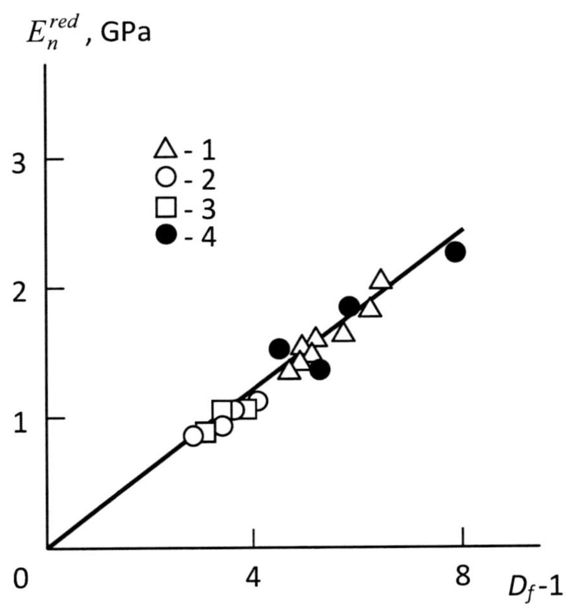


Figure 2. The dependence of reduced elasticity modulus $E_n^{red}$ of nanocomposites, calculated according to the equation (4), on fractal dimension $D_f$. Conventional signs are the same that in Figure 1.

The adduced in Figure 2 correlation $E_n^{red}$ ($D_f$-1), corresponding to the equation (4), confirms its correctness. Let us note, that the indicated equation allows to determine the upper limiting value of nanocomposite elasticity modulus at the condition that values $E_{pol}$ and $E_{PP}$ are known (it is naturally that as a basic one not only PP can be used, but any other polymer). As it is known [17], for the real physical objects the maximum value $d_f$=2.95. From the equation (1) it is followed that corresponding to its maximum value $D_f$=21. For example, if for PVC $E_{pol}$=1150 MPa [12], then this means according to the equation (4), that maximum attainable for nanocomposites PVC/$CaCO_3$ value $E_n$ is equal to ~ 5300 MPa. Therefore, particulate-filled polymer nanocomposites obey to general laws, obtained earlier for particulate-filled microcomposites.

The authors [19] obtained the following equation, showing the dependence of reinforcement degree $E_n/E_m$ for particulate-filled nanocomposites on their main characteristics:

$$\frac{E_n}{E_m} = \frac{0.19W_f l_{st}}{D_p^{1/2}}, \tag{5}$$

where $E_m$ is elasticity modulus of matrix polymer, $W_f$ is nanofiller mass contents, $l_{st}$ is length of statistical segment of polymeric matrix, $D_p$ is nanofiller particles diameter.

The value $l_{st}$ can be determined according to the formula [20]:

$$l_{st} = C_\infty l_0, \tag{6}$$

where $C_\infty$ is characteristic ratio, $l_0$ is length of main chain skeletal bond. For PP one should accept: $C_\infty$=6, $l_0$=1.54 Å [21].

In Table 1 the comparison of the obtained experimentally ($E_n/E_m$) and calculated according to the equation (5) $(E_n/E_m)^T$ values of reinforcement degree for nanocomposites PP/$CaCO_3$ is adduced. As one can see, the good enough correspondence of theory and experiment is obtained (the average discrepancy of ($E_n/E_m$) and $(E_n/E_m)^T$ makes up 8.4 %), but for all nanocomposites the condition $(E_n/E_m)<(E_n/E_m)^T$ is fulfilled. According to the equation (5) such effect can be given by nanoparticles aggregation, i.e. $D_p$ increasing. Using experimental values of reinforcement degree, real sizes of nanofiller particles aggregates $D_p^{agr}$ can be calculated, the dependence of which on filling mass degree $W_f$ in Figure 3 is adduced. The data of this Figure demonstrate strong enough aggregation of nanofiller particles at small $W_f$ (≤10 mass. %) and its small degree at $W_f$≥20 mass. %. Such unexpected enough correlation (it is usually considered that the stronger aggregation the higher filler contents [16]) can be explained by nanocomposites melt viscosity η change, since in this state only nanofiller particles aggregation is possible. In Figure 4 the dependence of melt relative viscosity for nanocomposites PP/$CaCO_3$ $\eta/\eta_0$ ($\eta_0$ is neat matrix polymer melt viscosity), calculated according to Einstein's formula [22]:

$$\frac{\eta}{\eta_0} = 1 + \frac{2,5W_f}{1-W_f}, \tag{7}$$

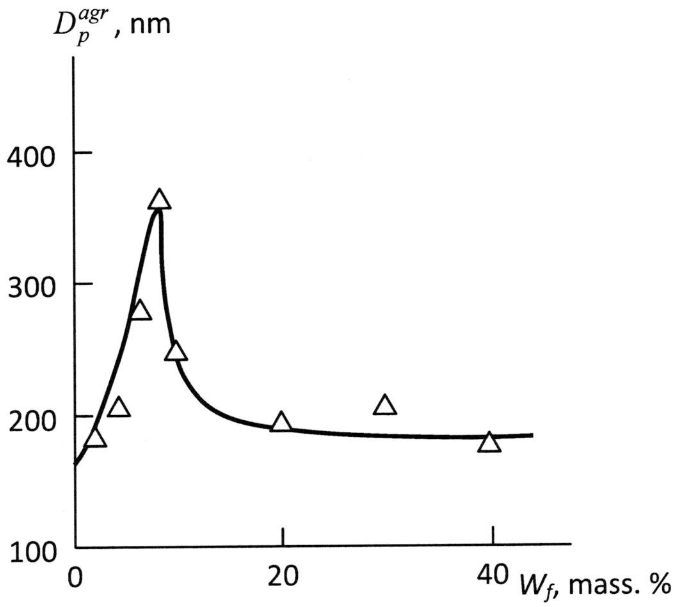


Figure 3. The dependence of diameter of nanofiller particles aggregates $D_p^{agr}$ on mass contents of $CaCO_3$ $W_f$ for nanocomposites PP/$CaCO_3$.

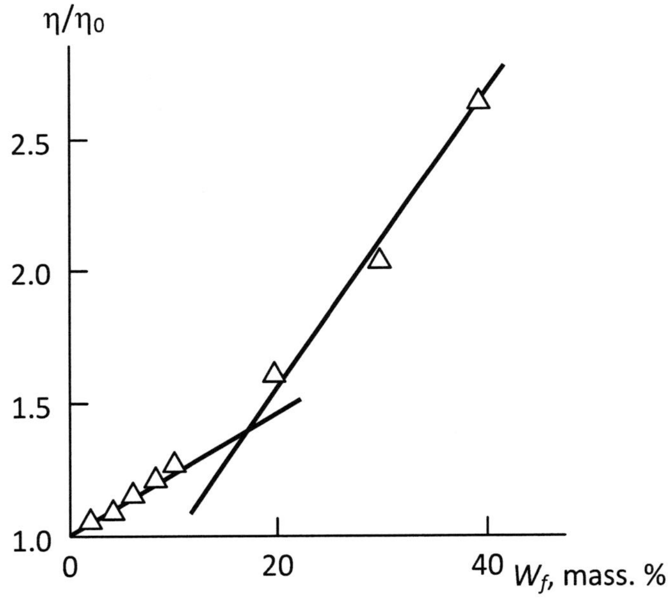


Figure 4. The dependence of melt relative viscosity $\eta/\eta_0$, calculated according to the equation (7), on $CaCO_3$ mass contents $W_f$ for nanocomposites PP/$CaCO_3$.

on $CaCO_3$ contents. As it follows from Figure 4 plot, at $W_f \approx 10$ mass. % relative viscosity gradient increases essentially (about twice) for large $CaCO_3$ contents. This means that relatively low nanocomposites PP/$CaCO_3$ melt viscosity at small $W_f$ allows high enough nanofiller particles mobility degree in melt that facilitates their aggregation, since nanoparticles in general have high specific surface, promoting to their strong agglomeration [10, 23]. A melt viscosity enhancement at $W_f$>10 mass. % reduces nanofiller particles mobility in melt, making difficult their aggregation (compare Figures 3 and 4). The similar anomalous effect of melt viscosity change as a function of $W_f$ for nanocomposites PP/$CaCO_3$ was observed in paper [15]. It is significant that, as it has been expected, nanofiller particles aggregation degree influences on nanocomposites elasticity modulus $E_n$. In Figure 5 the dependence $E_n(W_f)$ is adduced, which corresponds completely to Figures 3 and 4 data. As one can see, in the range of $W_f$=0-10 mass. %, where nanofiller particles aggregation is relatively high (Figure 3) and melt relative viscosity is small (Figure 4), $E_n$ growth at $W_f$ increase occurs much slower than in the range of $W_f$>10 mass. %.

Let us consider further the reasons of yield stress $\sigma_Y$ of relatively small increasing of particulate-filled nanocomposites in comparison with their elasticity modulus $E_n$. So, for nanocomposites PP/$CaCO_3$ at $E_n$ increasing on 50 % $\sigma_Y$ enhancement only on 10 % is observed [9] and for nanocomposites HDPE/$CaCO_3$ $W_f$ increasing $W_f$ up to 20 mass. % results to $\sigma_Y$ reduction in comparison with such nanocomposites with $CaCO_3$ smaller contents [5, 6]. The similar picture is observed for nanocomposites PVC/$CaCO_3$ [12]. Within the frameworks of the cluster model of polymers amorphous state structure yield stress $\sigma_Y$ is determined by the following equation [16, 24]:

$$\sigma_Y = \frac{Gb\rho_d^{1/2}}{2\pi}, \tag{8}$$

where $G$ is shearing modulus, $b$ is Burgers vector, $\rho_d$ is structure linear defects density (analog of dislocations for crystalline lattices).

Let us consider parameters estimation methods, which are included in the equation (8). The value $G$ is determined as follows [17]:

$$G = \frac{E_n}{d_f}, \tag{9}$$

and Burgers vector $b$ – according to the empirical equation [16, 24]:

$$b = \left(\frac{62.5}{C_\infty}\right)^{1/2}, \text{Å}. \tag{10}$$

For polymers as structure linear defects segments of macromolecules, included in densely-packed regions, which for nanocomposites with semicrystalline matrix are crystallites, local order regions (clusters) and interfacial regions, are accepted [23]. Relative fraction of the last $\varphi_{if}$ can be estimated with the aid of the equation [23]:

**Table 1. The structural and mechanical characteristics of nanocomposites PP/$CaCO_3$**

| $W_f$, mass. % | $E_n/E_m$ | $(E_n/E_m)^T$ | $d_f$ | $\varphi_{cl}$ | $G$, MPa | $\sigma_Y$, MPa | $\sigma_Y^T$, MPa |
|---|---|---|---|---|---|---|---|
| 0 | 1 | 1 | 2.787 | 0.194 | 498 | 35.4 | 36.8 |
| 2 | 1.025 | 1.040 | 2.792 | 0.189 | 509 | 35.4 | 37.1 |
| 4 | 1.047 | 1.060 | 2.795 | 0.183 | 519 | 35.6 | 37.3 |
| 6 | 1.060 | 1.120 | 2.796 | 0.181 | 525 | 35.8 | 37.5 |
| 8 | 1.071 | 1.160 | 2.796 | 0.181 | 531 | 36.1 | 37.3 |
| 10 | 1.108 | 1.200 | 2.800 | 0.174 | 549 | 36.4 | 38.0 |
| 20 | 1.243 | 1.400 | 2.816 | 0.148 | 612 | 37.5 | 39.2 |
| 30 | 1.347 | 1.600 | 2.826 | 0.132 | 661 | 38.2 | 39.2 |
| 40 | 1.504 | 1.800 | 2.840 | 0.112 | 735 | 39.2 | 39.4 |

$$\frac{E_n}{E_m} = 1 + 11\left(W_f + \varphi_{if}\right)^{1.7}. \tag{11}$$

The estimations according to the equation (11) have shown practical absence of interfacial regions in nanocomposites PP/$CaCO_3$. These estimations are confirmed experimentally by the data of atomic force microscopy on the example of nanocomposites HDPE/$CaCO_3$ [5]. In itself the fact of practical absence of interfacial regions in nanocomposites PP/$CaCO_3$ speaks for low enough level of interfacial adhesion nanofiller-polymeric matrix despite coupling agent usage [23]. The authors [25] showned that in case of ideal adhesion between filler and polymeric matrix the dependence $E_n/E_m(W_f)$ was described by Kerner's equation, which can be approximated by the following relationship:

$$\frac{E_n}{E_m} = 1 + 11.6W_f - 44.4W_f^2 + 96.3W_f^3. \tag{12}$$

The other similar relationship (the Guth's equation), using for the estimation of composites reinforcement degree in case of ideal adhesion, has the form [12]:

$$\frac{E_n}{E_m} = 1 + 2.5W_f + 14.1W_f. \tag{13}$$

The equations (12) and (13) give close values $E_n/E_m$ at $W_f$=40 mass. % (or $W_f$=0.4) for nanocomposites PP/$CaCO_3$: 4.70 and 4.46, that corresponds to elasticity modulus $E_n$=6.52 and 6.18 GPa, respectively. Since the experimental $E_n$ value at the indicated $CaCO_3$ content is equal to 2.09 GPa [9], then this comparison confirms the conclusion about low level of interfacial adhesion for these nanocomposites, which was made above.

Therefore, for the studied nanocomposites as the structure linear defects one should accept segments, included in crystallites and clusters with relative fractions $K$ (crystallinity degree) and $\varphi_{cl}$, respectively. $K$ value for PP is equal to 0.50 [9] and the value $\varphi_{cl}$ can be estimated with the aid of the equation [24]:

$$d_f = 3 - 6\left(\frac{\varphi_{cl}}{C_\infty S}\right)^{1/2}, \tag{14}$$

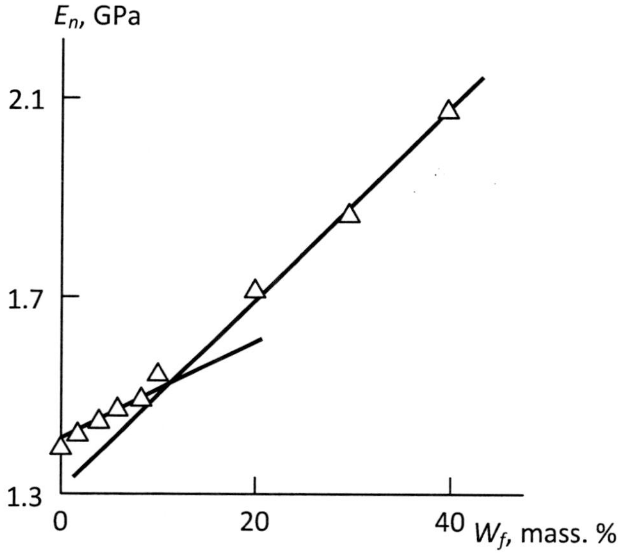


Figure 5. The dependence of elasticity modulus $E_n$ on $CaCO_3$ mass contents $W_f$ for nanocomposites PP/$CaCO_3$.

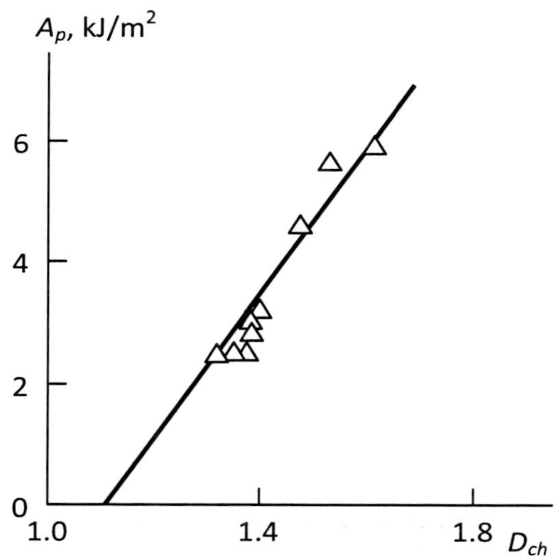


Figure 6. The dependence of impact toughness $A_p$ on fractal dimension $D_{ch}$ of chain part between clusters for nanocomposites PP/$CaCO_3$.

where $S$ is cross-sectional area of macromolecule, which is equal to 34.3 Å$^2$ for PP [26].

Further the value $\rho_d$ is calculated according to the following formula [16]:

$$\rho_d = \frac{K + \varphi_{cl}}{S}. \tag{15}$$

In table 1 the comparison of experimental $\sigma_Y$ and calculated according to the equation (8) $\sigma_Y^T$ yield stress values for nanocomposites PP/$CaCO_3$ is adduced and also necessary for such calculation values $d_f$, $\varphi_{cl}$ and $G$ are given. As one can see, a good correspondence of theory and experiment is obtained (the average discrepancy of $\sigma_Y$ and $\sigma_Y^T$ makes up 3.6 %). Let us note the polymeric matrix structure, characterized by dimension $d_f$, small change – at $W_f$=40 mass. % it makes up on the fractional part of $d_f$ (which gives the main information about the structure) only 6.7 %. For comparison, in case of composites polyhydroxiether/graphite at the same filler content $d_f$ change is equal to 32.8 %, i.e. approximately in 5 times more [16]. Antibate change of parameters $G$ and $\varphi_{cl}$, defining the value $\sigma_Y$ in the equation (8) is also significant. Hence, $\sigma_Y$ weak increase at $W_f$ growth for nanocomposites PP/$CaCO_3$ is due to a compensating effect of antibate change of parameters $G$ and $\varphi_{cl}$ (or $\rho_d$).

And let us consider in conclusion nanocomposites PP/$CaCO_3$ plasticity increasing effect at $W_f$ growth on the example of impact toughness $A_p$, which at $W_f$=40 mass. % raises in 2.4 times in comparison with neat PP. As Kausch shows [27], polymer molecular mobility increasing is always accompanied by its plasticity (impact toughness) growth. Within the frameworks of fractal analysis molecular mobility level can be estimated with the aid of fractal dimension $D_{ch}$ of chain part between its fixation points (cross-linking nodes, physical entanglements, clusters, etc.) [28]. The value $D_{ch}$ can be calculated with the aid of the equation [28]:

$$\frac{2}{\varphi_{cl}} = C_\infty^{D_{ch}}. \tag{16}$$

In Figure 6 the dependence $A_p(D_{ch})$ is adduced, which has an expected form: enhancement of molecular mobility level, characterized by dimension $D_{ch}$, results to the growth of nanocomposites plasticity, characterized by their impact toughness $A_p$. Let us note, that $D_{ch}$ and, consequently, $A_p$ increase is due to local order level reduction in amorphous phase of polymeric matrix, i.e. $\varphi_{cl}$ decrease at $W_f$ growth (table 1), in other words, by polymeric matrix structure "disturbance" at nanofiller introduction.

The fractal analysis application allows to estimate maximum attainable $A_p$ value. So, from Figure 6 plot it follows, that for nanocomposites PP/$CaCO_3$ at maximum $D_{ch}$=2 limiting magnitude of impact toughness is equal to 10.4 kJ/m$^2$.

## Conclusions

Therefore, performed in the present paper quantitative analysis of particulate-filled polymer nanocomposites main features within the frameworks of fractal analysis and cluster model of polymers amorphous state structure allows to elucidate the following structural causes of the indicated features:

The relatively small reinforcement degree of polymer nanocomposites, filled with calcium carbonate, is due to low level of interfacial adhesion polymer-$CaCO_3$, despite coupling agent using. Nanofiller particles aggregation plays in this aspect a definite role.

The weak increasing of nanocomposites yield stress at $CaCO_3$ contents growth is due to compensating effect of shear modulus change and relative fraction of polymeric matrix local order regions.

The melts viscosity anomalous behaviour of nanocomposites, filled with $CaCO_3$, is due to strong tendency to aggregation of nanofiller particles in virtue of their high specific surface at small nanofiller contents and, as consequence, nanoparticles high mobility. $CaCO_3$ contents increase raises essentially melt viscosity, reduces nanofiller particles mobility and decreases their aggregation degree.

The plasticity, characterized by impact toughness, increasing for nanocomposites PP/$CaCO_3$ is due to polymeric matrix structure "disturbance" at nanofiller introduction and, as consequence, to growth of molecular mobility level.

## References

[1] Suwanprateeb J. Composites. Part A. *Appl. Sci. and Manufact.*, 2000, v. 31A, № 3, p. 353-359.
[2] Osman A.M., Atallah A., Suter U.W. *Polymer,* 2004, v. 43, № 3, p. 1177-1183.
[3] Misra R.D.K., Nerikar P., Bertrand K., Murphy D. *Mater. Sci. Engng.,* 2004, v. 384A, № 2, p. 284-298.
[4] Parsons E.M., Boyce M.C., Parks D.M., Weinberg M. *Polymer,* 2005, v. 46, № 6, p. 2257-2265.
[5] Tanniru M., Misra R.D.K. *Mater. Sci. Engng.,* 2005, v. 405A, № 2, p. 178-193.
[6] Tanniru M., Misra R.D.K. *Mater. Sci. Engng.,* 2007, v. 424A, № 1, p. 53-70.
[7] Deshmane C., Yuan Q., Misra R.D.K. *Mater. Sci. Engng.,* 2007, v. 452-453A, № 3, p. 592-601.
[8] Zebarjad S.M., Sajjadi S.A. *Mater. Sci. Engng.,* 2008, v. 475A, № 2, p. 365-367.
[9] Yang K., Yang Q., Li G., Sun Y.S., *Feng D. Mater. Lett.,* 2006, v. 60, № 7, p. 805-809.
[10] Ma C.G., Mai Y.L., Rong M.Z., Ruan W.H., Zhang M.Q. *Composites Sci. Techn.,* 2007, v. 67, № 9, p. 2997-3005.
[11] Chen N., Wan C., Zhang Yo, Zhang Yi. *Polymer Test.,* 2004, v. 23, № 2, p. 169-174.
[12] Xie X.-L., Liu Q.-X., Li R.K.-Y., Zhou X.-P., Zhang Q.-X., Yu Z.-Z., Mai Y.-W. *Polymer,* 2004, v. 45, № 19, p. 6665-6673.
[13] Jin F.-L., Park S.-J. *Mater. Sci. Engng.,* 2008, v. 478A, № 3, p. 406-408.
[14] Phillips D.C., Harris B. In book: Polymer Engineering Composites. Ed. Richardson M.O.W. London, *Applied Science Publishers LTD,* 1978, p. 50-146.

[15] Sultonov N.Zh., Dzhangurazov B.Zh., Mikitaev A.K. Mater. *International Forum by Nanotechnologies "Rusnanotech-08"*. Moscow, 3-5 December 2008, p. 527-528.

[16] Mikitaev A.K., Kozlov G.V. *The Fractal Mechanics of Polymer Materials*. Nal'chik, Publishers of KBSU, 2008, 312 p.

[17] Balankin A.S. *Synergetics of Deformable Body*. Moscow, Publishers of Ministry Defence SSSR, 1991, 404 p.

[18] Kozlov G.V., Sanditov D.S. Anharmonic Effects and Physical-Mechanical Properties of Polymers. Novosibirsk, *Nauka*, 1994, 261 p.

[19] Aphashagova Z.Kh., Kozlov G.V., Burya A.I., Mikitaev A.K. *Materialovedenie, 2007*, № 9, p. 10-13.

[20] Wu S. J. *Polymer Sci.: Part B: Polymer Phys.*, 1989, v. 27, № 4, p. 723-741.

[21] Aharoni S.M. *Macromolecules,* 1983, v. 16, № 9, p. 1722-1728.

[22] Mills N.J. *J. Appl. Polymer Sci.,* 1971, v. 15, № 11, p. 2791-2805.

[23] Malamatov A.Kh., Kozlov G.V., Mikitaev M.A. Reinforcement Mechanisms of Polymer Nanocomposites. Moscow, *Publishers of Mendeleev RKhTU*, 2006, 240 p.

[24] Kozlov G.V., Novikov V.U. *Uspekhi Fizicheskikh Nauk*, 2001, v. 171, № 7, p. 717-764.

[25] Tugov I.I., Shaulov A.Yu. Vysokomolek. *Soed.* B, 1990, v. 32, № 7, p. 527-529.

[26] Aharoni S.M. *Macromolecules,* 1985, v. 18, № 12, p. 2624-2630.

[27] Kausch H.H. *Polymer Fracture*. Berlin, Heidelberg, New York, Springer-Verlag, 1978, 428 p.

[28] Kozlov G.V., Zaikov G.E. Structure of the Polymer Amorphous State. Utrecht-Boston, *Brill Academic Publishers,* 2004, 465 p.

In: Chemical and Biochemical Reactions
Editors: G. E. Zaikov, L. Liu et al.
ISBN 978-1-61122-860-1


*Chapter 15*

# THE NANOADHESION EFFECT ON STRUCTURE AND PROPERTIES FORMATION IN PARTICULATE-FILLED NANOCOMPOSITES

***G. V. Kozlov[1], Z. Kh. Aphashagova[1], Yu. G. Yanovskii[2] and G. E. Zaikov[3]***
[1]Kabardino-Balkarian State University,
[1]73Chernishevsky st., 360004, Nalchik, Russia
[2]Institute of Applied Mechanics of Russian Academy of Sciences,
32 a Leninskii pr., 119991 Moscow, Russia
[3]Institute of Biochemical Physics, Russian Academy of Sciences,
4 Kosygin st., 119991, Moscow, Russia

## ABSTRACT

The results of the present paper have shown the existence in polymer particulate-filled nanocomposites nanoadhesion effect, which consists in enhanced level of polymeric matrix-nanofiller interaction. This effect has purely dimensional origin and was realized for nanofiller particles with diameter less than 80 nm. The nanoadhesion effect influences strongly on macroscopic properties of those nanocomposites, in which it was realized. So, essential thermal stability enhancement and considerable thermal expansion coefficient reduction of nanocomposites on the base of phenylone at introduction 1 mass. % aerosil only are due to the nanoadhesion effect. Appreciation of nanoadhesion effect allows qualitatively and quantitatively to describe exactly nanocomposites reinforcement degree.

**Keywords:** Polymer nanocomposite, interfacial adhesion, thermal expansion, nanoadhesion, structure, thermal stability

## INTRODUCTION

At present polymer nanocomposites are some of the most perspective engineering materials [1]. Moreover the main attention is paid to nanocomposites filled by layered silicates (organoclays) [2] or nanotubes [3], whereas particulate-filled nanocomposites were studied to a far smaller extent. Nevertheless, possibilities of the last class of polymer nanocomposites are far from being exhausted [4].

As it is well known [5], for multiphase materials the interaction level between phases plays a defining role in properties formation of these materials. The said above relates completely to polymer composites too (nanocomposites). So, the authors [6] have demonstrated that the filler contents increase in polymer composites results in good interfacial adhesion case of polymeric matrix-filler to modulus elasticity enhancement, but interfacial adhesion level determination by experimental methods in polymer composites case comes up on definite difficulties. So, the adhesional strength value determined by standard methods can differ sufficiently from the same for composites in virtue of the existence of various kinds of nonaccounted effects in real composites: filler particles aggregation [7], polymeric matrix structure change due to filler introduction [8] and a number of other factors. The indicated above importance of the interfacial phenomena consists in their influence practically on all polymer composites properties. Therefore the most suitable from practical point of view method of the interfacial adhesion level determination in bulk composites is the estimation of this factor according to composite macroscopic properties, for which a number of techniques (for example, [9]) was elaborated. The purpose of the present paper is to determine the interfacial adhesion level and the factors, influencing on it, and the influence of the indicated level on thermal and mechanical properties of particulate-filled nanocomposites on the base of phenylone as well.

## EXPERIMENTAL

As polymer binding of the studied nanocomposites the linear heterochain copolymer phenylone S-2 of the following chemical constitution was used:

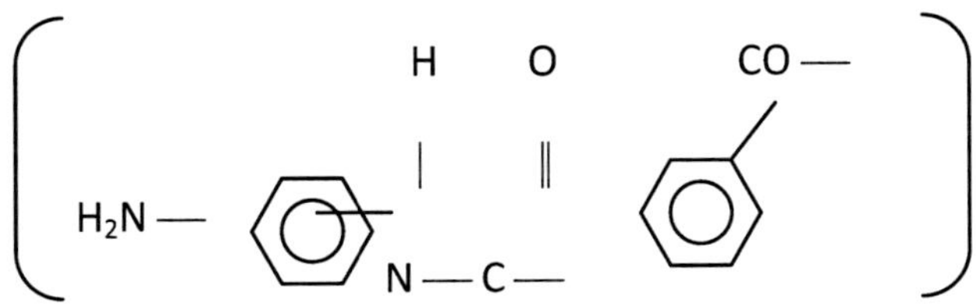


As particulate nonafiller the aerosil (A) with particles average diameter 15 nm and specific surface 170 $m^2/g$ and oxinitride silicium-yttrium (OSY) $Si_3N_4$-$Y_2O_3$ with average diameter of particles 64 nm an specific surface 43 $m^2/g$ were used. The contents of the first from the indicated nanofillers makes up to 0.3, 0.5 and 1.0 mass. %, the second one is varied within the range 0.2-10 mass. %.

The nanofiller introduction in polymeric matrix was realized in rotating electromagnetic field with the aid of nonequiaxed ferromagnetic particles with the ratio of length to diameter 4-5; the ferromagnetic particles volume, loading in an apparatus reactor, was within the range 0.04-0.05 from electromagnetic field action volume. The electromagnetic induction value of this field was within the range 0.08-0.12 Tl. At the indicated parameters it was found out experimentally that the optimal processing duration of nanocomposites in electromagnetic field was 270-300 s. Nanocomposites samples preparation was realized by compressional pressing method at temperature 537-616 K, pressure 40-100 MPa and exposure at the indicated parameters at a rate of 1 min/mm per product thickness.

The processing of powder-like compositions phenylone/A and phenylone/OSY in rotating electromagnetic field allows to inhibite (or decrease) nanofiller particles aggregation. For checking of the achievement of this effect the phenylone/A nanocomposite samples were processed by simple mechanical blending. Further the samples of nanocomposites, filled by aerosil and obtained with rotating electromagnetic field application, will be designated as phenylone/A-MF and obtained by mechanical blending – as phenylone/A. Nanocomposites filled by oxinitride silicium-yttrium were designated as phenylone/OSY.

The linear coefficient of thermal expansion of nanocomposites was determined on dilatometer DKV-SAM, intended for automatic registration of dilatometric curves of different solids, within the temperature range 298-498 K. For the tests the prismatic (6×4 mm) samples with the length 50±3 mm were used. The nonparallelism of sample grinded butt-ends consists of not more than ±0.02 mm. The number of samples for tests for each nanocomposite was no less than three.

For the surface structure study the three-dimensionally reflecting surface structure analyzer Zygo NewView 5022 was used.

The nanocomposites study by method of thermogravimetric analysis (TGA) was made on derivatograph Q-1500 of Paulik-Erdei system of the firm MOM (Hungary). The studies were carried out within the temperature range 298-873 K with the temperature rise rate 10 K/min, mass of a sample was equal to 100 mg. The curves recording was made with maximal sensitivity.

The mechanical tests on compression were made on testing machine FP-100 at temperature 293 K and strain rate $10^{-3}$ $s^{-1}$. The testing samples have diameter 10±0.5 mm and height 15±0.5 mm. For each nanocomposite five samples were tested.

## RESULTS AND DISCUSSION

In Figure 1 the curves 1-3 represent three main types of the dependences of linear coefficient of thermal expansion of the studied nanocomposites $\alpha_n$ on nanofiller volume contents $\varphi_n$. The straight line 1 illustrates the case, when between two phases of nanocomposites adhesion is absent and at $\alpha_m>\alpha_n$ ($\alpha_m$, $\alpha_n$ are thermal expansion coefficients of polymeric matrix and nanofiller, respectively) and the absence in interfacial layers residual compression strain polymeric matrix at heating will be expanded irrespective of filler particles and in this case $\alpha_n=\alpha_m$ [10]. The straight line 2 corresponds to a simple rule of mixtures [10]:

$$\alpha_n^m = \alpha_m(1-\varphi_n) + \alpha_n\varphi_n, \tag{1}$$

where $\alpha_n^m$ is the thermal expansion coefficient of mixture, $\varphi_n$ is the volume fraction of filler.

The rule (1) is correct for the ideal case only, when each phase expands independently one from another. And lastly, the straight line 3 corresponds to Terner equation [10]:

$$\alpha_n^T = \frac{\alpha_m(1-\varphi_n)K_m + \alpha_n\varphi_n K_n}{(1-\varphi_n)K_m + \varphi_n K_n}, \tag{2}$$

where $K_m$ and $K_n$ are a bulk elasticity modulus of polymeric matrix and nonofiller, respectively.

At the equation (2) derivation the strains equality method was used for calculation of mixtures thermal expansion coefficient depending on density, elasticity modulus, thermal expansion coefficient and mass relationship of composing components. If the made assumptions are correct, then the formula (2) will also be applicable for polymer nanocomposites.

Besides, in Figure 1 values $\alpha_n$ determined experimentally are shown by points for two series of the studied nanocomposites including phenylone, for which $\alpha_n$=$\alpha_m$. Let's note, that in Terner equation Young modulus $E$ magnitudes were used instead of the bulk elasticity modulus $K$. As it was known [10], such replacement allows to obtain a lower boundary of $\alpha_n$ value for polymer composites. From the data of Figure 1 it follows that $\alpha_n$ values for both nanocomposites series, filled by aerosil, are disposed substantially lower than the curve calculated according to Terner equation, i.e. they are substantially smaller than the lower limiting values $\alpha_n$ for polymer composites. Besides $\alpha_n$ values for nanocomposites phenylone/A-MF are significantly smaller than the corresponding values for nanocomposites phenylone/A. In other words, inhibition (or weakening) of aerosil particles aggregation process substantially decreases $\alpha_n$ value. In case of nanocomposites phenylone/OSY the values $\alpha_n$ at small $\varphi_n$ are somewhat smaller than the ones calculated according to Terner equation, although they are close to them, and at $\varphi_n$>0.05 the values $\alpha_n$ for them are disposed between the curves, calculated according to the mixtures rule and Terner equation, i.e. they have $\alpha_n$ values intermediate on absolute value and typical for polymer composites [10].

Let's consider physical significance of the described above $\alpha_n$ changes for nanocomposites, studied in this paper. It is possible to estimate polymeric matrix-filler interaction level (interfacial adhesion level) with the aid of the parameter $b$, determined according to the equation [10]:

$$\alpha_n = \alpha_n^m - b(\alpha_n^m - \alpha_n^T). \tag{3}$$

The larger the value $b$ the higher interfacial adhesion level. For a large enough number of polymer composites with different matrixes and fillers $b$ value varies from -0.19 to 1.39 [10]. Calculation according to the equation (3) for the studied nanocomposites shows that for phenylone/A-MF $b\approx12$, phenylone/A $b\approx6$ and for phenylone/OSY value $b$ varies within the range ~ 5.40-0.35, decreasing at $\varphi_n$ increase (Figure 2). Therefore, polymeric matrix-filler interaction level in the studied nanocomposites can exceed substantially (on one order of magnitude) this value for microcomposites. Following to the authors [11] we called this effect nanoadhesion. Let's remind that in paper [11] the adhesion level for two-layered films of polystyrene on micro- and nanolevel was measured. It turns out, that in the last case adhesional strength is substantially higher in comparison with the first one. The authors explained [11] the indicated effect of polymers nanoadhesion by different degree of interfacial layer formation completion between two films of polystyrene.

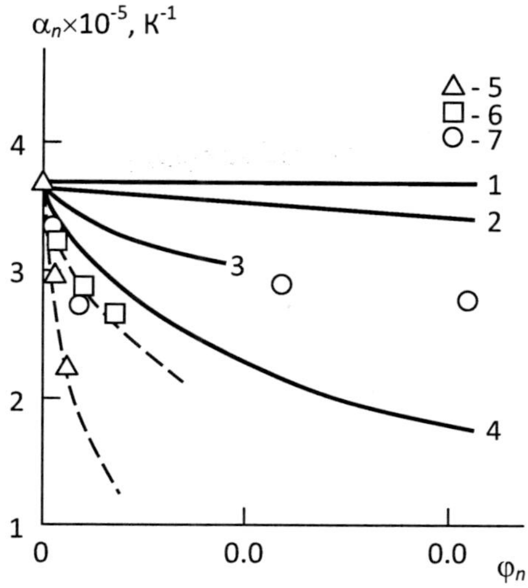


Figure 1. The dependences of linear coefficient of thermal expansion $\alpha_n$ on nanofiller volume contents $\varphi_n$. 1 – adhesion absence on interfacial boundary; 2 – mixtures rule; 3 – the Terner equation; 5-7 – experimental data for nanocomposites phenylone/A-MF (1, 2, 3, 5), phenylone/A (1, 2, 3, 6) and phenylone/OSY (1, 2, 4, 7).

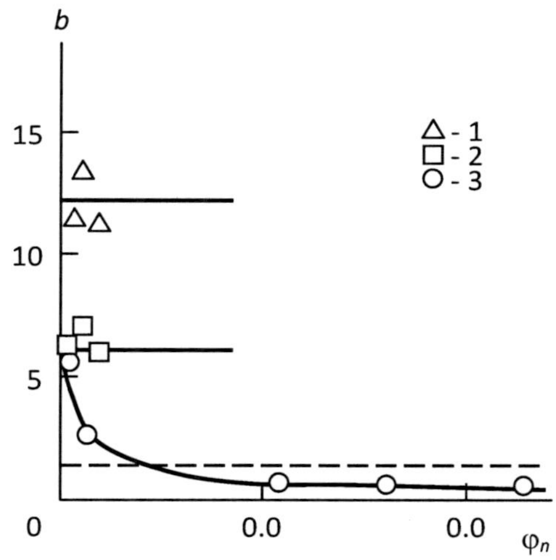


Figure 2. The dependences of parameter $b$ on nanofiller volume contents $\varphi_n$ for nanocomposites phenylone/A-MF (1), phenylone/A (2) and phenylone/OSY (3). The horizontal shaded line indicates upper limiting value $b$ for microcomposites.

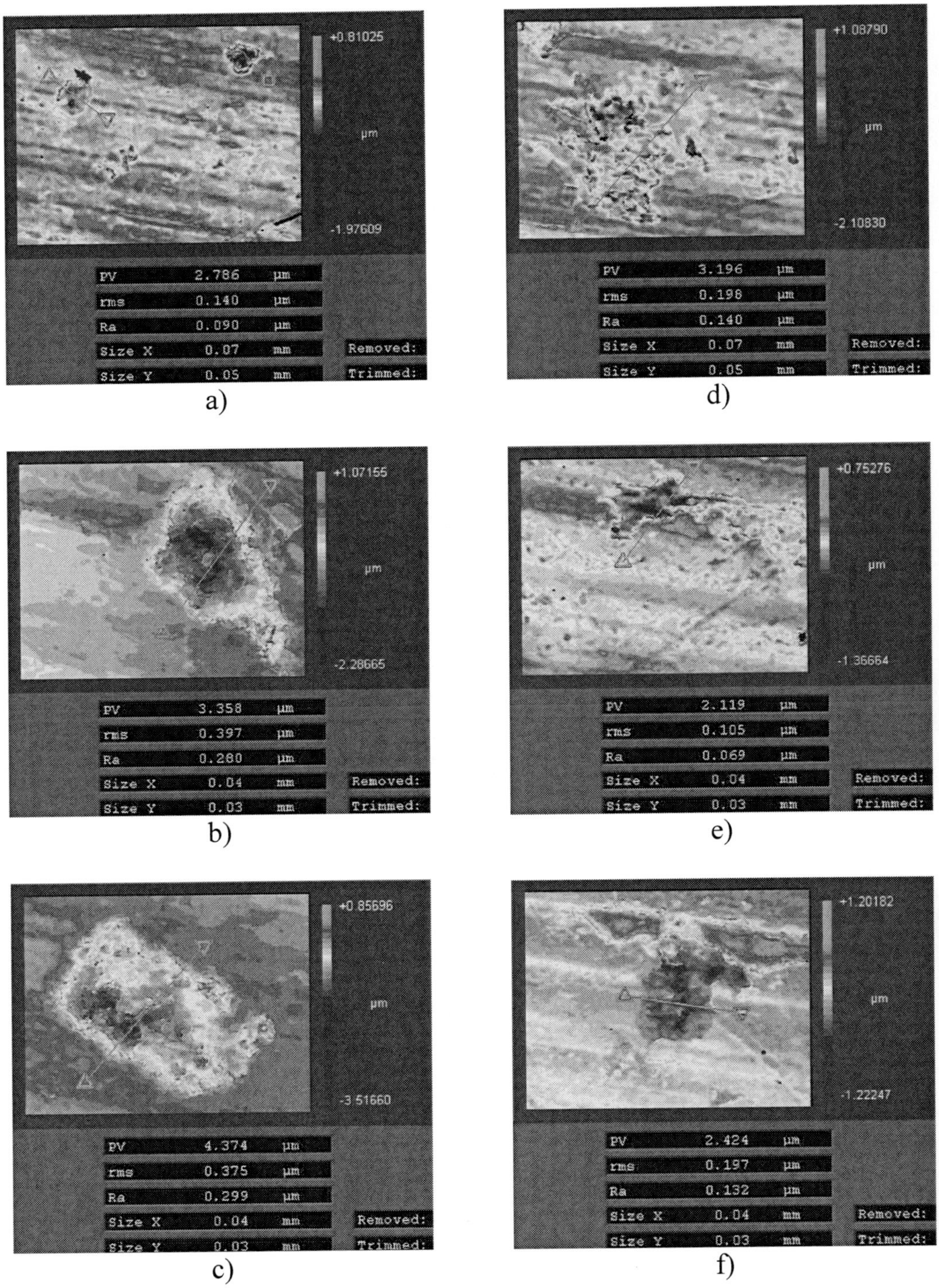


Figure 3. Electron micrographs of surfaces of nanocomposites phenylone/A (a, b, c) and phenylone/A-MF (d, e, f).

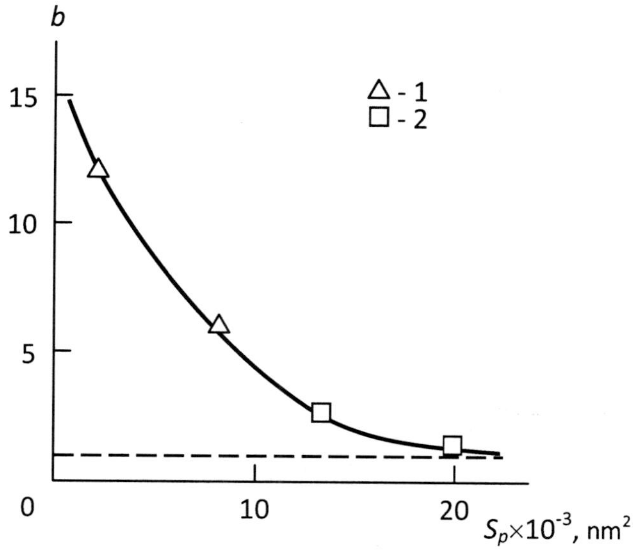


Figure 4. The dependence of parameter *b* on average area of nanofiller particles surface $S_p$ for nanocomposites filled by aerosil (1) an oxinitride silicium-yttrium (2). The horizontal shaded line indicates perfect microadhesion level ($b$=1).

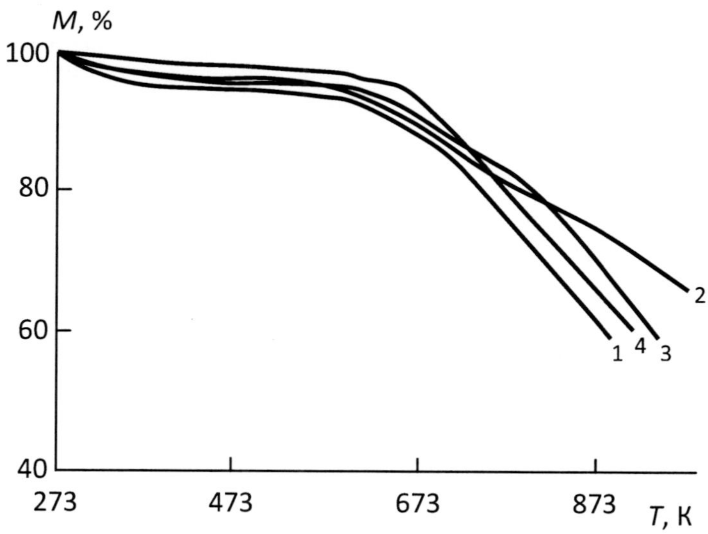


Figure 5. The TGA curves of phenylone (1) and nanocomposites phenylone/A with aerosil contents 0.3 (2), 0.5 (3) and 1.0 (4) mass. %.

In the nanocomposites case the nanoadhesion cause is a small contact area of polymeric mztrix-nanofiller. From the comparison of aerosil and OSY particles size and Figure 2 data one can make a conclusion that two factors as minimum influence on value b: nanofiller particles size and their aggregation level, i.e. in the long run nanofiller aggregates size as well. So, value *b* for aerosil is higher than for oxinitride silicium-yttrium (particles size are 15 and 64 nm, respectively) and for nanocomposites phenylone/A-MF value *b* is substantially larger (twice) than for phenylone/A as that was mentioned above, processing of phenylone and aerosil powder-like blend in rotating electromagnetic field reduces aerosil particles aggregation level. Besides, the effect of $b$ decrease at $\varphi_n$ increase for nanocomposites phenylone/OSY (Figure 2) is also due to enhancement of oxinitride silicium-yttrium particles aggregation level at its contents rise.

In Figure 3 the micrographs of the surface of phenylone/A (Figure 3a, b, c) and phenylone/A-MF (Figure 3d, e, f) with aerosil content 1 mass. % samples are shown, from which one can see the aggregation process inhibition for the last nanocomposites series. Figure 2d is particularly outstanding in this sense, where one can see “cloud” of aerosil particles, which could not unite in an aggregate owing to the application of processing in electromagnetic field. And on the contrary, Figure 2a, b, c clearly demonstrate substantial aggregation of aerosil particles in large enough aggregates. The obtained data assume that the studied nanoadhesion effect is purely dimensional effect or true nanoeffect [12]. It is well known that in adhesional strength measurements of different material pairs it is impossible to achieve the full contact of their surfaces. That is why the indicated contact is realized according to some hillocks only, that decreases contact area and, correspondingly, adhesional strength [13]. At very small (nanometer scale) contact area sizes the existence of large hillocks and depressions on the contact surface is scarcely probable that results to sharp adhesion level enhancement. This condition can be expressed more specifically: the nanoadhesion is achieved at the same scale of its participant sizes – nanoparticle and macromolecular coil.

In Figure 4 the dependence of parameter $b$ on nanofiller particles surface average area $S_p$ for nanocomposites phenylone/aerosil and phenylone/OSY is shown. As one can see, the decrease $b$ or adhesion level reduction at the increase $S_p$ (nanofiller particles size) is observed and at $S_p \approx 2\times10^4$ nm$^2$, that corresponds to particles diameter ~ 80 nm, this dependence tries to attain $b\approx1$, i.e. to perfect microadhesion level ($\alpha_n = \alpha_n^T$, that follows from the equation (3)).

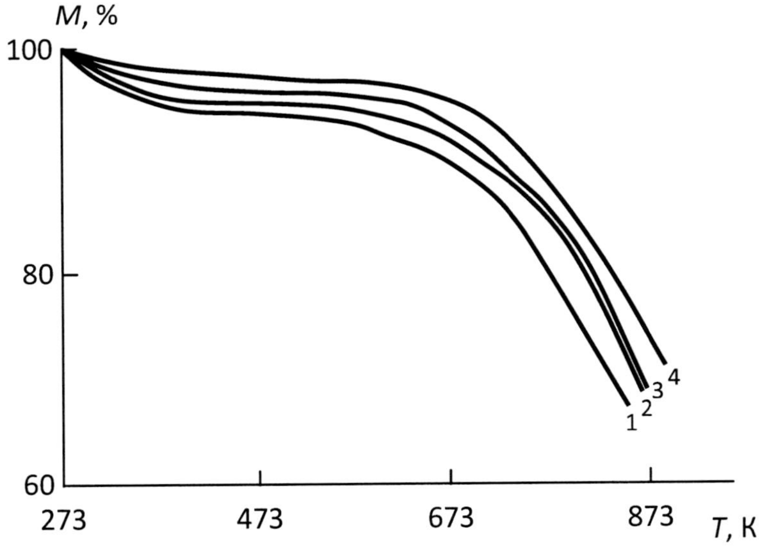


Figure 6. The TGA curves of phenylone (1) and nanocomposites phenylone/A-MF with aerosil contents 0.3 (2), 0.5 (3) and 1.0 (4) mass. %.

The nanoadhesion effect influences strongly on polymer nanocomposites macroscopic properties. As follows from the data of Figure 1, the introduction 1 mass. % of aerosil in phenylone results to $\alpha_n$ decrease in about 3 times (for nanocomposites phenylone/A-MF). In case of microcomposites such $\alpha_n$ decrease requires the introduction of filler with volume contents $\varphi_f$=0.3-0.7 [10], i.e. practically on two orders more. Further as an example let's consider nanoadhesion influence on thermal and mechanical properties of the studied nanocomposites.

At present a large number of kinetic equations for the description of thermogravimetric analysis (TGA) data is known [14, 15]. However, all the indicated models do not take into consideration the polymer melt structure and the character of diffusive processes proceeding in thermooxidative degradation course. The last are important in the respect of controlling of oxidizer (for example, oxygen) access to reactive centers of polymeric macromolecules [16]. As a rule, the temperature of thermal degradation beginning in TGA tests (for example, temperature of 20 %-th sample mass loss $T_{20\%}$) is higher than temperature of the so-called transition "liquid 1 – liquid 2" $T_{ll}$, which can be estimated as follows [17]:

$$T_{ll} \approx (1.20 \pm 0.05) T_g, \tag{4}$$

where $T_g$ is glass transition temperature of polymer.

At $T_{ll}$ there occurs a transition of the polymer melt from "liquid with fixed structure" (where residual structural ordering is observed [17]) to true liquid state or "structureless liquid". Nevertheless, melt "structureless" at $T>T_{ll}$ relates to the absence of the supermolecular structure, but the structure of a macromolecular coil in melt remains an important structural factor (in essence, unique at $T>T_{ll}$) [16]. Therefore structural analysis of thermal stability of polymer nanocomposites on the base of phenylone, filled by aerosil, will be given fur ther.

In Figures 5 and 6 the TGA curves for nanocomposites phenylone/A and phenylone/A-MF are shown, respectively. As the plots of these figures show, on the first stage for the studied materials within the temperatures range 293-383 K sample mass gradual decrease on 1.5-2.0 is observed, connected with moisture loss. Then, up to $T$=573 K sample mass remains practically nonchanged. The intensive thermooxidative degradation of phenylone and nanocomposites on its base, accompanied by a considerable mass loss, begins in the air above 663 K.

The experimentally determined temperatures of sample mass 20 %-th loss $T_{20\%}$ for all studied materials are listed in Table 1. As one can see, the introduction of only one percent of aerosil into phenylone rises its thermal stability on 76-80 K. Let's consider the physical fundamentals of this strong enough effect.

As it was known [18], within the framework of a strange (anomalous) diffusion conception on fractal object it's possible to select two its main types: slow and fast diffusion. The dependence of mobile displacement $s$ on time $t$ is assumed as a basis of such division [18]:

$$s \sim t^{\beta}, \tag{5}$$

**Table 1. The comparison of experimental $T_{20\%}$ and calculated according to the equation (8) $T_{20\%}^{T}$ temperatures of sample 20 %-th mass loss for nanocomposites phenylone/aerosil**

| Material | The aerosil contents, mass. % | $T_{20\%}$, K | $T_{20\%}^{T}$, K, the equation (16) | $T_{20\%}^{T}$, K, the equation (14) |
|---|---|---|---|---|
| Phenylone S-2 | - | 773 | 799 | 799 |
| Phenylone/A | 0.3 | 775 | 803 | 805 |
| | 0.5 | 798 | 810 | 805 |
| | 1.0 | 849 | 825 | 804 |
| Phenylone/A-MF | 0.3 | 798 | 809 | 805 |
| | 0.5 | 818 | 823 | 805 |
| | 1.0 | 853 | 837 | 804 |

where for classical case β=1/2, for slow diffusion β<1/2 and for the fast one - β>1/2.

The most precisely the macromolecular coil structure, which is fractal object, can be characterized with the aid of its fractal (Hausdorff) dimension $\Delta_f$, describing the coil elements distribution in space [19]. Earlier within the framework of fractional derivatives theory the interrelation $\Delta_f$ and β was shown, which is expressed analytically as follows [18]:

$$\beta = \frac{\Delta_f - 1}{4} \tag{6}$$

for slow diffusion and

$$\beta = \frac{\Delta_f - 1}{\Delta_f} \tag{7}$$

for the fast one. The structural boundary between the mentioned kinds of diffusion one can consider the value $\Delta_f$=2.5 at its general variation $2.0 \le \Delta_f < 3.0$: at $\Delta_f < 2.5$ (less compact macromolecular coils) the fast diffusion of oxidizer (oxygen) is realized, at $\Delta_f > 2.5$ – a slow one.

For theoretical estimation of value $T_{20\%}$ ($T_{20\%}^{T}$) the following equation was used [20]:

$$\Delta_f = c\left(T_{20\%}^{T} - T_g\right)^{\beta}, \tag{8}$$

where $c$ is constant, accepted equal to 0.093 for fast diffusion and 0.305 – for slow one [20].

The equation (8) defines three factors, influencing on polymer materials thermal stability: chemical constitution of a polymer, characterized by $T_g$ value, polymer melt structure, characterized by dimension $\Delta_f$, and a type (intensity) of oxidizer diffusion, connected with the structure and characterized by exponent $\beta$ [20].

To carry out the estimation of dimension $\Delta_f$ is possible with the aid of a number of methods. However, in the present case the most obvious one will be the following. As it is known [21, 22], the introduction of filler in polymer can change its effective fractal dimension $\Delta_f$ (let's remind, that for polymer materials melt the condition $\Delta_f = d_f$ was accepted), which can be calculated according to the equation [23]:

$$d_f = 3 - 6\left(\frac{\varphi_{den}}{SC_\infty}\right)^{1/2}, \tag{9}$$

where $\varphi_{den}$ is relative fraction of nanocomposites densely-packed regions, in which relaxation processes are inhibited, $S$ is a cross-sectional area of a macromolecule (for phenylone $S$=17.9 Å$^2$ [24]), $C_\infty$ is characteristic ratio, which is equal to 3 for phenylone [24].

To densely-packed regions of nanocomposites clusters in polymeric matrix, nanofiller and interfacial regions with relative fractions $\varphi_{cl}$, $\varphi_n$ and $\varphi_{if}$, respectively can be attribute. The value $\varphi_{cl}$ for phenylone can be obtained according to the equation (9) at $d_f$=2.447 [24]. The relation between $\varphi_{if}$ and $\varphi_n$ can be obtained as follows. As it is known [7] both polymeric matrix and aerosil particles surface, interacting at interfacial regions formation, are fractal objects. At interaction of such objects there is the only linear scale $l$, defining their interpenetration distance [25]. As in polymer nanocomposites nanofiller elasticity modulus, as a rule, is larger than the corresponding parameter for polymeric matrix, then it is assumed [7], that in this case aerosil penetration in polymeric matrix occurs and then $l$ is equal to interfacial layer thickness $l_{if}$. Further it can be written [25]:

$$l_{if} \sim a\left(\frac{r_n}{a}\right)^{2(d-d_{surf})/d}, \tag{10}$$

where $a$ is a lower linear scale of polymeric matrix fractal behavior, $r_n$ is a nanofiller particles radius, $d$ is dimension of Euclidean space, in which fractal is considered (obviously, in our case $d$=3), $d_{surf}$ is fractal dimension of nanofiller particles surface. The proportionality coefficient in the relationship (10) is accepted equal to 1.0 [22].

For polymers the value $a$ is accepted equal to statistical segment length $l_{st}$ [7], determined according to the formula [26]:

$$l_{st} = l_0 C_\infty, \tag{11}$$

where $l_0$ is the length of skeletal bond of the main chain, which is equal to 1.40 Å for phenylone [24].

The value $d_{surf}$ can be determined according to the following equation [27]:

$$S_u = 400 r_n^{d_{surf}-d}, \tag{12}$$

where specific surface of aerosil particles $S_u$ is given in $m^2/g$ and $r_n$ – in nm.

The equations (11) and (12) give $l_{st}$=0.42 nm, $d_{surf}$=2.95 and then $l_{if}$=2.74. Further the following equation can be used [22]:

$$\varphi_{if} = \varphi_n \left[ \left( \frac{r_n + l_{if}}{r_n} \right)^3 - 1 \right], \tag{13}$$

which at $r_n$=7.5 nm and $l_{if}$=2.74 nm gives the following relationship between $\varphi_{if}$ and $\varphi_n$ for aerosil:

$$\varphi_{if} = 1.55\varphi_n . \tag{14}$$

Let's note, that for the layered silicates (organoclays) the equation (14) has a form [21, 22]:

$$\varphi_{if} = 1.91\varphi_n . \tag{15}$$

As nanocomposites reinforcement degree is defined by sum ($\varphi_n$+$\varphi_{if}$) (see the equation (18)), then the comparison of the equations (14) and (15) shows, that by its reinforcing action nonaggregated aerosil is close to exfoliated organoclays. The equations (14) and (15) were obtained for perfect adhesion polymeric matrix-nanofiller, i.e. for the condition $b$=1. At nanoadhesion effect availability ($b$>1) the value $l_{if}$ is proportional to $b$ [11] and at the first approximation the equation (14) is necessary to write as follows:

$$\varphi_{if} = 1.55\varphi_n b . \tag{16}$$

Then for the value $\varphi_{den}$ in the equation (9) in case of nanocomposites phenylone/aerosil let's obtain finally:

$$\varphi_{den} = \varphi_{cl} + \varphi_n + 1.55\varphi_n b . \tag{17}$$

Further one can calculate theoretical values $T_{20\%}^T$, using the equations (7), (8), (9) and (17). The comparison of experimental $T_{20\%}$ and calculated by the indicated method $T_{20\%}^T$ values of sample 20 %-th mass loss temperatures in TGA tests is shown in table 1. As one can see, the good enough correspondence of theory and experiment is obtained (the average discrepancy values $T_{20\%}$ and $T_{20\%}^T$ makes up 2.2 %). In this table the similar calculation of $T_{20\%}^T$ without accounting of nanoadhesion effect is given, i.e. at the calculation the equation (15) was used instead of the equation (17). In this case theory and experiment correspondence grows substantially worse. This means, that the indicated above $T_{20\%}$ enhancement at the introduction of particulate nanofiller (aerosil) small amounts by 76-80 K is due to nanoadhesion effect. At the indicated effect absence such enhancement would not exceed 32 K.

The equation (8) gives a strict and physically substantiated definition of temperature $T_{20\%}$. However, the thermal stability structural analysis can be made by empirical more obvious method. In Figure 7 the dependence $T_{20\%}$ on value $\varphi_{den}$ is shown, which demonstrates linear increase $T_{20\%}$ at $\varphi_{den}$ enhancement. The extrapolation of linear plot $T_{20\%}(\varphi_{den})$ to limiting values $\varphi_{den}$=0 and $\varphi_{den}$=1.0 gives the values $T_{20\%}$=616 and 1040 K, respectively. Theoretically these values $T_{20\%}$ can be calculated according to the equation (8) at the condition $\Delta_f$=3 (slow diffusion) and $\Delta_f$=2 (fast diffusion), respectively. Such calculation gives the following values $T_{20\%}^T$: 645 and 1040 K, respectively, i.e. close enough to the one obtained by extrapolation. Hence, as follows from the equations (16) and (17), such empirical treatment assumes also nanoadhesion effect influence on the value $T_{20\%}$.

Let's consider briefly nanoadhesion influence on nanocomposites phenylone/aerosil mechanical properties, namely, on their reinforcement degree. Compression tests have shown, that the elasticity modulus $E_m$ of initial matrix phenylone is equal to 3.05 GPa and nanocomposite phenylone/A-MF with aerosil content 0.3 mass. % - 3.29 GPa. As it is known [22], nanocomposites reinforcement degree is defined by the equation:

$$\frac{E_n}{E_m} = 1 + 11\left(\varphi_n + \varphi_{if}\right)^{1.7}, \tag{18}$$

where $E_n$ is nanocomposite elasticity modulus.

By using in the equation (18) the estimation $\varphi_{if}$ according to the formula (14) let's obtain theoretical value $E_n$ ($E_n^T$) equal to 3.06 GPa and by using formula (16) $E_n^T$=3.33 GPa. It is not difficult to see, that the second value $E_n^T$, obtained with nanoadhesion effect accounting, corresponds to the experiment much better (the discrepancy of $E_n$ and $E_n^T$ makes 1 %), than obtained without taking into consideration the indicated effect (the discrepancy is about 10 %). Therefore, nanoadhesion effect results to nanocomposites elasticity modulus increase by 0.263 GPa even at insignificant aerosil content of 0.3 mass. %.

Figure 8 gives the obvious notion of nanoadhesion effectivity action, where the curves of stress-strain ($\sigma$-$\varepsilon$) for initial phenylone and nanocomposite phenylone/A-MF with aerosil content 0.3 mass. % are shown. As one can see, the introduction of aerosil indicated amount changes not only mechanical characteristics of matrix polymer, but its fracture type.

Let's consider in more detail nanoadhesion effect influence on nanocomposites reinforcement degree on the example of phenylone/OSY systems. For the indicated nanocomposites the equations (10)-(14) give the following relationship between $\varphi_{if}$ and $\varphi_n$:

$$\varphi_{if} = 0.102\varphi_n. \tag{19}$$

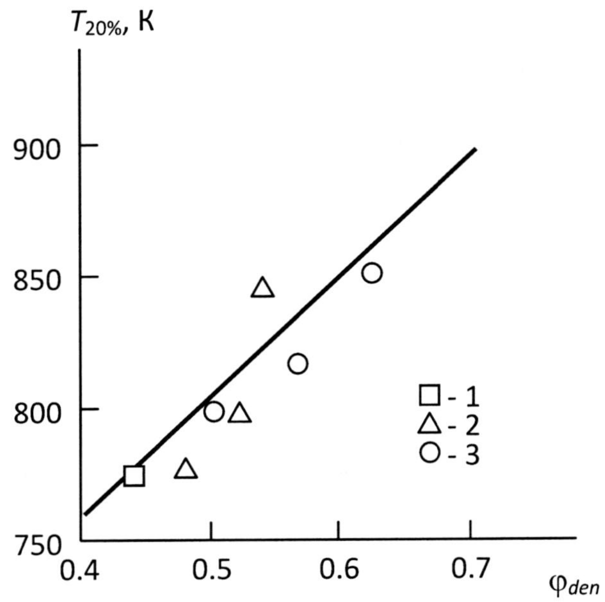


Figure 7. The dependence of sample 20 %-th mass loss temperature $T_{20\%}$ on relative fraction of densely-packed regions $\varphi_{den}$ for phenylone (1) and nanocomposites phenylone/A (2), phenylone/A-MF (3).

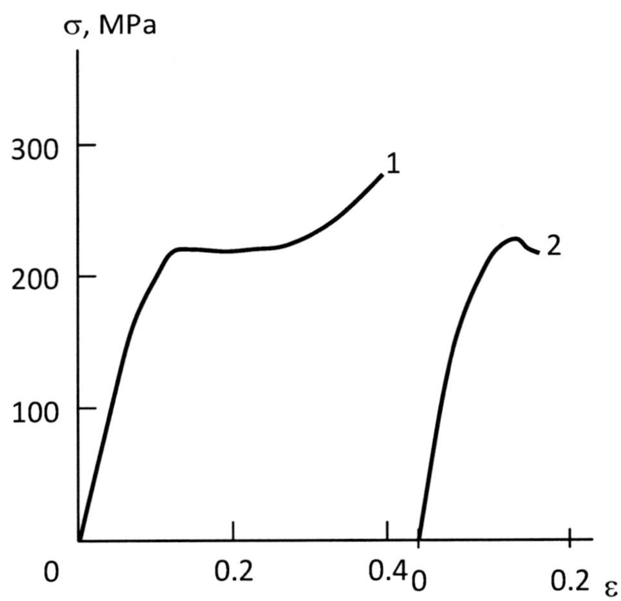


Figure 8. The diagrams stress-strain (σ-ε) for phenylone (1) and nanocomposite phenylone/A-MF with aerosil content 0.3 mass. % (2) in compression tests.

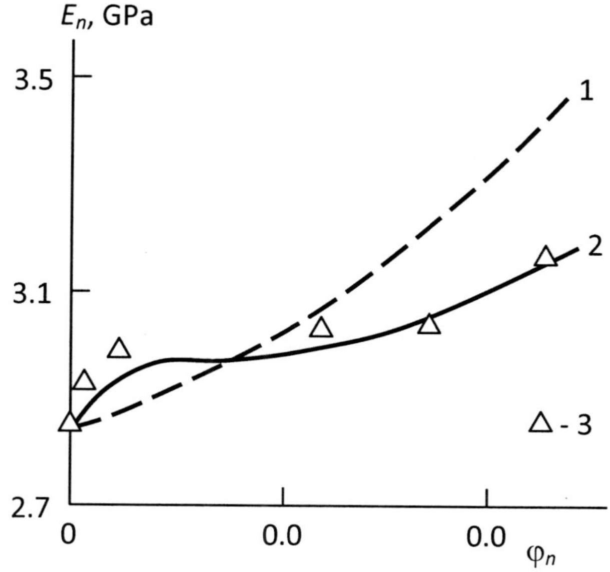


Figure 9. The dependences of elasticity modulus $E_n$ on nanofiller volume contents $\varphi_n$ for nanocomposites phenylone/OSY. 1 – the calculation according to the equations (18) and (19); 2 – according to the equations (18) and (20); 3 – experimental data.

In Figure 9 the comparison of dependences $E_n(\varphi_n)$ calculated according to the equations (18) and (19) is adduced and is also obtained experimentally. As one can see, the indicated curves $E_n(\varphi_n)$ correspondence leaves much to be desired. Therefore let's use the equation (19) modification similar to the formula (16):

$$\varphi_{if} = 0.102\varphi_n b. \tag{20}$$

In Figure 9 the comparison of the dependences $E_n(\varphi_n)$, obtained experimentally and calculated according to the equations (18), (20) is also shown. As follows from Figure 9 data, in this case much better correspondence of theory and experiment (their average discrepancy makes 1.9 %) is obtained. But the main merit of the equation (20) does not even consist in this (using of the equations (18) and (19) gives also acceptable average discrepancy of theory and experiment, namely, 4.2 %). Firstly, the calculation with due regard to the nanoadhesion effect (the equation (20)) allows to reproduce the main qualitative features of the experimental curve $E_n(\varphi_n)$, namely, to increase ("hump") of $E_n$ at small $\varphi_n$ ($b>1$, nanoadhesion effect) and slower enhancement of $E_n$ at $\varphi_n \geq 0.023$ ($b<1$ – adhesion level reduction due to OSY particles aggregation). Secondly, the equations (16) and (20) allow quantitative accounting interfacial adhesion level on nanocomposites reinforcement degree.

## CONCLUSIONS

Therefore, the present paper results have shown the existence in polymer particulate-filled nanocomposites nanoadhesion effect, which consists in the enhanced level of polymeric matrix-nanofiller interaction. This effect has purely dimensional origin and is realized for nanofiller particles with diameter less than 80 nm. The nanoadhesion effect influences strongly on macroscopic properties of those nanocomposites, in which it is realized. So, the essential thermal stability enhancement and the considerable thermal expansion coefficient reduction of nanocomposites on the base of phenylone at the introduction only 1 mass. % aerosil are due to the nanoadhesion effect. Appreciation of the nanoadhesion effect allows qualitatively and quantitatively an exact description of nanocomposites reinforcement degree.

## REFERENCES

[1] Pul C., Owens F. *The Nanotechnologies.* Moscow, Technosfera, 2005, 336 p.

[2] Sheng N., Boyce M.C., Parks D.M., Rutledge G.C., Abes J.I., *Cohen R.E. Polymer,* 2004, v. 45, № 2, p. 487-506.

[3] Buchachenko A.L. *Uspekhi Khimii,* 2003, v. 72, № 5, p. 419-437.

[4] Okhlopkova A.A., Vinogradov A.V., Pinchyk L.S. *Plastics Filled by Ultradisperse Inorganic Particles.* Gomel, IMMS NASB, 1999, 164 p.

[5] Lipatov Yu.S. The Interfacial Phenomena in Polymers. Kiev, *Naukova Dumka,* 1980, 260 p.

[6] Khunyants N.N., Lyapunova M.A., Manevich L.I., Oshmyan V.G., Shaulov A.Yu. *Mekhanika Kompozithykh Materialov,* 1986, v. 22, № 2, p. 231-234.

[7] Kozlov G.V., Yanovskii Yu.G., Lipatov Yu.S. *Mekhanika Kompozitsionnykh Materialov I Konstruktsii, 2002*, v. 8, № 1, p. 111-149.

[8] Kozlov G.V., Lipatov Yu.S. *Composite Interfaces,* 2002, v. 9, № 6, p. 509-527.

[9] Kubat J., Rigdahl M., Welander M. *J. Appl. Polymer Sci.,* 1990, v. 39, № 5, p. 1527-1539.

[10] Holliday L., Robinson J. In book: Polymer Engineering Composites. Ed. Richardson M.O.W. London, *Applied Science Publishers LTD*, 1978, p. 241-283.

[11] Tanaka K., Nagamura T. *Mater. Intern. Conf. on Polymer and Advanced Materials.* POLYMEX-2006, Huatulko, Mexico, November 5-9, 2006, Session 1, p. 51.

[12] Andrievskii R.A. *Rossiiskii Khimichesk. Zhurnal*, 2002, v. 46, № 5, p. 50-56.

[13] Wu S. Polymer Interface and Adhesion. New York, *Marcel Dekker,* 1982, 492 p.

[14] Wendlandt U. Thermal Methods of Analysis. Moscow, *Mir,* 1978, 526 p.

[15] Zuru A.A., Whitehead R., Griffiths D.L. *Thermochim. Acta,* 1990, v. 164, № 3, p. 285-305.

[16] Kozlov G.V., Zaikov G.E. The Structural Stabilization of Polymers: Fractal Models. Leiden-Boston, *Brill Academic Publishers,* 2006, 345 p.

[17] Bershtein V.A., Egorov V.M. Differential Scanning Calorymetry in Physics-Chemistry of Polymers. L*eningrad, Khimiya,* 1990, 256 p.

[18] Shogenov V.Kh., Akhkubekov A.A., Akhkubekov R.A. Izvestiya VUZov, *Severo-Kavkazsk. region, estestv. nauki,* 2004, № 1, p. 46-50.

[19] Vilgis T.A. Physica A, 1988, v. 153, № 2, p. 341-354.
[20] Dolbin I.V., Burya A.I., Kozlov G.V. *Teplofizika Vysokikh Temperatur,* 2007, v. 45, № 3, p. 355-358.
[21] Kozlov G.V., Malamatov A.Kh., Antipov E.M., Karnet Yu.N., Yanovskii Yu.G. *Mekhanika Kompozitsionnykh Materialov I Konstruktsii,* 2006, v. 12, № 1, p. 99-140.
[22] Malamatov A.Kh., Kozlov G.V., Mikitaev M.A. The reinforcement Mechanisms of Polymer Nanocomposites. Moscow, *Publishers RKhTU*, 2006, 240 p.
[23] Kozlov G.V., Novikov V.U. Uspekhi Fizichesk. Nauk, 2001, v. 171, № 7, p. 717-764.
[24] Kozlov G.V., Burya A.I., Dolbin I.V. *Prikladnaya Fizika,* 2006, № 1, p. 14-18.
[25] Hentschel H.G.E., Deutch J.M. *Phys. Rev.* A, 1986, v. 29, № 12, p. 1609-1611.
[26] Wu S. J. Polymer Sci.: Part B: *Polymer Phys.,* 1989, v. 27, № 4, p. 723-741.
[27] Kozlov G.V., Shustov G.B. *Khimicheskaya Technologiya*, 2006, № 1, p. 24-26.

In: Chemical and Biochemical Reactions
Editors: G. E. Zaikov, L. Liu et al.

ISBN 978-1-61122-860-1


*Chapter 16*

# ON PROCESSES OF THE CHARGE TRANSFER IN THE ELECTRICAL CONDUCTING POLYMER MATERIALS

***T. Marsagishvili[1] and J. Aneli[2*]***
[1]Institute of the Inorganic Chemistry and Electrochemistry, Mindeli str.11, Tbilisi 0186, Georgia
[2]Institute of Machine Mechanics, Mindeli str.10, Tbilisi 0186,Georgia

## ABSTRACT

The analysis of charge transfer processes in the electrical conducting solid polymer systems has been carried out. The processes in these systems are divided on two types. First type- the processes of charge transfer between electrodes and particles, second one – the processes of the charge transfer between conductive particles.

The describing of medium is carried out with using of Green temperature functions of polarization operators for the medium molecules. It permits taking into account the effects of frequency and space dispersion.

The analytical expressions for kinetic parameters of the charge transfer processes from electrodes to particles and between particles in condensed matter have been obtained. The comparison of the general theoretical dependence of the current in ECPC on the average distance between conducting particles with analogical dependence for some real ECPC are presented in the paper.

**Keywords:** Charge transfer, vector-potential, Green function, polymer composite, structure, electro-conductivity, filler content, inter- and intra-phase interactions

* E-mail: jimaneli@yahoo.com

## BACKGROUND

In recent time so-called nontraditional electrical conducting materials, particularly composites, containing a dielectric basis (ceramics, polymers) and conducting dispersive fillers (carbon-graphite and metallic powders), display a growing competition to traditional conductors (e.g. metals, classical semiconductors). This situation is due to many positive characteristics of these materials, among which are the high corrosion stability, accessible technology of production and low cost. Now there are many scientific and technical works in the sphere of the noted materials with both theoretical and experimental character (see for example [1-5]).

Growth of ECPC conductivity with the increase of conducting filler content is the rule without exclusions [1-4, 6]. The specific feature of this dependence is an increase of conductivity $\gamma$ or, which is the same, a decrease of $\rho$ at definite (for a particular composite) threshold filler concentration, induced by an insulator-conductor transition. This transition conforms to the so-called threshold of proceeding, or percolation. In this case $\gamma$ value jump, which may reach several decimal degrees, is stipulated by formation of a continuous chain of filler particles in the polymer matrix - the infinite cluster [7,8].

At present the problem of the conductivity mechanism of ECPC is still to be discussed. As to the opinion of some investigators the charge transfer is conducted by chains, consisted of filler particles having direct electric contact [9,10]. On the opinion of other authors conductivity of ECPC is caused by thermal emission of electrons through spaces between particles [11,12]. It is another opinion that current exists in ECPC with air gaps of polymer films between filler particles. In this case electrons, which energy is lower than the potential barrier value may be tunneled through it, if their own wave-length is comparable with space width of insulating film [13-15]. The Wessling model [16] considers formation of conductive chains as the process, based on the non-equilibrium thermodynamics. At present the percolation theory [17] is widely used for calculations of $\gamma$ for conducting composites (with both organic and inorganic binders). Accordingly to this theory $\gamma$ of composites consisted of non-interacting phases. Experimental and theoretical studies of composite conductivity [18, 19] were conducted in superfine gaps between graphite particles. In this case a significant meaning was devoted to the polymer molecule state in the gap, if the filler particles were of a hypothetic form of a truncated cone.

The analysis of works on the investigations of electrically conducting properties of ECPC induces one general conclusion: despite a variety of the above considered models of electrically conducting ECPC, no one could unfortunately pretend for the versatility. Each model includes one or several approximations and suppositions, which aggravate the correctness of estimations of ECPC conducting properties. That is why the comparison of theoretically calculated data with the experimental results usually gives deviations, which reach several orders in some cases. The coincidence is rarely reached at definite concentrations of conducting filler and specific conditions of the composite production.

In the presented work a new approach to charge transfer processes in the polymer materials contained the conducting fillers has been proposed.

# 1. Electromagnetic Interactions in the Irregular System. Models for Green Functions of the Operators of Electromagnetic Vector-potential

The processes in the systems with electrical conducting polymer media may be divided on several types. First of all it is the process of the electron transfer from the electrode to the electrical conducting admixture and second one –the electron transfer from one conducting particle to another in the polymer matrix. Third type of the processes – electron transfer from conducting particle to the electrode in the polymer.

## 1.1. The System Hamiltonian

The processes of charge transport from the electrode to particle may be described analytically at choosing of the concrete system for which it is possible to write the system Hamiltonian. A principal difficulty of investigation of charge transfer processes is connected with necessity of using of quantum approach. In the frame of this approach there is no models generally accepted for the processes of charge transfer in the irregular systems. The using of such apparatus of mathematical physics as the apparatus of Green temperature function (GF) allows a describing of the complex condensed system and processes of charge transfer in them. This technique allows the unification of theoretical approaches, using widely the different models for describing of the effects of the system of frequency and space dispersion [20-23].

In the unrelativistic approximation a Hamiltonian of the condensed system may be presented as

$$\breve{H} = H^{T}_{rad} + H_{C} - \int d\vec{r} A^{T}(\vec{r})\ j^{T}(\vec{r}),$$

where $H^{T}_{rad}$ - the Hamiltonian of transverse photons, $H_c$ –Hamiltonian of the “Coulomb” medium (the particles interacting between them by Coulomb law); $A^T$ – field vector-potential, J – current density of the “Coulomb” medium.

In the frame of linear response theory the quantum-statistical average value of the vector-potential has the view:

$$\langle A_\alpha(\vec{r},t)\rangle_{ex} = -\int d\vec{r} \int_{-\infty}^{\infty} dt' D^{R}_{\alpha\beta}(\vec{r},\vec{r}';t-t') j^{ex}_{\beta}(\vec{r}',t') + \int dr' \int_{-\infty}^{\infty} dt' G^{R}_{A_\alpha\phi}(\vec{r},\vec{r}';t-t')\rho^{ex}(\vec{r}',t'),$$

where $D^{R}_{\alpha\beta}(\vec{r},\vec{r}';t-t')$ - retarded of GF of vector –potential operators equaled to:

$$D^{R}_{\alpha\beta}(\vec{r},\vec{r}';t-t') = -i\theta(t-t')\langle[A_\alpha(\vec{r},t)A_\beta(\vec{r}',t')]\rangle$$

The GF of operators $\vec{A}$ and the scalar potential $\Phi$ $G^R_{A_\alpha\phi}$.is defined analogically. For the quantum-statistical average operators of the electromagnetic field strength and system polarization the relations for Fourier components have the view:

$$G^E_{\alpha\beta}(\vec{k},\vec{k}';\omega) = i\omega D^R_{\alpha\beta}(\vec{k},\vec{k}';\omega) - ik'_\beta G^R_{A_\alpha\phi}(\vec{k},\vec{k}';\omega) - ikG^R_{\phi A_\beta}(\vec{k},\vec{k}';\omega) + i\frac{k_\alpha k'_\beta}{\omega} G^R_{\phi\phi}(\vec{k},\vec{k}';\omega)$$

$$G^P_{\alpha\beta}(\vec{k},\vec{k}';\omega) = \frac{i}{\omega} G^R_{j_\alpha A_\beta}(\vec{k},\vec{k}';\omega) - \frac{k'_\beta}{\omega} G^R_{j_\alpha\phi}(\vec{k},\vec{k}';\omega)$$

From these correlations for complex susceptibility of the system the following correlation is obtained:

$$\chi = G^P (G^E)^{-1},$$

where $(G^E)^{-1}$ - inverse operator.

The closed equation for GF $D^{TT}$ and G may be obtained only for model system. For example, GF may be calculated when the characteristic frequencies are sufficiently low.

In case of photon frequencies lower than ones corresponding to the interzone transfers of the medium electrons the system Hamiltonian may be presented as:

$$\breve{H} = H_f^T + T_n + T_e + V_{ee} + V_{nn} + V_{en} + V_{ef} + V_{nf},$$

where $T_n$ is the operator of kinetic energy of the atoms nuclei of Coulomb medium, $T_e$ – the operator of medium electrons kinetic energy, $V_{ee}$, $V_{nn}$ and $V_{en}$ the operators of interactions between valence electrons and medium nuclei, $V_{ef}$ и $V_{nf}$ - Hamiltonians of the interactions between electrons and nuclei with transverse photons of electromagnetic field.

If instead of free photon Hamiltonian in the vacuum the effective photon Hamiltonian including the medium optical polarization will be included the system Hamiltonian of type

$$\breve{H} = \tilde{H}^T_f + H_C + \tilde{H}^T_{int},$$

will be obtained, where

$$\tilde{H}^T_{int} = -\int \vec{P}(\vec{r})\vec{E}^T(\vec{r})d\vec{r} = -\int \vec{J}(\vec{r})\vec{A}^T(\vec{r})d\vec{r}$$

Considerable simplification is appeared for local, homogeneous and isotropic electron subsystem, where the simple dependence between GF operators of electron polarization and high frequency dielectric permittivity $\varepsilon_\infty$ exists. In this case for the photon energy in the medium we have:

$$\frac{1}{8\pi}\int\left[\vec{E}^{T}(\vec{r})\right]^{2}d\vec{r}-\Delta\breve{F}^{e}=\frac{1}{8\pi\varepsilon_{\infty}}\int\left[\vec{E}^{T}(\vec{r})\right]^{2}d\vec{r}$$

By means of the same approximation it maybe find between different polarizations. The average electron polarization of the medium will be defined by electrical fields both nuclei and electrons:

$$\vec{P}_{0}^{e}=\frac{\varepsilon_{\infty}-1}{4\pi}\left(-4\pi\vec{P}^{n}-4\pi\vec{P}^{e}\right).$$

From this expression we have

$$\vec{P}_{0}^{e}=-\left(1-\frac{1}{\varepsilon_{\infty}}\right)\vec{P}^{n};\ \vec{P}^{n}=\varepsilon_{\infty}\vec{P}$$

At last it must be noted that in simple case considered above, the substitute of $H_{f}^{T}$ instead of $\widetilde{H}_{f}^{T}$ leads to taking in account of the fact that the transverse photons in the medium are propagate in $\sqrt{\varepsilon_{\infty}}$ times slower than in vacuum. At choosing of the model for processing of experimental data or for interpretation of the experimental results it must be remembered that the correlation of different polarization modes of medium molecules is obtained in the frame of definite and essential restrictions.

## 1.2. Equation of Dyson for Green Function of the Electromagnetic Field Vector Potential. Approximation for Multi-particle Green Function

The problem of transverse photons interacting with Coulomb medium may be solved for definite approximation. For this aim temperature GF of transverse photons may be written as:

$$D_{\alpha\beta}^{TT}(\vec{r},\vec{r}';\tau-\tau')=-\left\langle T_{\tau}A_{\alpha}(\vec{r},\tau)A_{\beta}(\vec{r}',\tau'S)\right\rangle_{0}/\left\langle S\right\rangle_{0}, \tag{1}$$

where average corresponds to states of Hamiltonian $\breve{H}_{0}=H_{C}+H_{f}^{T}$ and S-matrix has a view:

$$S=T_{\tau}\exp\left[\int_{0}^{\beta}d\tau\int d\vec{r}\,\vec{j}(\vec{r},\tau)\vec{A}(r,\tau)\right];\ \beta=(k_{B}T)^{-1},$$

where $k_B$ is Boltzmann constant, and T is absolute temperature.

After decomposition of the exponential function in the equation (1) we will obtain the infinite sum of quantum-mechanical averages containing after action of the operator $T_\tau$, operators $\vec{A}$ and $\vec{j}$:

$$D_{\alpha\beta}^{TT}(\vec{r},\vec{r}';\tau-\tau') = -\frac{1}{\langle S\rangle_0}\sum_{n=0}^{\infty T}\frac{1}{n!}\Big\langle T_\tau \int d\vec{r}_1 ... d\vec{r}_n \int_0^\beta d\tau_1 ... \times$$

$$\int_0^\beta d\tau_n \cdot A_\alpha^T(\vec{r},\tau) A_\beta^T(\vec{r}',\tau') A_{i_1}^T(r_1,\tau_1) j_{i_1}^T(r_1,\tau_1) .. A_{i_n}^T(\vec{r}_n,\tau_n) j_{i_n}^T(\vec{r}_n,\tau_n) \Big\rangle_0$$

Quantum-statistical average is calculated by eigenstates of Hamiltonian $\breve{H}_0$, which is a square form of the vector-potential of the operators $\vec{A}$. In accordance with Wick theorem each average of the operators $\vec{A}$ may be presented in view of multiplication of any couples of averages.

By such way it can be obtained the dependence between $D^{TT}$ and $G_{Aj}$ in the operator form

$$D^{TT} = d^{TT} - d^{TT} G_{jA}. \tag{2}$$

On this stage this dependence has exact character. If the deviation of polarization from equilibrium magnitude depends linearly on the deviation of nuclei coordinates from their equilibrium values and the oscillation of nuclei is presented in the harmonic approximation then the Hamiltonian of such Coulomb medium may be presented in square form in the operators of fluctuations of polarization $(\vec{P})$ and current density $(\vec{j})$. In this case the quantum-statistical average of the operators j over eigenstates of Hamiltonian $H_c$ scattered in pairs for obtaining of closed equation for GF $D^{TT}$.

The unlinking of average of many operators on the pairs is often based on different "heuristic" methods, such as approach of superposition in the classic field theory and self-consistent field and approach of random phases in the quantum field theory. There is not an exact criterion in these models. Although the application of such criterions allows to obtain the analytical solutions for many tasks.

At last we have

$$G_{jA} = -g_{jj} D^{TT}, \tag{3}$$

where

$$g_{j_\alpha j_\beta}(\vec{r},\vec{r}';\tau-\tau') = -\langle T_\tau j_\alpha(\vec{r},\tau) j_\beta(\vec{r}',\tau')\rangle_0$$

is the temperature GF operators of the current density on the states of Hamiltonian $H_c$.

From (2) and (3) we get

$$D^{TT} = d^{TT} + d^{TT} g_{jj} D^{TT}. \tag{4}$$

This is equation of Dyson for GF of electromagnetic field vector-potential. In this equation the role of mass operator plays GF of current density operators of Coulomb medium $g_{jj}$. The equation GF (4) has a solution for $D^{TT}$

$$D^{TT} = \left[\left(d^{TT}\right)^{-1} - g_{jj}\right]^{-1}, \tag{5}$$

where index "-1" means inverse operator. Eq. (5) is also the solution of task for transverse polariton. When GF $g_{jj}$ is known (from model calculations), the definition of GF $D^{TT}$ reduced to finding of reverse operator. In connection with it the task is simplified essentially for homogeneous system, for which GF $g_{j_\alpha j_\beta}$ received the view.

$$g_{j_\alpha j_\beta}\left(\vec{k}, \omega_n\right) = g^L_{jj}\left(\vec{k}, \omega_n\right)\frac{k_\alpha k_\beta}{k^2} + g^T_{jj}\left(\vec{k}, \omega_n\right)\left[\delta_{\alpha\beta} - \frac{k_\alpha k_\beta}{k^2}\right],$$

where the indexes L and T are longitudinal and transverse components, respectively. Using of GF $D^{TT}$ view (which is differed from vacuum GF operators of cross-section component of the vector of electromagnetic field by renormalization of light rate in the vacuum)

$$d^{TT}_{\alpha\beta}\left(\vec{k}, \omega_n\right) = \frac{4\pi}{\omega_n^2 + k^2/\varepsilon_\infty}\left[\delta_{\alpha\beta} - \frac{k_\alpha k_\beta}{k^2}\right].$$

We get following expression for GF $D^{TT}$

$$D^{TT}_{\alpha\beta}\left(\vec{k}, \omega_n\right) = \frac{4\pi\left[\delta_{\alpha\beta} - \frac{k_\alpha k_\beta}{k^2}\right]}{\omega_n^2 + k^2/\varepsilon_\infty - 4\pi g^T_{jj}\left(\vec{k}, \omega_n\right)}. \tag{6}$$

The eq.(6) defines the behavior of transverse photons in irregular condensed system.

The following modeling of spreading of transverse photons in the medium is possible by modeling of GF of Coulomb medium.

By means of presented equations and known GF $D^{TT}$ we can to define GF $G_{jA}$. Using the same approximation for solution of eq.(4) we can also obtain the closed equations for GF $G_{Fj}$ and $G_{jj}$.

$$G_{Aj} = -d^{TT} G_{jj} = -D^{TT} g_{jj}$$

$$G_{jj} = g_{jj} + g_{jj} d^{TT} G_{jj}$$

The formal solution of these functions receives a view:

$$G_{jj} = \left(g_{jj}^{-1} - d^{TT}\right)^{-1}$$

$$G_{Aj} = -d^{TT}\left(g_{jj}^{-1} - d^{TT}\right)^{-1} = \left(1 - g_{jj}^{-1}\left(d^{TT}\right)^{-1}\right)^{-1}$$

More suitable form of GF $D^{TT}$ may be got if this function will be expressed by microscopic dielectric permeability instead of GF of the operators of current density $G_{jj}$.

### 1.3. Dielectric Permittivity for Irregular Condensed System

To express of the dielectric permittivity by means of properties of Coulomb medium that is by means of different components $g_{jj}$ GF contenting the operators of scalar potential $G_{j\phi}, G_{A\phi}, G_{\phi A}$ and $G_{\phi\phi}$ must be also expressed by means of this function. Now let us consider the function

$$G^{R}_{j_\alpha\phi}\left(\vec{r},\vec{r}';t-t'\right) = -i\theta\left(t-t'\right)\left\langle\left[j_\alpha\left(\vec{r},t\right),\phi\left(\vec{r}',t'\right)\right]\right\rangle$$

Two times differentiation of both sides in respect to t′ gives

$$\Delta_{\vec{r}'}G^{R}_{j_\alpha\phi}\left(\vec{r},\vec{r}';t-t'\right) = -i\theta\left(t-t'\right)\left\langle\left[j_\alpha\left(\vec{r},t\right),-4\pi\rho\left(\vec{r}',t'\right)\right]\right\rangle,$$

where ρ is the density of the system charge. By further differentiation accordingly this t' and using of the continuity equation we get

$$\frac{\partial}{\partial t'}\Delta_{\vec{r}'}G^{R}_{j_\alpha\phi}\left(\vec{r},\vec{r}';t-t'\right) = -4\pi i\delta\left(t-t'\right)\left\langle\left[j_\alpha\left(\vec{r},t\right),\rho\left(\vec{r}',t'\right)\right]\right\rangle + 4\pi\frac{\partial}{\partial r'_\beta}G^{R}_{j_\alpha j_\beta}\left(\vec{r},\vec{r}';t-t'\right)$$

The commutator in the first component from right side of this equation is constant. For definition of its value we present the Fourier coefficients in terms of t, $\vec{r}$ and $\vec{r}'$.

$$-i\omega k'^2 G^{R}_{j_\alpha\phi}\left(\vec{k},\vec{k}';\omega\right) = -4\pi i\left\langle\left[j_\alpha\left(\vec{k}\right),\rho^*\left(\vec{k}'\right)\right]\right\rangle - 4\pi i k'_\beta G^{R}_{j_\alpha j_\beta}\left(\vec{k},\vec{k}';\omega\right).$$

At $\omega=0$ and using a fact that GF hasn't poles for $\omega=0$, we get

$$\left\langle\left[j_\alpha\left(\vec{k}\right),\rho^*\left(\vec{k}'\right)\right]\right\rangle = -k'_\beta G^{R}_{j_\alpha j_\beta}\left(\vec{k},\vec{k}';\omega=0\right)$$

and

$$G^{R}_{j_\alpha\phi}\left(\vec{k},\vec{k}';\omega\right) = \frac{4\pi}{\omega k'^2}k'_\beta G^{R}_{j_\alpha j_\beta}\left(\vec{k},\vec{k}';\omega\right) - 4\pi k'_\beta G^{R}_{j_\alpha j_\beta}\left(\vec{k},\vec{k}';\omega=0\right).$$

The equations for FG $G^R_A$, and $G^R$ may be obtained analogically

$$G^R_{A_\alpha\phi}\left(\vec{k},\vec{k}';\omega\right)=\frac{4\pi}{\omega k'^2}G^R_{A_\alpha j_\beta}\left(\vec{k},\vec{k}';\omega\right)-4\pi k'_\beta G^R_{j_\alpha j_\beta}\left(\vec{k},\vec{k}';\omega=0\right)$$

$$G^R_{\phi A_\alpha}\left(\vec{k},\vec{k}';\omega\right)=-\frac{4\pi}{\omega k'^2}k_\beta G^R_{j_\beta A_\alpha}\left(\vec{k},\vec{k}';\omega\right)+4\pi k_\beta G^R_{j_\beta A\alpha}\left(\vec{k},\vec{k}';\omega=0\right)$$

$$G^R_{\phi\phi}\left(\vec{k},\vec{k}';\omega\right)=\frac{(4\pi)^2 k_\alpha k_\beta}{\omega^2 k^2 k'^2}\left[\,G^R_{j_\alpha j_\beta}\left(\vec{k},\vec{k}';\omega\right)-G^R_{j_\alpha j_\beta}\left(\vec{k},\vec{k}';\omega=0\right)-\omega\left[\frac{\partial}{\partial\omega}G^R_{j_\alpha j_\beta}\left(\vec{k},\vec{k}';\omega\right)\Big|_{\omega=0}\right]\right]$$

The obtained equations express the complex permeability of the polar system by means of FG operators of current density of Coulomb medium $g_{jj}$. This dependence usually has too cumbersome character. Very simple results we get for homogeneous medium for which

$$G^E\left(\vec{k},\vec{k}';\omega\right)=\left\{i\omega D^R\left(\vec{k},\omega\right)\left[\delta_{\alpha\beta}-\frac{k_\alpha k_\beta}{k^2}\right]+i\frac{k^2}{\omega}G^R_{\phi\phi}\left(\vec{k},\omega\right)\frac{k_\alpha k_\beta}{k^2}\right\}\delta_{\vec{k},\vec{k}'}\,,$$

$$G^P_{\alpha\beta}\left(\vec{k},\vec{k}';\omega\right)=\left\{\frac{i}{\omega}G^{RT}_{jA}\left(\vec{k},\omega\right)\left[\delta_{\alpha\beta}-\frac{k_\alpha k_\beta}{k^2}\right]-i\frac{k^2}{\omega^2}G^{RL}_{j\phi}\left(\vec{k},\omega\right)\frac{k_\alpha k_\beta}{k^2}\right\}\delta_{\vec{k},\vec{k}'}\,,$$

where GF $D^R$ and $G^R$ depend on k.

For homogeneous medium it may be obtained

$$\chi_{\alpha\beta}\left(\vec{k},\vec{k}';\omega\right)=\left\{\chi^T\left(\vec{k},\omega\right)\left[\delta_{\alpha\beta}-\frac{k_\alpha k_\beta}{k^2}\right]-\chi^L\left(\vec{k},\omega\right)\frac{k_\alpha k_\beta}{k^2}\right\}\delta_{\vec{k},\vec{k}'}\,, \tag{7}$$

where

$$\chi^T(k,\omega)=\frac{1}{\omega^2}G^{RT}_{jA}(k,\omega)\left[D^R(k,\omega)\right]^{-1}, \tag{8}$$

$$\chi^L(k,\omega)=\frac{1}{k}G^{RL}_{j\phi}(k,\omega)\left[G^R_{\phi\phi}(k,\omega)\right]^{-1} \tag{9}$$

These expressions (7), (8) and (9) for components of complex permeability $\chi^T$ may be simplified by means of presented above expressions. Moreover, eq.(9) may be obtained only for temperature GF. The expression for $\chi^T$ may be obtained only in the point $i|\omega_n|$, where $\omega_n = 2\pi n k_\beta T$ and n = 0, +1, 2, ...

$$\chi^{T}\left(k, i\left|\omega_{n}\right|\right)=-\frac{1}{\omega_{n}^{2}} g_{jj}^{T}\left(k, \omega_{n}\right)$$

For GF operators of the vector-potential $D^{TT}(k,\omega_n)$ we therefore have

$$D_{\alpha\beta}^{TT}\left(\vec{k}, \omega_{n}\right)=\frac{4\pi\left[\delta_{\alpha\beta}-\frac{k_{\alpha}k_{\beta}}{k^{2}}\right]}{k^{2}/\varepsilon_{\infty}+\omega_{n}^{2}\varepsilon^{T}\left(\vec{k}, i\left|\omega_{n}\right|\right)},$$

where $\varepsilon^{T}$ is a transverse component of microscopic dielectric permittivity, that is

$$\varepsilon^{T}=1+4\pi\chi^{T}.$$

If GF $g_{jj}^{T}(k,\omega_n)$ will be calculated in the frames of some model for Coulomb medium than by means of analytical form of this function $\omega_n$ it isn't difficult the calculation of corresponding dependence of time retarded GF $g_{jj}^{RT}(\omega)$ on the frequency $\omega$ and, consequently, similar dependence on $\varepsilon^{T}(\omega)$ and $D^{TT}(\omega)$. Taking into account that for k – 0 a longitudinal component of the complex permeability $\chi^{L}$ coincides with transverse component $\chi^{L}$ and by such way we get the function $\chi(k=0;\omega)$ for long-wave limit.

The results obtained above present the basis for analysis of spreading of the electromagnetic waves in the rather general form.

The computational technique and dependences presented in this part, give the possibility to consider the polymer systems in pure quantum-mechanical style, to calculate all possible GF and complex dielectrical permittivity of the polymer systems and describe the charge transfer processes in them.

## 1.4.Modeling of Green Function for Describing of the Effects of the Spread Dispersion of Irregular Condensed System

The effects of the frequency dispersion of the medium may be taken into account using different model functions for GF $g_{ik}(\vec{r},\vec{r}')$. The model function may be selected by some general considerations. For example, for GF operators of the system scalar potential it is known that for their definition it may be used the results of solution of electrostatic problems, with accuracy in one electron charge GF $g_{\varphi\varphi}(\vec{r},\vec{r}')$ coincides with amount of electrostatic potential in the point $\vec{r}$, if the point charge will be placed in point $\vec{r}'$. In this case the behavior of GF $g_{\varphi\varphi}(\vec{r},\vec{r}')$ must be of type:

$$g_{\varphi\varphi}(r,r') = C(r)\frac{1}{|r-r'|}.$$

Here the behavior of FG $g_{\varphi\rho}$ and $g_{\rho\varphi}$ will be of type:

$$g_{\varphi\rho}(\vec{r},\vec{r}\,') = g_{\rho\varphi}(\vec{r},\vec{r}\,') = -f(\vec{r})\delta(\vec{r}-\vec{r}\,')$$

The Fourier-component of function $f(\vec{r})$ maybe connected with longitudinal component of dielectric penetration $\varepsilon^l(k)$ of medium. After standard manipulation we get:

$$f(\vec{r}) = \frac{1}{(2\pi)^{3/2}}\int e^{i\vec{k}\vec{r}} f(\vec{k})$$

and

$$f(\vec{k}) = \frac{1}{4\pi}\left(1 - \frac{1}{\varepsilon^l(k)}\right)$$

For Fourier-component we have correlation:

$$g_{\varphi\rho}(\vec{k},\vec{k}\,') = -\frac{1}{4\pi}\left(1 - \frac{1}{\varepsilon^l(k)}\right)\delta_{\vec{k}\vec{l}\,'}$$

In the case of local homogeneous medium from the above formula we get:

$$g_{\varphi\rho}(\vec{k},\vec{k}\,') = -\frac{1}{4\pi}\left(1 - \frac{1}{\varepsilon^l(k)}\right)\delta_{\vec{k}\vec{l}\,'} = -\frac{C_0}{4\pi}\delta_{\vec{k},k'}$$

In the space of $\vec{r}$ the last formula has a view:

$$G_{\varphi\rho}(\vec{r},\vec{r}\,') = -\frac{1}{4\pi}C_0\delta(\vec{r}-\vec{r}\,') \qquad (10)$$

For account of the effects of the spatial dispersion $\delta$- function in the formula (10) may be “smeared”.

For it let us introduce the function

$$g_{\varphi\rho}^{(\lambda)}(\vec{r},\vec{r}\,') = -\frac{C_0}{4\pi}\Delta_\lambda\left(|\vec{r}-\vec{r}\,'|\right).$$

As $\Delta_\lambda(|\vec{r}-\vec{r}'|)$ function we select the exponentially decreasing function normalized to unity of view

$$\Delta_\lambda(|\vec{r}-\vec{r}'|)=\frac{1}{8\pi\lambda^3}e^{-\frac{|\vec{r}-\vec{r}'|}{\lambda}}$$

The application of this model function is useful for systems, in which the medium molecules have more or less round form and at calculation of the kinetic parameters it is enough to take into account mainly the nearest to conducting particle surrounding.

Analogically it may be considered the model of system in which the effects of the spatial dispersion of medium is described by more complex function, for example, the damped oscillations:

$$\Delta_\lambda(|\vec{r}-\vec{r}'|)=\frac{\nu(\nu^2-3\mu^2)}{8\pi}e^{-\nu|\vec{r}-\vec{r}'|}\cos(\mu|\vec{r}-\vec{r}'|).$$

Here the parameter $\frac{1}{\mu}$ is approximately equals to the molecular diameter of medium. The using of this model function is applicable to systems, in which medium molecules have approximately round form and at calculations of the kinetic parameters it is necessary to take into account many layers of molecules from surrounding of an electrical conducting particle. For describing of the effects of spatial dispersion of medium it may be used the function of type:

$$\Delta_\lambda(|\vec{r}-\vec{r}'|)=\frac{1}{\pi^{3/2}\lambda^3}\exp\left(-\frac{|\vec{r}-\vec{r}'|^2}{\lambda^2}\right)$$

The application of this model function is applicable for systems having large size admixture molecules (conducting particles) and medium molecules having nearly round forms and at calculation of the kinetic parameters it is enough to take into account mainly the nearest surrounding to admixture particles.

As a function describing the effects of spatial dispersion of medium $\Delta_\lambda$ it may be selected also a stepwise function:

$$\Delta_\lambda(|\vec{r}-\vec{r}'|)=\frac{3}{4\pi\lambda^3}\theta(|\vec{r}-\vec{r}'|)\theta(\mu-(|\vec{r}-\vec{r}'|))$$

The using of last model function is applicable for systems, in which the medium is considered as a dielectrical background.

### 1.5. Modeling of Green Functions for Describing of Effects of Frequency Dispersion Irregular Condensed System

For showing of the possibility of taking into account of the particles frequency spectrum in the medium we will use an poles approximation, which includes Debay and resonance function, for Green function of the operators of medium polarization fluctuation. In this case we have

$$U(\omega_n) = \sum_{i=1}^{m} \frac{u_i}{|\omega_i - \omega_n|}, \ \omega_n = 2\pi nkT, n = 0, \pm 1 \pm 2 \pm 3...$$

where $u_i$ – experimental constants, m- the number of Green function poles.

## 2. Processes of the Charge Transfer from Metal or Semiconductor Electrode to the Admixed Particle in the Polymer

At creation of the polymer conducting system some problems is arisen, which are connected with peculiarities of using of obtained material. First of all the working electrode may be both metal and semiconductor. In both cases it is possible the correct description of the system for caring out of analytical calculations of kinetic parameters of charge transfer process.

The next question – the material of conducting particle, its geometric form, distance between particles and orientation of particle relatively to the electrode.

It is essential also the possibility of adsorption of polymer molecules on the surfaces of electrically conducting metal particle and electrode, the “solvatation” effects of conducting particle and electrode by polymer molecules. Similar details, which are the characteristics of the system are mainly defined by method of preparing of the conducting polymer material, properties of the components used at production of conducting material (composite). In dependence on an obtained system it must be used any model for calculation of kinetic parameters. For example, if the polymer molecule is adsorbed chemically on the surface of the conducting particle, then instead of electron transfer process between electrical conducting particles located in the polymer matrix it must be considered the process of charge transfer with participation of adsorbed particles. In this case the adsorbed particle become the essential element of the process. In all kinetic parameters it will be participated intramolecular reorganization of adsorbed particles. All analytical calculations are essentially complicated however their conducting in a rather correct form is possible.

Our calculations will be carried out for metal particles and consequently the “solvatation” effects may be neglected.

## 2.1. The System Hamiltonian

The Hamiltonian of initial state we present in the view:

$$H^i = H_e^i + H_m^i + H_p^i + H_{p,m}^{\text{int}} + H_{e,m}^{\text{int}} + H_{e,p}^{\text{int}} ,$$

Where $H_e^i$ is the Hamiltonian of semiconductor or metal electrode in the initial state; $H_p^i$ is the Hamiltonian of particle in the initial state; $H_m^i$ is the Hamiltonian of medium; $H_{e,m}^{\text{int}}$ - interaction of the electrode with medium; $H_{p,m}^{\text{int}}$ - interaction of the particle with medium; $H_{e,p}^{\text{int}}$ -Hamiltonian of the interaction of the particle with an electrode. The particle here may be both polarized and "solvated".

Analogically Hamiltonian in the final state has the view:

$$H^f = H_e^f + H_m^f + H_p^{f*} + H_{p,m}^{\text{int}} + H_{e,m}^{\text{int}} + H_{e,p}^{\text{int}} + e\eta ,$$

Where η – over-voltage in the point of particle finding.

For semiconductive electrodes the distribution of electrons by energy levels of electrode is described by Fermi distribution

$$f(\varepsilon) = (1 + \exp((E - E_F)/kT)^{-1} ,$$

where $E_F$ – Fermi energy, k – Boltzmann constant, T – temperature.

However if the energy of the level essentially differs from the Fermi level energy the distribution of energetic levels transfers to the Boltzmann distribution. Besides this it is necessary to solve one more problem for semiconductive electrode to find out the number of quantum states in the volume unit of the substance $\rho(\varepsilon)$. The conducting of exact calculation for such magnitudes is difficult, however some estimation may be made for the crystals.

## 2.2. Current Density for the Process of Charge-transfer from Metal or Semiconductor Electrode to the Conducting Particles in the Polymer Matrix

The calculation of the current density connected with process of charge-transfer from electrode to particle may be carried out on the basis of quantum-mechanical calculations for process of transition between two electron states.

The density of cathode current may be presented as

$$i_c = e\frac{N_0}{S}\int d\varepsilon \rho(\varepsilon)\rho(\varepsilon_1) n_F(\varepsilon) n_F(\varepsilon_1) W_c(\varepsilon, \varepsilon_1) ,$$

where e – the electron charge, S – electrode surface, $N_0$ – number of particles in the system volume of particles $V_0$, $\rho(\varepsilon)$ - the density of single electron states in the electrode, $\rho(\varepsilon_1)$ - the density of single electron states in the particle, $n_F(\varepsilon)$ - Fermi function of the electron distribution , $W_c(\varepsilon,\varepsilon_1)$ - probability of the electron transfer from given energetic level of the electrode ε on the given energetic level of particle $\varepsilon_1$.

The electron transfer may be both electron non-adiabatic and electron adiabatic. Usually in the wide range of the parameters the process of charge transfer of the electron is electron-nonadiabatic transfer and below first of all we will consider namely electron-nonadiabatic transfer of the electron.

At integration on the energy it must be noted that usually the contribution to the integral on energies posses small range of energies and formal integration on energies may be carried out in the infinite limits.

The probability of the electron-nonadiabatic transfer of the electron is expressed as

$$W_c(\varepsilon,\varepsilon_1)=\frac{\beta}{i}\exp(\beta F_i)\int d\theta Sp\left[\exp((-\beta(1-\theta)H^i)L\exp(-\beta\theta H^f)L\ \right]$$

Omitting the cumbersome calculations we present the quantum expression for density of cathode current of heterogeneous process with participation of metallic or semiconductor electrode with Fermi distribution

$$i_c=e\pi\left|L_{fi}(\vec{R}^*,\psi^*)\right|^2\int d\varepsilon d\varepsilon' kT\rho(\varepsilon)\rho(\varepsilon')\exp(-2\ln(\sin\pi\theta^*))\Phi(\vec{R}^*,\psi^*)U(\vec{R}^*,\Psi^*)$$
$$\exp\left\{-\beta\theta^* e\eta-\beta\theta^*\Delta F-\Psi^m(\vec{R}^*,\psi^*;\theta)\ \right\},\beta=1/kT \tag{11}$$

ΔF Here is free energy of the process. The star in the designation marks means the value of this coordinate in the point of maximum at calculation of corresponding integration by given coordinate, as a rule, by the saddle-point method.

The saddle-point $\theta^*$ maybe found from equation:

$$e\eta+\beta\Delta F+\frac{\partial\psi^m(R^*,\psi^*;\theta)}{\partial\theta}+2\pi ctg(\pi\theta)=0,$$

In these formulae $L_{fi}$ is the resonance integral from interaction of particle with surface of semiconductor or metal electrode. Matrix element is calculated by using of wave functions in the frames of concrete model for particle. The resonance integral $L_{fi}$ may be considered as some phenomenological parameter. Arguments of this resonance integral characterize the geometrical characteristic of process, distance to surface (R), spatial orientation of particle (Ψ) at transfer of charge. Function $U(R^*,\psi^*)$ is calculated for concrete processes allowing for the geometry of electrode and particles. Function $\Phi(\vec{R}^*,\psi^*)$ presents the function of distribution of admixture particles. This function may be model one, connected with particles concentration. Function $\Psi^{m}(\vec{R}^*,\psi^*;\theta)$ is one of reorganization of the polymer medium. Its formal expression has the view:

$$\Psi^{m}(\vec{R}^*,\psi^*,\theta)=\frac{1}{\pi}\int d\vec{r}\,d\vec{r}'\Delta E_i(\vec{r},\vec{R}^*,\psi^*)\Delta E_k(\vec{r}',\vec{R}^*,\psi^*)$$

$$\int_{-\infty}^{\infty}d\omega\,\mathrm{Im}\,g_{ik}^{R}(\vec{r},\vec{r}';\omega)\frac{sh\frac{\beta\omega(1-\theta)}{2}sh\frac{\beta\omega\theta}{2}}{\omega^2 sh\frac{\beta\omega}{2}},$$

Where ΔE is the change of the electrical field voltage of admixture particle and electrode at process of charge transfer.

This function in general describes both the processes of tunnelling of an electron and classic reorganization of medium at charge transfer.

Let us introduce the energy of reorganization of the polymer medium by means of relation

$$E_r^m(\vec{R},\psi)=-\frac{1}{2}\int d\vec{r}\,d\vec{r}'\Delta E_i(\vec{r};\vec{R},\psi)g^{R}_{ik}(\vec{r},\vec{r}';\omega=0)\Delta E_k(\vec{r}';\vec{R},\psi)$$

In the factorization approximation for function $g^R$ over spatial and time coordinates the medium reorganization function may be presented in view:

$$\Psi^{m}(\vec{R}^*,\psi^*,\theta)=E_r^m\frac{2}{\hbar}\int_{-\infty}^{\infty}d\omega f(\omega)\frac{sh\frac{\beta\omega(1-\theta)}{2}sh\frac{\beta\omega\theta}{2}}{\omega^2 sh\frac{\beta\omega}{2}}$$

At integration over r and r′ it is necessary to take into account the structure of medium and the situation, when both effects of spatial dispersion of medium (function g(r, r′)) and the effects of it frequency dispersion ( function f(ω)) will be described by different model functions allowing for existing of definite modes of polymer polarization.

The carried out calculations show that complete fulfilling of the analytical calculations is impossible and it is necessary to carry out of the numerical integration.

The activation energy may be defined by formula:

$$E_a=-2\ln(\sin\pi\theta^*)+\theta^*(1-\theta^*)E_r^m-\theta^* e\eta-\Delta F\theta^*$$

The above presented in section II relations are valid also for processes of the charge transfer of the electron from particle to electrode at corresponding change of marks of energetic parameters.

The calculations of the kinetic parameters for processes of electron-adiabatic processes of electron may be considered in detail analogically with nonadiabatic processes. However it may be used the another method of estimation of the parameters for adiabatic processes.

The probability of transfer may be presented in view:

$$W_{ad} = A_{ad} \exp(-\beta E_a)$$

In this formula the pre-exponent multiplayer may be obtained from analogical expression for nonadiabatic process by substituting of electron resonance integral on the critical value by formula

$$L \to L_c / \sqrt{2}$$

Where critical value for single-frequency model of medium (with frequency $\omega_m$) has a view

$$L_c = \left(\frac{kTE_r^m \omega_m^2}{\pi^3}\right)^{1/4}$$

The activation energy of the process is defined by the way in which the adiabatic term is designed from potential energies of the channel terms of initial and final states $U_i$ and $U_f$.

$$U = 0.5(U_i + U_f) - (0.25(U_i - U_f)^2 + L^2)^{0.5}$$

Further on the points of minimum and the maximum of the system energy and the activation energy is defined as difference between values corresponding to maximum and minimum at initial state.

## 3. Processes of Charge Transfer between Admixture Particles in the Polymer

The processes of charge transfer in the conducting polymer systems are defined by a number of factors. First of all there are the geometrical parameters of conducting particles, their shape, mutual orientation of interacted particles, possibility of adsorption of polymer molecules on the surface of conducting particles, effects of "solvatation" of conducting particle with polymer molecules. Similar details of such systems are mainly defined by method of preparation of conducting polymer material, properties of components. In dependence on peculiarities of structure of composite material it is necessary to use one or another model for calculation of kinetic parameters. So, if the polymer molecule is chemically adsorbed on the surface of conducting particle actually instead of the charge transfer process between metal particles in the polymer matrix it is necessary to consider the process with participation of adsorbed particles. So, the adsorbed particle becomes the essential element of the process.

*In all kinetic parameters it will be presented intramolecular reorganization of adsorbed particles. All calculations will be essentially complicated, but execution of them in sufficiently correct form is possible.*

One of the serious problem at investigation of such systems – is the distribution of electromagnetic field near particle surfaces. This distribution often plays the defining role at calculation of kinetic parameters. The picture of distribution of the field essentially depends on "solvating" capability of the polymer molecules. For solvated particles also it is necessary the using of the model of particles with oscillation subsystem in which not at all cases it is possible to describe what namely chemical bonds participate in the oscillation, however it can be foreseen some generalized oscillated modes. There are possibility of analytical calculations, although in this case the definite number of parameters are arisen, which are the characteristics of the system and numerical estimation of which are very approximate.

## 3.1. System Hamiltonian

The process of charge transfer between admixture particles in the polymer is considered as process of transfer between two metal electrodes in the condensed medium. The model for such processes may be presented on the basis of above conducted calculations for processes of transfer between electrodes and particles. For conducting of calculations it is necessary to detail of the system Hamiltonian. It may be used the same model approximations for describing of particles (1 and 2) and medium, which were used in the last section. In result for initial state the Hamiltonian one get the view:

$$H^i = H^i_{p1} + H^i_m + H^i_{p2} + H^{\text{int}}_{p1,m} + H^{\text{int}}_{p2,m} + H^{\text{int}}_{p1,p2}$$

Analogical view has the Hamiltonian of the final state.

## 3.2. The Rate Factor of Charge Transfer Process between Conducting Particles

The current density for the process of electron transfer from particle 1 to particle 2 in the polymer matrix has a few

$$i_c = e\pi \left| V_{fi}(\vec{R}^*, \psi^*) \right|^2 \int d\varepsilon d\varepsilon' kT \rho_1(\varepsilon) \rho_2(\varepsilon') \exp(-2\ln(\sin \pi\theta^*)) \Phi(\vec{R}^*, \psi^*) A(\vec{R}^*, \Psi^*)$$
$$\exp\left\{-\beta\theta^* e\eta_{12} - \beta\theta^* \Delta F_{12} - \Psi_{12}{}^m(\vec{R}^*, \psi^*; \theta)\right\}$$

Here the parameters have sense analogical to formula (11) and critical point for $\theta^*$ is defined from equation:

$$e\eta_{12} + \beta\Delta F_{12} + \frac{\partial\psi^m(R^*,\psi^*;\theta)}{\partial\theta} + 2\pi ctg(\pi\theta) = 0,$$

The preexponent multiplayer is calculated for concrete particles allowing for the particles geometry, space orientation and the distance of electron transfer.

For activation energy of transfer process we have:

$$E_a = -2\ln(\sin\pi\theta^*) + \theta^*(1-\theta^*)E_r^m - \theta^* e\eta_{12} - \Delta F_{12}\theta^*$$

The presented expressions allow one to conduct simple estimations of process kinetic parameters in frames of strongly simplified models. In this way one could calculated the electron kinetic parameters using of average magnitudes of admixture concentrations and, consequently, to get average magnitudes of parameters.

At estimation of kinetic parameters it must be foreseen that direct charge transfer takes place really in the limits of 30 Å, and that because of electron tunnelling. The charge transfer through intermediate state, naturally, is possible, but the probability of the final process will be equal to the product of probabilities of all elementary acts of electron transfer.

## 4. Scheme for Carrying out of Quantitative Estimation for the Processes

For estimation of kinetic parameter values of different processes it is below presented the schemes of calculations:

1. For calculation of the kinetic parameters first of all it is necessary to make more exact, namely what particles react and what is the reaction medium, what electrodes are used at real measurements. It is necessary also to have maximal information about electrode, its zone structure, a bending of zones near the surface, etc.
2. To introduce the degrees of freedom of the reaction and divide them on classic and quantum for medium. Classical degrees of freedom may be ones, which satisfy the condition

   $$th\left(\frac{\omega_i(1-\theta^*)}{2kT}\right) \approx \frac{\omega_i(1-\theta^*)}{2kT} << 1, \ th\frac{\omega_f\theta^*}{2kT} \approx \frac{\omega_f\theta^*}{2kT} << 1$$

   At fulfilling of preliminary calculations the number 0.5 may be taken as value of $\theta^*$.
3. To fulfil the estimation of the medium reorganization energy by simple formulas, e.g. [4]. However it is necessary to take into account here, that in the dependence of medium only definite modes of polarization are reorganized and, consequently, instead of static dielectric penetration it must be used the values of the dielectric permittivity from left and right sides of corresponding pick of absorption. The allowing for the effects of frequency and spatial dispersion may be conducted lately

at exact calculations. After preliminary estimation for final calculations it must be selected a model for describing of spatial dispersion of medium.

4. To define or estimate the electron resonance integral or dipole moment of transition using quantum-chemical methods, sums law, fulfilled calculations for similar systems. It must be foreseen that quantum-chemical calculations allow one to conduct calculations either in vacuum or in medium with static-dielectric background, but not in medium with complex structure. Therefore, at final calculations of the kinetic parameters of processes it is often necessary to introduce the corresponding corrections in the results, e.g. in the characteristic value of decay of the resonance integral.
5. To define the character of charge transfer process if it is electron-nonadiabatic or adiabatic. For electron-nonadiabatic process it must be used above given methods and for adiabatic ones – the transfer rule, in which is the method of calculation of pre-exponent, rate factor and definition of activation energy of the process after calculation of transition configuration and the finding of the coordinates of minimum of initial system state and maximum in the transition state. The calculation of the Landau-Zener parameter may be carried out by using of formula:

   $$\gamma_e = 2\pi L^2 /(|v||\Delta U)|$$

   Where $v$ is the rate of movement over the reaction coordinate of the system near transition configuration, $\Delta U$ is the difference of inclinations of terms of potential energy near point of their crossing. If the condition $\gamma_e >> 1$ is fulfilled then the process has an electron-adiabatic character. At calculation of the kinetic parameters it must be used the adiabatic terms [23]. The transmission factor in this case is $\kappa \approx 1$. At fulfilling of reciprocal condition the process has electron-nonadiabatic character.
6. To define of transition configuration. The definition of transition configuration over spatial and rotational coordinates of reaction is more complex task, there are no general methods. It is necessary to have information about distribution of the reacting particles in the medium for using of right function for their distribution. The information is necessary about possible mutual orientation of reacted particle. At writing of the potentials of reagent interactions all this information is necessary for introduction of reaction coordinates and calculation of transition configuration by them.
7. To calculate the transition coefficient for heterogeneous reaction $\theta^*$.
8. To estimate of the activation energy for classic degree of freedoms and tunnelling factor – for quantum ones.
9. After conducting of preliminary calculations to make more exact the models of processes for each stage of electron transfer process, to co-ordinate the models of separate stages and fulfil of the exact calculations on the full scheme for all system.

As it was above noted, experimental data treatment requires the exacting of parameters entered to the results of analytical calculations obtained in the frames of theoretical models. However, some conclusions may be made on the basis of semi-classical estimations of the parameters. We will provide estimations for metallic electrode and electrical conducting particles.

One of the defining factors at calculation of the process rate constant is the distance of transition. In the same time this parameter defines often the character of process if it is adiabatic or non-adiabatic. It may be shown qualitatively that if the character of the charge transfer process is adiabatic the rate constant weakly depends on transfer distance because the main dependence in the process rate constant on the distance is presented through electron resonance integral. In the same time, for adiabatic processes it is necessary to change the electron resonance integral on its critical value, which is not presented by the function of the transfer distance. The calculations in different systems show that in the condensed medium the electron transfer process has adiabatic character only on rather small distances. Consequently, on small distances the rate constant of the process will not be dependent on distance. Moreover, the value of rate constant will mainly be dependent on the medium properties and for very rough model of homogeneous, isotropic local medium in the frames of single mode model it will be presented by frequency of this mode $\omega_m$ and by energy of the reorganization of medium $E_r^m$ and the square of the electron resonance integral must be changed on the following value:

$$\frac{(kTE_r^m)^{1/2}}{2\pi^{3/2}}\omega_m .$$

In the last expression the energy of medium reorganization depends on the electron transfer distance, but this dependence has relatively week character. At variation of distance of transfer on 60-70% the reorganization energy is changed on about 25-40%, in dependence on the model.

The electron resonance integral for adiabatic processes of the electron transfer is exponentially decreasing function of the distance.

$$L(R) = L_0 \exp(-R/\Delta),$$

For many processes of nonadiabatic electron transfer the characteristic distance of the decay of electron resonance electron $\Delta$ (it is proposed that the distance R is measured in angstroms).

Qualitatively the current density in the polymer system with electrical conducting filler is the function of distance between conducting particles and graphically corresponding dependence has the graphical view presented in Figure 1 ($\Delta = 0.4$ Å).

For comparison of the theoretically calculated dependences of electrical current density on the average distance between conductive particles in the polymer matrix with experimentally obtained results for three polymer composites, containing the nanopowders of nickel and silver, all curves are presented on the one diagram (Figure 1), showing that characters of these curves are qualitatively the same. The deviation between numerical values of the theoretical and experimental data is defined by inhomogeneous distribution of conducting particles in the polymer matrix.

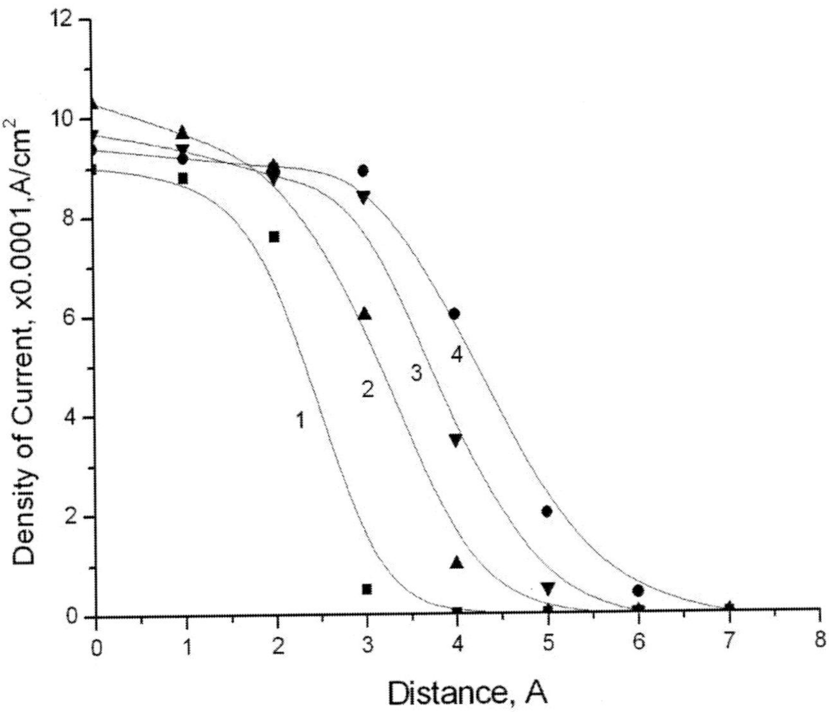


Figure 1. Dependence of density of current on the distance between conducting particles in is ECPM. 1-theoretical calculated ; 2 – ECPM based on polydimethylvinilsiloxan with silver nanoparticles (50-100 nm); 3 – ECPM based on polydimethylvinilsiloxan with silver nanoparticles (200-250nm); 4 – ECPM based on polyvinilalcohol with nickel nanoparticles.

## CONCLUSION

Comparison of theoretically calculated values of the density of electrical current in dependence on the average distance between conducting particles with conductivity of the real systems shows that the considered approach to the analysis of processes of the charge transfer in electrical conducting polymer composites describes satisfactorily the conducting properties of these materials.

## REFERENCES


[1] R.N. Norman. *Conductive Rubber and Plastics,* Amsterdam, Elsevier, (1970).

[2] A. Donnet, A. Voet: *Carbon Black, Marcell Decker,* New-York-Basel, (1976).

[3] V.E. Gul', L.Z. Shenfil: *Electrically Conductive Polymeric Composites*, Moscow, Khimia, 240 (1984) (in Russian).

[4] *Carbon Black Polymer Composites: The Physics of Electrically Conducting Composites,* Marcell Decker, New-York, (1982).

[5] J.N.Aneli, L.M.Khananashvili, G.E.Zaikov, Structuring and Conductivity of Polymer Composites, *Novo-Sci. Publ.,* New-York, 1998.

[6] K.A. Pechkovskaya. Carbon Black as Rubber Reinforcement, Moscow, *Khimia,* 304 (1968) (in Russian).

[7] B.I. Shklovski, A.L. Efros. Electronic Properties of Alloyed Semiconductors, Moscow, *Nauka*, 416 (1979) (in Russian).

[8] G. Clero, T. Giroult, I.Russank. *Acad. Sci. Comptes Rendus, Ser.B,* 281 (13), 227 (1975).

[9] Z.Petrovich, B.Martinovich,V.Diviankovich, *J.Budinski-Simenovich, J.Appl.Pol. Sci.* 49, 1659 (1993)

[10] L.Benguigu, J.Jakubovich, M.Narkis, *J.Pol.Sci., Part B: Polymer Physics,* 25,127(1987).

[11] F.F. Koshelev, A.E. Kornev, E.M. Spiridonova. *Electrically Conductive Polymeric Materials,* Moscow, CBTI, 61 (1961) (in Russian).

[12] L.K. Van Beek, B.I. *Van Pul. Carbon,* 2, 121 (1964).

[13] G.Beaucage, S.Rane, D.W.Shaffer, G.Long, D.Fisher, *J.Pol.Sci.,Part* B,37,1105 (1999).

[14] B. Lee: *Polymer Eng.Sci.,* 32(1), 36 (1992).

[15] W.Bridge. B, M.J. Folkes, B.R. Wood: *J.Phys.D.,* 23(7), 890(1990).

[16] A.K.Zainutdinov, A.A.Kasimov, M.A.Magrupov. *Pisma v zhurnal eqsperimentalnoi i tekhnicheskoi fiziki,* 18(2),29(1992) (in Russian).

[17] B. Wessling. *Macromol.Chem.,* 185, 1265 (1984).

[18] B.I. Shklovski A.L. Efros. *Uspekhi Fizicheskikh Nauk,* 117(3), 401 (1975) (in Russian).

[19] N.S. Enikolopian, S.G. Gruzova, N.M. Galashina, E.N. Sklyarov, L.N. Grigorov: *Dokl. AN USSR*, 274(6), 1404 (1984) (in Russian).

[20] L.N. G rigorov, N.M.Galashina, N.S. *Enikolopyan: Dokl. AN USSR,* 274 (4), 840 (1984) (in Russian).

[21] A.A.Abrikosov, L.P.Gorkov, I.E.Dzialoshinskii. *Methods of quantum theory of field in the statistic physics.*Moscow, Physmatgiz,1962, 444(in Russian).p.

[22] F.Platsman, P.Wolf. The waves and interactions in the plasma of solid state.Moscow, *Mir,*1975, 435 p.(in Russian).

[23] R.R.Dogonadze, T.A.Marsagishvili. *The chemical physics of solvation.* Part A.Amsterdam,Elsevier,1985, p.39.

[24] R.R.Dogonadze, A.M.Kuznetsov, T.A.Marsagishvili. Present state of the theory of charge transfer processes in condensed phase. *Electrochimica Acta,*1980, V.25,N1, p.1-28.

In: Chemical and Biochemical Reactions
Editors: G. E. Zaikov, L. Liu et al.
ISBN 978-1-61122-860-1


*Chapter 17*

# CALCULATIONS OF BOND ENERGY IN CLUSTER AQUEOUS NANOSTRUCTURES

**G. A. Korablev[1], N. V. Khokhriakov[1] and G. E. Zaikov[2]**
[1]Izhevsk State Agricultural Academy, Studencheskaya St. 11, Izhevsk, 426000, Russia
[2]N.M. Emanuel Institute of Biochemical Physics, Kosygina 4, Moscow, 119991, Russia

## ABSTRACT

The bond energy in some cluster aqueous structures has been calculated using spatial-energy notions and compared with quantum-mechanical methods. It is demonstrated that the variety and specifics of aqueous solutions are determined by the changes in structural energy characteristics of these cluster compounds.

**Keywords:** bond energy, cluster, nanostructures, water, quantum-mechanical methods

Based on modified Lagrangian equation for relative movement of two interacting material points the notion of spatial-energy parameter (P-parameter), which is a complex characteristic of important atomic values responsible for interatomic interactions and directly connected with electron density in atom [1].

The value of relative difference of P–parameters of interacting atoms – components of $\alpha$ -coefficient of structural interactions was used as the main quantitative characteristic of structural interactions in condensed media:

$$\alpha = \frac{P_1 - P_2}{(P_1 + P_2)/2} 100\% \quad (1)$$

The nomogram of dependence of structural interaction degree upon the coefficient $\alpha$ (the same for a wide range of structures) was obtained applying the reliable experimental data. This approach allowed evaluating the degree and direction of structural interactions of phase-formation, isomorphism and solubility in multiple systems, including the molecular ones. In particular, the peculiarities of cluster-formation in the system $CaSO_4 - H_2O$ were studied [2].

To evaluate the direction and degree of phase-formation processes the following equations were used [1]:

1) To calculate the initial values of P-parameters:

$$\frac{1}{q^2/r_i} + \frac{1}{W_i n_i} = \frac{1}{P_э}\ ; \quad \frac{1}{P_0} = \frac{1}{q^2} + \frac{1}{(Wrn)_i} \quad P_э = P_0 / r_i \qquad (2, 3, 4)$$

here: $W_i$ - orbital energy of electrons [3]; $r_i$ – orbital radius of i-orbital [4]; $q = Z^*/n^*$ - [5, 6]; $n_i$ – number of electrons of the given orbital, $Z^*$ and $n^*$ - nucleus effective charge and effective main quantum number. The value $P_0$ will be called spatial-energy parameter (SEP), and $P_E$ - effective P-parameter.

**Table 1. P-parameters of atoms calculated via the bond energy of electrons**

| Atom | Valent electrons | W (eV) | $r_i$ (Å) | $q^2$ (eVÅ) | $P_0$(eVÅ) | R (Å) | $P_E=P_0/R$ (eV) |
|---|---|---|---|---|---|---|---|
| H | $1S^1$ | 13.595 | 0.5295 | 14.394 | 4.7985 | 0.5295<br>0.28 | 9.0624<br>17.137 |
| C | $2P^1$ | 11.792 | 0.596 | 35.395 | 5.8680 | 0.77<br>0.69 | 7.6208<br>8.5043 |
| | $2P^2$ | 11.792 | 0.596 | 35.395 | 10.061 | 0.77 | 13.066 |
| | $2S^1$ | 19.201 | 0.620 | 37.240 | 9.0209 | 0.77 | 11.715 |
| | $2S^2$ | | | | 14.524 | 0.77 | 18.862 |
| | $2S^2+2P^2$ | | | | 24.585 | 0.77 | 31.929 |
| | $\frac{1}{2}(2S^2+2P^2)$ | | | | | | 15.964 |
| O | $2P^1$ | 17.195 | 0.4135 | 71.383 | 4.663 | 0.66 | 9.7979 |
| | $2P^2$ | 17.195 | 0.4135 | 71.383 | 11.858 | 0.66<br>0.59 | 17.967<br>20.048 |
| | $2P^4$ | 17.195 | 0.4135 | 71.383 | 20.338 | 0.66 | 30.815 |

**Table 2. Structural $P_S$-parameters**

| Radical, molecules | $P_1$ (eV) | $P_2$ (eV) | $P_3$ (eV) | $P_4$ (eV) | $P_S$ (eV) | Orbitals of oxygen atom |
|---|---|---|---|---|---|---|
| OH | 17.967 | 17.137 | | | 8.7712 | $2P^2$ |
| OH | 9.7979 | 9.0624 | | | 4.7080 | $2P^1$ |
| $H_2O$ | $2\times 17.138$<br>$2\times 9.0624$ | 17.967<br>17.967 | | | 11.788<br>9.0226 | $2P^2$<br>$2P^2$ |
| $C_2H_5OH$ | $2\times 15.964$ | $2\times 9.0624$ | 9.7979 | 9.0624 | 3.7622 | $2P^1$ |

The calculation results by equations [2,3,4] for some elements are given in Table 1, where we can see that for hydrogen atom the values of $P_E$–parameters considerably differ at the distances of orbital ($r_i$) and covalent radii (R). The hybridization of valent orbitals of carbon atom were evaluated as an averaged value of P-parameters of $2S^2$ and $2P^2$-orbitals.

2) To calculate the value of $P_S$-parameter in binary and complex structures:

$$\frac{1}{P_S} = \frac{1}{N_1 P_1} + \frac{1}{N_2 P_2} + \ldots\ldots.. \qquad (5)$$

where N – number of homogeneous atoms in each subsystem.

The results of these calculations for some systems are given in Table 2.

3) To determine the bond energy (E) in binary and more complicated structures:

$$\frac{1}{E} \approx \frac{1}{P_E} = \frac{1}{P_1 (N/K)_1} + \frac{1}{P_2 (N/K)_2} + \ldots\ldots\ldots. \qquad (6)$$

Here (as applied to cluster systems): $K_1$ and $K_2$ – number of subsystems forming the cluster system; $N_1$ and $N_2$ – number of homogeneous clusters [7].

Thus for $C_{60}(OH)_{10}$ $k_1 = 60$, $k_2 = 10$.

It is assumed that the stable aqueous cluster ($H_2O$) can have the same static number of subsystems (k) as the number of subsystems in the system interacting with it [8]. For example, aqueous cluster of N $(H_2O)_{10}$ type interacts with fullerene $[C_6OH]_{10.}$

Apparently, cluster $[(C_2H_5OH)_6 - H_2O]_{10}$ can be formed similarly to cluster $[C_6OH)10$, that assumes the structural interaction of subsystems $(C_2H_5OH)_{60}$ – $(H_2O)_{10.}$ And the interaction of aqueous clusters can be considered as the interaction of subsystems $(H_2O)_{60}$ – $N(H_2O)_{60}$.

Based on such notions and assumptions the bond energy in corresponding systems was calculated by the equation (6), the results are given in Table 3.

The calculation data obtained by N.V. Khokhriakov based on his quantum-mechanical technique [9] are given here for comparison.

Both techniques present comparable bond energy values (eV). Transfer multiplier: (1 $\frac{kcal}{mol}$ =0.04336 eV). Besides, the technique of P-parameter allows explaining why the energy value of cluster bonds of water molecule with fullerene $C_{60}(OH)_{10}$ two times exceeds the bond energy between the molecule in cluster water (Table 3).

In accordance with the nomogram the structure phase-formation can take place only with the relative difference of their P-parameters below 25%-30%, and the most stable structures are formed at $\alpha < (6\text{-}7)\%$.

Table 4 gives the values of coefficient $\alpha$ in systems H-C, H-OH and H-$H_2O$, which are within 0.44 – 7.09(%).

But in the system H-C for carbon and hydrogen atoms the interactions at the distances of covalent radii were considered, but for other systems – at the distances of orbital radius.

**Table 3. Calculations of bond energy – E (eV)**

| System | $C_{60}$ | $(OH)_{10}$ | $(H_2O)_{10}$ | | $P_E E$ (calculation) | |
|---|---|---|---|---|---|---|
| | $P_1/\kappa_1$ | $P_2/\kappa_2$ | $P_3/\kappa_3$ | $N_3$ | Equation (6) | Quantum-mechanical |
| $C_{60}(OH)_{10}$ – - $N(H_2O)_{10}$ | 15.964/60 | 8.7712/10 | 11.788/10 | 1 | 0.174 | 0.176 |
| | | | | 2 | 0.188 | 0.209 |
| | | | | 3 | 0.193 | 0.218 |
| | | | | 4 | 0.196 | 0.212 |
| | | | | 5 | 0.197 | 0.204 |
| $(H_2O)_{60}$– - $N(H_2O)_{60}$ | $P_1/\kappa_1$ | $P_2/\kappa_2$ | $N_2$ | | | |
| | 9.0226/60 | 9.0226/60 | 1 | | 0.0768 | 0.0863 |
| | | | 2 | | 0.1020 | 0.1032 |
| | | | 3 | | 0.1128 | 0.1101 |
| | | | 4 | | 0.1203 | 0.1110 |
| | | | 5 | | 0.1274 | 0.115 |
| $(C_2H_5OH)_{60}$ – - $(H_2O)_{10}$ | $P_1/\kappa_1$ | $P_2/\kappa_2$ | | | | |
| | 3.7622/60 | 9.0226/10 | | | 0.0586 | 0.0607 |
| $(C_2H_5OH)_{10}$ – - $(H_2O)_{60}$ | $P_1/\kappa_1$ | $P_2/\kappa_2$ | | | | |
| | 3.7622/10 | 9.0226/60 | | | 0.1074 | ≈ 0,116 |

Thus, the interaction in the system H-C at the distances of covalent radius plays a role of fermentative action, which results in the transition of dimensional characteristics of water molecules from orbital radius to the covalent one, i.e. to the formation of the system $C_{60}$ $(OH)_{10} - N(H_2O)_{10}$ with bond energy between the main components two times exceeding the one between water molecules themselves.

The broad possibilities of aqueous clusters in changing their spatial-energy characteristics apparently explain all other water properties with its different names: mineral, holly, live, spring, radioactive, etc.

**Table 4. Spatial-energy interactions in the system H-R, where R= C, (OH), $H_2O$**

| System | $P_1$(eV) | $P_2$(eV) | $\alpha = \frac{\Delta P}{<P>} 100\%$ | Type of spatial bond |
|---|---|---|---|---|
| H-C | 17.137 | 15.964 | 7.09 | Covalent |
| H-OH | 9.0624 | 8.7712 | 3.27 | Orbital |
| $H-H_2O$ | 9.0624 | 9.0226 | 0.44 | Orbital |

## CONCLUSIONS

Structural interactions in the bond H-C at the distances of covalent radius play the role of fermentative action, which results in the transition of dimensional characteristics of water molecules from orbital radius to covalent one, i.e. to the system: $C_{60}(OH)_{10}$ - $N(H_2O)_{10}$.

Broad possibilities of aqueous clusters in the change of their spatial-energy characteristics apparently explain all other unique properties of water with different names: mineral, holly, live, spring, radioactive, etc.

## REFERENCES

[1] Korablev G.A. *Spatial-Energy Principles of Complex Structures Formation,* Leiden, the Netherlands, Brill Academic Publishers and VSP, 2005, 426 pages (Monograph).

[2] Korablev G.A., Yakovlev G.I., Kodolov V.I. Some peculiarities of cluster-formation in the system $CaSO_4$-$H_2O$. *Chemical physics and mesoscopy,* vol. 4, №2, 2002, p.188-196.

[3] Fischer C.F. Average-Energy of Configuration Hartree-Fock Results for the Atoms Helium to Radon.//*Atomic Data,*-1972, № 4, p. 301-399.

[4] Waber J.T., Cromer D.T. Orbital Radii of Atoms and Ions.//*J. Chem. Phys* -1965, -V 42, -№12, p. 4116-4123.

[5] Clementi E., Raimondi D.L. Atomic Screening constants from S.C.F. Functions, 1.//*J.Chem. Phys.*1963, v.38, №11, p. 2686-2689.

[6] Clementi E., Raimondi D.L. Atomic Screening constants from S.C.F. Functions, 2.//*J.Chem. Phys.*-1967, v.47, №4, p. 1300-1307.

[7] Korablev G.A., Zaikov G.E. Energy of chemical bond and spatial-energy principles of hybridization of atom orbitals.//*J. Applied Polymer Science.* USA, 2006,V.101, n.3, p.2101-2107.

[8] Hodges M.P., Wales D.J. Glolal minima of protonated Water clusters. *Chemlocal Physies Letters,* 324, (2000), p.279-288.

[9] Khokhriakov N.V., Melchor Ferrers. Electron properties of contacts in ideal carbon nanotubes. // *Chemical physics and mesoscopy.* Vol. 4, №2, 2002, p. 261-263.

In: Chemical and Biochemical Reactions
Editors: G. E. Zaikov, L. Liu et al.
ISBN 978-1-61122-860-1


*Chapter 18*

# MATHEMATICAL MODELLING DEPENDENCE OF THERMODYNAMIC CHARACTERISTICS UPON SPATIAL-ENERGY PARAMETER OF FREE ATOMS

***G. A. Korablev***[1,*]**, *N. G. Korableva***[1] ***and G. E. Zaikov***[2,†]
[1]Izhevsk State Agricultural Academy, Rizhevsk 426000, Russia
[2]N.M. Emanuel Institute of Biochemical Physics, RAS, 4 Kosygina St., Moscow 119991, Russia

## ABSTRACT

The dependence of some thermodynamic characteristics upon initial spatial-energy parameters of free atoms has been analyzed. The corresponding equations have been obtained, the dissociation energies of binary molecules and enthalpy of single-atom gas formation have been calculated based on them.

**Keywords:** Spatial-energy parameter, formation enthalpy, sublimation, dissociation energy.

## INTRODUCTION

Thermodynamic parameters (enthalpy, entropy, thermodynamic potential) allows describing various physical, chemical and other processes and explicitly assess the possibility of their flow without vividly using physical models. The application of reliable values of formation enthalpies is required for searching new perspective materials and compounds,

* Doctor of Chemical Science, Professor, Head of Department of Physics at Izhevsk State Agricultural Academy. Izhevsk 426052, 30 let Pobedy St., 98-14, tel.: +7(3412) 591946. E-mail: korablev@udm.net; biakaa@mail.ru, korablev@udm.net
† Doctor of Chemical Science, Professor of N.M. Emanuel Institute of Biochemical Physics, RAS, Russia, Moscow 119991, 4 Kosygina St., tel.: +7(495)9397320, e-mail: chembio@chph.ras.ru

assessment of molecule kinetic properties, analysis mechanisms of rocket propellant combustion, etc. [1,2,3].

The establishment of connection between the structure and thermochemical parameters, as well as kinetic and thermochemical characteristics of interacting systems is of great importance [4,5]. But the analysis of dependences between main parameters of chemical thermodynamics and spatial-energy characteristics of free atoms is still topical.

For this purpose the methodology of spatial-energy parameter (P-parameter) has been used in this research [6].

## RESEARCH TECHNIQUE

The comparison of multiple regularities of physical and chemical processes allows assuming that in many cases the principle of adding reciprocals of volume energies or kinetic parameters of interacting structures is fulfilled.

Some examples are: ambipolar diffusion, cumulative rate of topochemical reaction, change in the light velocity when moving from vacuum into the given medium, resultant constant of chemical reaction rate (initial product – intermediary activated complex – final product).

Lagrangian equation for the relative motion of isolated system of two interacting material points with masses $m_1$ and $m_2$ in the coordinate $x$ with acceleration $\alpha$ can be as follows:

$$\frac{1}{1/(m_1 a\Delta x)+1/(m_2 a\Delta x)} \approx -\Delta U \quad \text{or} \quad \frac{1}{\Delta U} \approx \frac{1}{\Delta U1}+\frac{1}{\Delta U2} \tag{1}$$

where $\Delta U_1$ and $\Delta U_2$ – potential energies of material points on elemental interaction section, $\Delta U$ – resultant (mutual) potential energy of these interactions.

The atom system is formed by oppositely charged masses of nucleus and electrons. In this system the energy characteristics of the subsystems are: orbital energy of electrons and effective energy of nucleus considering screening effects (by Clementi). Either the bond energy of electrons (W) or the ionization energy of an atom ($E_i$) can be used as the orbital energy.

Therefore, assuming that the resultant interaction energy in the system orbital-nucleus (responsible for interatomic interactions) can be calculated following the principle of adding reciprocals of some initial energy components, the introduction of P-parameter [6,7] as averaged energy characteristic of valent orbitals based on the following equations has been substantiated:

$$\frac{1}{q^2/r_i}+\frac{1}{W_i n_i}=\frac{1}{P_э} \tag{2}$$

$$P_E=\frac{P_0}{r_i} \tag{3}$$

$$\frac{1}{P_0} = \frac{1}{q^2} + \frac{1}{(wrn)_i} \quad (4)$$

$$q = \frac{Z^*}{n^*} \quad (5)$$

where: $W_i$ – bond energy of an electron [8]; $n_i$ – number of elements of the given orbital; $r_i$ – orbital radius of *i*–orbital [9]; Z* and n* – effective charge of a nucleus and effective main quantum number [10,11].

The value of $P_0$ will be called spatial-energy parameter, and the value of $P_E$ – effective P-parameter. The effective $P_E$-parameter has a physical sense of some averaged energy of valent electrons inside the atom and is measured in energy units, e.g. in electron-volts (eV).

The calculations demonstrated that the values of $P_E$-parameters numerically equal (in the limits of 2%) the total energy of valent electrons (U) by statistic model of an atom. Using the known relationship between the electron density (β) and interatomic potential by statistic model of an atom, the direct dependence of $P_E$-parameter upon the electron density on the distance $r_i$ from the nucleus can be obtained:

$$\beta_i^{2/3} = \frac{AP_0}{r_i} = AP_E$$

where A – constant.

Based on the equations (2-5) $P_E$ and $P_0$-parameters of free atoms for the majority of elements of the periodic system have been calculated [6,7]. Some of these calculations are given in Table 1.

Modifying the rules of addition for reciprocals of energy values of subsystems with reference to complex structures, the formula for calculating $P_C$-parameter of a complex structure is obtained:

$$\frac{1}{P_C} = \left(\frac{1}{NP_E}\right)_1 + \left(\frac{1}{NP_E}\right)_2 + \ldots \quad (6)$$

where $N_1$ and $N_2$ – number of homogeneous atoms in subsystems.

## CALCULATION OF DISSOCIATION ENERGY OF BINARY MOLECULES VIA AVERAGE VALUES OF $P_o$-PARAMETERS

The application of methods of valent bond and molecular orbitals to complex structures faces significant obstacles for prognosticating energy characteristics of the bonds formed.

Based on the equation (6), the application of the formula for calculating the dissociation energy has turned out to be practically suitable [12]:

$$\frac{1}{D_0} = \frac{1}{P_C} = \frac{1}{\left(P_E \frac{N}{K}\right)_1} + \frac{1}{\left(P_E \frac{N}{K}\right)_2} \quad (7),$$

where N – bond order, K – maxing or hybridization coefficient that usually equals the number of valent electrons considered, and the value of $P_E$ ( N/K) has a physical sense of the averaged energy of spatial-energy parameter accrued to one valent electron of the orbitals registered. For complex structures $P_E$-parameter was being averaged by all main valent orbitals.

For binary molecules the dissociation energy ($D_o$) corresponds to the value of chemical bond energy: $D_o$=$E$.

The calculation results of dissociation energy by the equation (7), given ion Table 2, demonstrated that $P_C$=$Д_0$. For some molecules containing such elements as F, N and O, the values of ion radius have been applied to register the bond ionic character for calculating $P_E$-parameter (in Table 2 marked with *). For such molecules as $C_2$, $N_2$, $O_2$ the calculations have been made by multiple bonds. In other cases the average values of bond energy have been applied. The calculated data do not contradict the experimental ones [2,3].

## Equation of Dependence of Thermodynamic Characteristics upon Spatial-Energy Parameters of Free Atoms

The dissociation energy (E) of the molecule breakage into two parts numerically equals the difference of the heat produced during the formation of dissociation products and during the initial molecule formation:

$$E = D_0 = \left[\Delta H_0(R_1) + \Delta H_0(R_2)\right] - \Delta H_0(R_1 R_2) \quad (8)$$

where $D_0$ – dissociation energy of a molecule (rupture energy of its bond), $\Delta H_0(R_1)$ and $\Delta H_0(R_1)$ – formation enthalpy at 0°K respectively of dissociation products $R_1$ and $R_2$, $\Delta H_0(R_1 R_1)$ – initial energy formation enthalpy.

Since for binary molecules $E$=$D_0$, the equation (7) gives:

$$E = \frac{\left(P_E \frac{N}{K}\right)_1 \left(P_E \frac{N}{K}\right)_2}{\left(P_E \frac{N}{K}\right)_1 + \left(P_E \frac{N}{K}\right)_2} \quad (7a)$$

From the formulas (8) and (7a) we have:

$$\frac{\left(P_E\frac{N}{K}\right)_1\left(P_E\frac{N}{K}\right)_2}{\left(P_E\frac{N}{K}\right)_1+\left(P_E\frac{N}{K}\right)_2}=\Delta H_0(R_1)+\Delta H_0(R_2)-\Delta H_0(R_1R_2) \quad (9)$$

This is the equation of direct dependence of thermodynamic values and initial spatial-energy characteristics of a free atom.

The relationship obtained (9) allows defining the corresponding dependences with the experimental thermodynamic characteristics of chemical reactions.

Thus in thermodynamic experimental methods for determining the formation enthalpy the following equation is used:

$$\Delta H_0 = T\left(\Delta\Phi_T^* - R\ln K_E\right)$$

where: $K_E$ – equilibrium constants of chemical reaction; T – thermodynamic temperature of the process; $\Delta\Phi_T^*$ – change in the reaction of considered reduced thermodynamic potential; R – universal gas constant.

## EVALUATION OF SINGLE-ATOM GAS FORMATION ENTHALPY ($\Delta H_G^0$)

In the experimental evaluation method of single-atom gas formation enthalpy from solids, a known thermodynamic identity can be used:

$$\Delta H_G^0 \equiv \Delta H^0_{SD} + \Delta H_{S,298} \quad (10)$$

where: $\Delta H_G^0$ – formation enthalpy of gaseous substance;

$\Delta H^0_{SD}$ – formation enthalpy of solid in nonstandard state;

$\Delta H_{S,298}$ – sublimation enthalpy (substance transition from solid into gaseous state).

The heats of formation of chemical elements in standard state are taken as zero. The list of such elements is known, e.g. [13] and mainly incorporates elements in solid state. Thus the single-atom gas formation enthalpy in these cases equals the sublimation enthalpy and is determined by physical and chemical criteria of the process. The sublimation comes to diffusion movements of heated particles into the surface layer with further effusion (outflow). The diffusion activation energies in external and internal regions are quite different [14].

It is usually believed that sublimation flows by the particle migration from more strongly bonded state with the biggest number of neighbors to less strongly bonded and further – to the adsorbed surface layer [15]. By analogy, the number of interacting molecules in liquid decreases, when molecules move from the lower part of the surface layer to its upper part.

The diffusion activation energy is defined by the values of electron densities of migrating particle and particles surrounding it at the distance of atom-molecular interaction radius (R), i.e. such interaction is carried out via the power field of particles evaluated by the values of their P-parameters [16]. In such a model the diffusion process has basically the similar nature in any aggregate state and its energy and directedness are defined by three main factors:

1) value of P-parameters of structures;
2) number of particles;
3) radius of atom-molecular inetarction.

**Table 1. P-parameters of atoms calculated via bond energy of electrons**

| Atom | Valent electrons | W (eV) | $r_i$ (Å) | $q^2{}_0$ (eVÅ) | $P_0$ (eVÅ) | R (Å) | $P_0/R$ (eV) | $P_0/$ $R(n^*+1)$ | $r_i$ (Å) | $P_0/r_i$ (eV) | $P_0/$ $r_i(n^*+1)$ |
|---|---|---|---|---|---|---|---|---|---|---|---|
| 1 | 2 | 3 | 4 | 5 | 6 | 7 | 8 | 9 | 10 | 11 | 12 |
| Li | $2S^1$ | 5.3416 | 1.506 | 5.86902 | 3.475 | 1.55 | 2.2419 | 0.7473 | 0.68 | 5.1103 | 1.7034 |
| Na | 3S1 | 4.9552 | 1.713 | 10.058 | 4.6034 | 1.69 | 2.4357 | 0.60892 | 0.96 | 4.6973 | 1.1743 |
| K | 4S1 | 4.0130 | 2.612 | 10.993 | 4.8490 | 2.36 | 2.0547 | 0.4372 | 1.33 | 3.6459 | 0.7757 |
| Rb | 5S1 | 3.7511 | 2.287 | 14.309 | 5.3630 | 2.48 | 2.1625 | 0.43250 | 1.49 | 3.5993 | 0.71986 |
| Cs | 6S1 | 3.3647 | 2.516 | 16.193 | 5.5628 | 2.68 | 2.0757 | 0.3992 | 1.65 | 3.3714 | 0.6483 |
| Mg | 3S1 | 6.8859 | 1.279 | 17.501 | 5.8568 | 1.60 | 3.6616 | 0.91544 | 0.74 | 7.9173 | 1.9793 |
| | 3S2 | 6.8859 | 1.279 | 17.501 | 8.7787 | 1.60 | 514667 | 1.3717 | 0.74 | 11.863 | 2.9658 |
| Ca | 4S1 | 5.3212 | 1.690 | 17.406 | 5.929 | 1.97 | 3.0096 | 0.64035 | 1.33 | | |
| | 4S2 | 5.3212 | 1.690 | 17.406 | 6.6456 | 1.97 | 4.902 | 0.95535 | 1.04 | 8.5054 | 1.8097 |
| Sr | 5S1 | 4.8559 | 1.836 | 21.224 | 6.790 | 2.15 | 2.9205 | 0.56409 | | | |
| | 5S2 | | | | 9.6901 | 2.15 | 4.5070 | 0.90140 | 1.20 | 8.0751 | 1.6150 |
| Ba | 6S1 | 4.2872 | 2.060 | 22.950 | 6.3768 | 2.21 | 2.6854 | 0.5549 | 1.38 | 7.2328 | 1.3909 |
| | 6S2 | | | | 9.9812 | 2.21 | 4.5164 | 0.8685 | | | |
| Sc | 4S1 | 5.7174 | 1.570 | 19.311 | 6.1279 | 1.64 | 3.7365 | 0.7950 | | | |
| | 4S2 | 5.7174 | 1.570 | 19.311 | 9.3035 | | | | | | |
| | 3d1 | 9.3532 | 0.539 | 81.099 | 4.7463 | 1.64 | 2.8941 | 0.6158 | | | |
| | 4S2+3d1 | 9.3532 | 0.539 | 81.099 | 14.050 | 1.64 | 8.5671 | 1.8228 | 0.83 | 16.928 | 3.6016 |
| Y | 5S1 | 6.3376 | 1.693 | 22.540 | 6.4505 | 1.81 | 3.5638 | 0.71276 | | | |
| | 5S2 | | | | 10.030 | | 5.5417 | 1.1083 | | | |
| | 4d1 | 6.7965 | 0.856 | 229.18 | 5.6756 | | 3.1357 | 0.62714 | | | |
| | 4d1+5S2 | | | | 15.706 | | 8.6771 | 1.7354 | 0.97 | 16.192 | 3.2384 |
| La | 6S1 | 4.3528 | 1.915 | 34.681 | 6.7203 | 1.87 | 3.5937 | 0.6911 | | | |
| | 6S2 | | | | 11.259 | | 6.0209 | 1.1579 | | | |
| | 4f1 | 10.302 | 0.4234 | 17870 | 4.2576 | | 2.2768 | 0.4378 | | | |
| | 5P1 | 25.470 | 0.827 | 145.53 | 18.400 | | 9.8396 | 1.8922 | | | |
| | 6S2+5P1 | | | | 29.659 | | 15.860 | 3.0501 | 1.04 | 28.518 | 5.4843 |
| | 6S2+4f1 | | | | 15.517 | | 8.2976 | 1.5957 | 1.04 | 14.920 | 2.8693 |
| Ti | 4S1 | 6.0082 | 1.477 | 20.879 | 6.2273 | 1.46 | 4.2653 | 0.9075 | | | |
| | 4S2 | 6.0082 | 1.477 | 20.879 | 9.5934 | 1.46 | 6.5708 | 1.3960 | 0.78 | 12.299 | 2.6169 |
| | 3d1 | 11.990 | 0.496 | 106.04 | 5.556 | 1.46 | 3.8053 | 0.80965 | | | |
| | 3d2 | 11.990 | 0.469 | 106.04 | 10.558 | | | | | | |
| | 4S2+3d2 | | | | 20.151 | 1.46 | 13.802 | 2.9366 | 0.64 | 31.486 | 6.6981 |
| | 4S1+3d1 | | | | | | | 1.7172 | | | |
| Zr | 5S1 | 5.5414 | 1.593 | 23.926 | 6.5330 | 1.60 | 4.0831 | 0.8166 | | | |
| | 5S2 | | | | 10.263 | | 6.4146 | 1.2829 | | | |
| | 4d1 | 9.1611 | 0.790 | 153.76 | 6.9121 | | 4.3201 | 0.8640 | | | |
| | 4d2 | | | | 13.229 | | 8.2681 | 1.6536 | | | |
| | 5S2+4d2 | | | | 23.492 | 1.6 | 14.683 | 2.9365 | 0.82 | 28.640 | 5.7298 |

In advanced researches carried out for solid solutions in the frames of generalized grid model it has been found out that "effective diffusion coefficient depends on local composition, own volumes of component atoms and potentials of paired interactions" [17]. At the same time two types of diffusion are also distinguished: "normal" and "bottom-up" [18], this apparently agrees with the notion of volume (internal) and surface diffusion.

In a liquid the radius of the sphere of molecular interaction $R \approx 3r$, where r – radius of a molecule. Liquids are mainly formed by the elements of first and second periods of the system. For the second period we can write down: $R \approx 3r = (n+1)r$, where n – main quantum number. For both periods (first and second) we have $R = (\langle n \rangle + 1)r \approx 2,5r$.

Let us assume that this principle with definite approximation can be extended to various aggregate states of elements of all other periods but taking into consideration screening effects introducing the value of effective main quantum number ($n^*$) instead of n. These values of $n^*$ and $n^*+1$ taken by Slater [19] are shown in Table 3.

**Table 2. Dissociation energies of two-atom molecules – $D_0 \left(\frac{KJ}{mol}\right)$**

| Structure | First atom | | | | Second atom | | | | $P_C$ (eV) | $D_0$ calcul. | $D_0$ experim. |
|---|---|---|---|---|---|---|---|---|---|---|---|
| | Orbitals | N/k | $P_E$ (eV) | $P_E \frac{N}{k}$ | Orbitals | N/k | $P_E$ (eV) | $P_E \frac{N}{k}$ | | | |
| 1 | 2 | 3 | 4 | 5 | 6 | 7 | 8 | 9 | 10 | 11 | 12 |
| CCl | $2P^1$ | 1/1 | 7.6208 | 7.6208 | $3P^1$ | 1/1 | 8.5461 | 8.5461 | 4.0209 | 388.9 | 393.3 |
| CBr | $2P^1$ | 1/1 | 7.6208 | 7.6208 | $4P^1$ | 1/1 | 8.0430 | 8.0430 | 3.9130 | 377.7 | 364 |
| CJ | $2P^1$ | 1/1 | 7.6208 | 7.6208 | $5P^1$ | 1/1 | 7.2545 | 7.2545 | 2.2523 | 217.4 | 209.2 |
| CN | $2P^2$ | 2/2 | 13.066 | 13.066 | $2S^22P^3$ | 2/5 | 47.413 | 18.965 | 7.7358 | 746.7 | 755.6 |
| CN | $2P^2$ | 2/2 | 14.581 | 14.581 | $2P^3$ | 2/3 | 25.127 | 16.751 | 7.796 | 752.5 | 755.6 |
| C-O | $2P^2$ | 1/2 | 13.066 | 6.533 | $2P^2$ | 1/2 | 17.967 | 8.984 | 3.782 | 365 | 356 |
| NO | $2P^1$ | 1/1 | 9.2839 | 9.2839 | $2P^2$ | 2/2 | 20.048 | 20.048 | 6.346 | 612.5 | 626.8 |
| CH | $2P^2$ | 1/2 | 13.066 | 6.533 | $1S^1$ | 1/1 | 9.066 | 9.066 | 3.7969 | 366.5 | 333±1 |
| OH | $2P^2$ | 1/2 | 17.967 | 8.9835 | $1S^1$ | 1/1 | 9.066 | 9.066 | 4.5118 | 435.5 | 423.7 |
| ClF | $3S^23P^5$ | 1/7 | 29.391* | 4.1987 | $2S^22P^5$ | 1/7 | 38.202* | 5.4574 | 2.5579 | 246.9 | 229.1 |
| ClO | $3S^23P^5$ | 1/7 | 29.391* | 4.1987 | $2P^2$ | 2/2 | 8.7191* | 8.7191 | 2.8337 | 273.5 | 264 |
| ClO | $3P^1$ | 1/1 | 4.7216* | 4.7216 | $2S^22P^4$ | 1/6 | 30.738* | 5.123 | 2.450 | 237.2 | 264 |
| FO | $2P^1$ | 1/1 | 4.9887* | 4.9887 | $2P^2$ | 1/2 | 8.7191* | 4.3596 | 2.327 | 224.6 | 219.2 |
| NF | $2P^3$ | 1/3 | 10.696* | 3.5653 | $2P^1$ | 1/7 | 38.202* | 5.4574 | 2.486 | 239.5 | 298.9 |
| NCl | $2P^3$ | 1/3 | 22.296 | 7.432 | $3P^1$ | 1/1 | 8.5461 | 8.5461 | 3.9751 | 383.7 | 384.9 |
| $H_2$ | $1S^1$ | 1/1 | 9.0624 | 9.0624 | $1S^1$ | 1/1 | 9.066 | 9.066 | 4.533 | 437.5 | 432.2 |
| $Li_2$ | $2S^1$ | 1/1 | 2.2419 | 2.2419 | $2S^1$ | 1/1 | 2.2419 | 2.2419 | 1.121 | 108.2 | 98.99 |
| $B_2$ | $2P^1$ | 1/1 | 5.4885 | 5.4885 | $2P^1$ | 1/1 | 5.4885 | 5.4885 | 2.744 | 264.9 | 276±21 |
| C-C | $2P^1$ | 1/1 | 7.6208 | 7.6208 | $2P^1$ | 1/1 | 7.6208 | 7.6208 | 3.810 | 367.8 | 376.7 |
| C=C | $2P^2$ | 2/2 | 13.066 | 13.066 | $2P^2$ | 2/2 | 13.066 | 13.066 | 6.533 | 630.6 | 611 |
| N-N | $2P^3$ | 1/3 | 10.696* | 3.5653 | $2P^3$ | 1/3 | 10.696* | 3.5653 | 1.783 | 172.1 | 161 |
| N=N | $2S^22P^3$ | 2/5 | 22.745* | 9.098 | $2S^22P^3$ | 2/5 | 22.745* | 9.098 | 4.549 | 439 | 418 |
| O-O | $2P^2$ | 1/2 | 8.7191 | 4.3596 | $2P^2$ | 1/2 | 8.7191 | 4.3596 | 2.1798 | 210.4 | 213.4 |
| O=O | $2S^22P^4$ | 2/6 | 30.738* | 10.246 | $2S^22P^4$ | 2/6 | 30.738* | 10.246 | 5.123 | 494.5 | 498.3 |

**Table 3. Effective Main Quantum Number**

| | | | | | | |
|---|---|---|---|---|---|---|
| n | 1 | 2 | 3 | 4 | 5 | 6 |
| $n^*$ | 1 | 2 | 3 | 3.7 | 4 | 4.2 |
| $n^*+1$ | 2 | 3 | 4 | 4.7 | 5 | 5.2 |

Thus we assume that the radius of sphere of atom-molecular interaction during the particle diffusion in the sublimation process is defined as follows:

$$R=(n^*+1)r,$$

where r – dimensional characteristic of atom structure. Total change of R is from 3r to 5,2r.

When forming single-atom gases, the sublimation process is accompanied by the rupture of paired bond of atoms of near surroundings. The averaged value of structural $P_S$-parameter of interacting atoms can be the assessment of formation enthalpy and numerically equals the value of $P_0$-parameter falling on the radius unit of atom-molecular interaction but taking into consideration the relative number of interacting particles by the equation:

$$P_S = \frac{P_0}{R}\gamma = \frac{P_0\gamma}{r(n^* + 1)} \approx \Delta H_\Gamma^o$$

where $\gamma$ - coefficient taking into consideration the relative number of interacting particles and equaled to (as the calculations demonstrated):

$$\gamma = \frac{N_0}{N}$$

Here $N_0$ – number of particles in the sphere volume of the radius R,

N – number of particles of realized interactions depending on the process type (internal or surface diffusion).

Inside the liquid below the top layer 2R thick, the resultant force of molecular interaction equals zero.

Applying the initial analogy to internal diffusion and sublimation we can consider that such equilibrium state corresponds to the equality $N_0$=N, and then $\gamma$=1.

On the top part of liquid surface layer the volume of the sphere of atom-molecular interaction and the number of particles in it is practically 2 times lower in comparison with internal layers under 2R, i.e. $\frac{N_0}{N} \approx \frac{1}{2}$ and $\gamma=\frac{1}{2}$ – for surface diffusion and sublimation.

In case of volume diffusion even a more extreme option is possible, when the number of particles of realized interactions 2 times exceeds $N_0$.

Thus two-valent elements magnesium and calcium form either two or even four covalent bonds (in chelate compounds): two covalent bonds and two – by donor-acceptor mechanism. For such and analogous cases with volume diffusion $\gamma$=2.

**Table 4. Calculation of formation enthalpy of single-atom gases – $\Delta H_{298}$**

1 eV = 96.525 $kJ/mol$

| Group | Atom | Orbitals | $P^o$ ($eVA^o$) | $\gamma$ | r ($A^o$) | $\Delta H = \frac{P_o \gamma}{r(n^*+1)}$, ($eV$) | $r^u$ ($A^o$) | $\Delta H = \frac{P_o \gamma}{r_u(n^*+1)}$, ($eV$) | $\Delta H$ calculation ($kJ/mol$) | $\Delta H^\circ$ 298 ref. data ($kJ/mol$) |
|---|---|---|---|---|---|---|---|---|---|---|
| 1 | 2 | 3 | 4 | 5 | 6 | 7 | 8 | 9 | 10 | 11 |
| 1a | Li | $2S^1$ | 3.4750 | 1 | | | 0.68 | 1.703 | 164.4 | 159.3 |
| | Na | $3S^1$ | 4.6034 | 1 | | | 0.98 | 1.1743 | 113.3 | 107.5 |
| | K | $4S^1$ | 4.8490 | 2 | 2.36 | 0.874 | | | 84.36 | 88.9 |
| | Rb | $5S^1$ | 5.3630 | 2 | 2.48 | 0.865 | | | 83.49 | 80.9 |
| | Cs | $6S^1$ | 5.5628 | 2 | 2.68 | 0.7984 | | | 77.07 | 77 |
| 2a | Be | $2S^2$ | 7.512 | ½ | | | 0.34 | 3.682 | 355.4 | 326.4 |
| | Mg | $3S^2$ | 8.7787 | ½ | | | 0.74 | 1.483 | 143.1 | 147.1 |
| | Ca | $4S^2$ | 8.8456 | 1 | | | 1.04 | 1.8097 | 174.7 | 177.8 |
| | Sr | $5S^2$ | 9.6901 | 1 | | | 1.20 | 1.6150 | 155.9 | 160.7 |
| | Ba | $6S^2$ | 9.9812 | 2 | 2.21 | 1.737 | | | 167.7 | 179.1 |
| 3a | Sc | $4S^2 3d^1$ | 14.050 | 1 | | | 0.74 | 4.040 | 389.96 | 379.1 |
| | Y | $5S^1 4d^2$ | 17.527 | 2 | 1.62 | 4.342 | | | 419.2 | 423 |
| | La | $6S^2 5p^1$<br>$6S^2 4f^1$ | 29.659<br>15.517 | 2<br>2 | 1.87<br>1.87 | 6.1002<br>3.1915 | | | <448.4> | 429.7 |

**Table 4. Continued**

| | | | | | | | | | | |
|---|---|---|---|---|---|---|---|---|---|---|
| 4a | Ti | $4S^1 3d^1$<br>$4S^2 3d^2$ | 11.7853<br>20.151 | 2<br>2 | 1.46<br>1.46 | 3.4344<br>5.8732 | | | <449.2> | 468.6 |
| | Ti | $4S^1 3d^{1^1}$<br>$4S^2 3d^2$ | 11.7853<br>20.151 | 1<br>1 | | | 0.78<br>0.64 | 3.214<br>6.6981 | <478.4> | 468.6 |
| | Zr | $5S^2 4d^2$ | 23.492 | 2 | 1.60 | 5.873 | | | 566.9 | 600 |
| | Hf | $6S^2 5d^2$ | 24.498 | 2 | $r_K = 1.44$ | 6.543 | | | 631.6 | 620.1 |
| 5a | V | $4S^2 3d^1$<br>$4S^1 3d^2$ | 15.776<br>17.665 | 1<br>1 | | | 0.67<br>0.67 | 5.6010<br>5.6090 | <512.5> | 514.6 |
| | Nb<br>($5S^2 4d^3$) | $5S^2 4d^2$ | 25.577 | 1 | | | 0.67 | 7.6349 | 736.9 | 722.6 |
| | Nb<br>($5S^1 4d^4$) | $5S^1 4d^2$<br>$5S^1 4d^4$ | 20.805<br>30.607 | 1<br>1 | | | 0.767<br>0.65 | 5.425<br>9.2748 | <709.4> | 722.6 |
| | Nb<br>(3, 4, 5) | | | 1 | | | | | <723.2> | 722.6 |
| | Ta | $6S^2 5d^2$<br>$6S^2 5d^3$ | 26.314<br>32.722 | 1<br>1 | | | 0.737<br>(0.66) | 6.8662<br>9.5344 | <791.5> | 786.1 |
| 6a | Cr | $4S^2$<br>$4S^2 3d^1$ | 10.535<br>17.168 | 1<br>1 | | | 0.83<br>0.64 | 2.7006<br>5.7074 | <406.1> | 397.5 |
| | Mo (2)<br>($5S^2 4d^4$) | $5S^1 4d^1$ | 19.574 | 1 | | | 0.737 | 5.312 | <658> | 656.5 |
| | Mo (4)<br>($5S^1 4d^4$) | $5S^1 4d^3$ | 28.293 | 1 | | | 0.68 | 8.3215 | | |
| | W(4)<br>W(5) | $6S^2 6S^1$<br>$6S^2 5d^3$ | 27.879<br>34.828 | 2 | 1.40<br>1.40 | 3.8295<br>4.7841 | | | <831.5> | 856.9 |
| 7a | Mn (2) | $4S^1 3d^1$ | 12.924 | 1 | | | 0.91 | 3.0217 | 291.7 | 284.5 |

| | | | | | | | | | | |
|---|---|---|---|---|---|---|---|---|---|---|
| | Tc (3) | $5S^1 4d^2$ | 23.866 | 2 | 1.36 | 7.0194 | | | 674.1 | 657 |
| | Re (4) | $6S^2 5d^2$ | 29.806 | 1 | | | 0.72 | 7.961 | 768.4 | 775.7 |
| | Re (6) | $6S^2 5d^4$ | 44.519 | ½ | | | 0.52 | 8.232 | 794.6 | |
| | Re (4,6) | | | | | | | | <781> | |
| 8 | Fe (2)<br>Fe (3) | $4S^1 3d^1$<br>$4S^2 3d^1$ | 12.717<br>16.664 | 1<br>1 | | | 0.80<br>0.67 | 3.3822<br>5.2918 | <418.6> | 417.1 |
| | Co (2)<br>Co (3) | $4S^1 3d^1$<br>$4S^2 3d^1$ | 12.707<br>16.680 | 1<br>1 | | | 0.78<br>0.64 | 3.4662<br>5.5785 | <436.5> | 428.4 |
| | Ni (2)<br>Ni (3) | $4S^1 3d^1$<br>$4S^1 3d^2$ | 12.705<br>16.897 | 1<br>1 | | | 0.74<br>0.60 | 3.6530<br>5.992 | <465.5> | 429.3 |
| | Ni (2) | $4S^1 3d^1$ | 12.705 | 2 | 1.24 | 4.360 | | | 420.8 | 429.3 |
| | Ru (3) | $5S^1 4d^2$ | 23.636 | 1 | | | 0.68 | 6.982 | 671 | 656.8 |
| | Rh (3) | $5S^2 4d^1$ | 21.114 | 1 | | | 0.75 | 5.6304 | 543.5 | 556.5 |
| | Pd (2) | $5S^2$ | 12.057 | 1 | | | 0.64 | 3.7681 | 363.7 | 372.3 |
| | Os (3) | $5d^3$ | 25.986 | 2 | $r_k = 1.26$ | 7.922 | | | 766 | 790.2 |
| 8 | Ir (2)<br>Ir (4) | $6S^1 5d^1$<br>$6S^2 5d^2$ | 17.631<br>30.790 | 2<br>2 | 1.35<br>1.35 | 5.023<br>8.772 | | | <665.8> | 669.5 |
| | Pt (2) | $6S^1 5d^1$ | 17.381 | 1 | | | 0.6 | 5.971 | 576.4 | 565.7 |
| 16 | Cu (2) | $4S^1 3d^1$ | 13.242 | 1 | | | 0.80 | 3.5218 | 339.9 | 337.6 |
| | Ag<br>Ag (1) | $4d^1$<br>$4d^1$ | 9.7843<br>9.7843 | 2<br>2 | $r_k = 1.25$<br>1.44 | 3.131<br>2.718 | | | <282.3> | 284.9 |
| | Au (3) | $6S^1 5d^2$ | 27.536 | 1 | 1.44 | 3.6774 | | | 355.0 | 368.8 |

**Table 4. Continued**

| | | | | | | | | | | |
|---|---|---|---|---|---|---|---|---|---|---|
| 2б | Zn (1)<br>Zn (2) | $3d^1$<br>$4S^2$ | 6.1153<br>11.085 | 1<br>1 | 1.39<br>1.39 | 0.9361<br>1.6968 | | | <127.1> | 130.5 |
| | Cd (2) | $5S^2$ | 11.839 | ½ | | | 0.99 | 1.1959 | 115.4 | 111.8 |
| | Cd (2) | $5S^1 4d^1$ | 17.145 | ½ | 1.56 | 1.0991 | | | 106.1 | 111.8 |
| | Cd (1, 2) | | | | | | | | <110.8> | 111.8 |
| | Hg (1) | $5d^1$ | 11.266 | ½ | 1.60 | 0.6771 | | | 65.35 | 61.40 |
| 3б | B | $2S^2 2p^1$ | 16.086 | 1 | 0.91 | 5.892 | | | 568.7 | 561 |
| | Al | $3S^2 3p^1$ | 18.093 | 1 | 1.43<br>$r_k = 1.25$ | 3.163<br>3.619 | | | <327.3> | 329.3 |
| 3б | Ga<br>Ga | $4S^2 4p^1$ | 20.760<br>8.8961 | 1<br>2 | 1.39<br>1.39 | 3.178<br>2.723 | | | <284.8> | 273.0 |
| | In | $5S^2 5p^1$ | 21.841 | 1 | 1.66 | 2.6314 | | | <241.5> | 238.1 |
| | | $5S^2 5p^1$ | 21.841 | ½ | | | 0.92 | 2.374 | | |
| | Tl | $6S^2 6p^1$ | 22.012 | ½ | | | 1.05 | 2.015 | 194.5 | 181.0 |
| 4б | C | $2p^2$<br>$2S^2 2p^2$ | 10.061<br>24.585 | 1<br>1 | 0.77<br>0.77 | 4.3554<br>10.643 | | | <723.9> | 716.7 |
| | Si | $3p^2$ | 10.876 | 2 | 1.17 | 4.648 | | | 448.6 | 452 |
| | Ge | $4p^2$<br>$4p^2$ | 12.072<br>12.072 | 2<br>2 | 1.39<br>1.24 | 3.3656<br>4.1428 | | | <378.3> | 376.6 |
| | Ge | $4p^2$ | 12.072 | 1 | | | 0.65 | 3.9516 | 381.43 | 376.6 |
| | Sn | $5p^2$ | 13.009 | 2 | 1.58 | 3.2934 | | | 317.9 | 302.1 |
| | Pb | $6S^2 6p^2$ | 32.526 | 1 | 1.50 | 2.0952 | | | 202.2 | 195.1 |
| | Pb | $6p^2$ | 13.460 | | | | 1.26 | 2.0543 | 198.3 | 195.1 |
| 5б | N | $2p_r^5$ | 21.966 | | | | 1.48 | 4.947 | 477.5 | 472.7 |

| | | | | | | | | | | |
|---|---|---|---|---|---|---|---|---|---|---|
| | P | $3p^1$ | 7.7864 | 2 | $r_K$ =1.16 | 3.3562 | | | 323.9 | 316.3 |
| | As (5)<br>As (3) | $4S^2 4p^3$<br>$4p^3$ | 40.232<br>18.645 | ½<br>1 | 1.40<br>$r_K$ = 1.21 | 3.057<br>3.2785 | | | 295.1<br>316.5 | 301.8 |
| | As (3 ,5) | | | | | | | | <305.8> | |
| 5б | Sb | $5p^3$<br>$5p^3$ | 20.509<br>20.509 | 1<br>1 | $r_K$ = 1.39<br>1.61 | 2.9509<br>2.5477 | | | <265.2> | 268.2 |
| | Bi | $6p^3$ | 21.919 | 1 | 1.82 | 2.3160 | | | <207.3> | 209.2 |
| | Bi | $6p^3$ | 21.919 | 1 | | | 2.13 | 1.9790 | | |
| 6б | O (2) | $2p^2$ | 11.858 | 1 | | | 1.40 | 2.823 | <253.1> | 249.2 |
| | O (4) | $2p^4$ | 20.338 | ½ | | | 1.40 | 2.421 | | |
| | S (2) | $3S^1 3p^1$ | 21.673 | 2 | $r_K$ =0.94 | 2.882 | | | <280.8> | 277.0 |
| | S (4) | $3p^4$ | 21.375 | 1 | | | 1.82 | 2.9360 | | |
| | Se (4) | $4p^4$ | 24.213 | ½ | $r_K$ =.17 | 2.2032 | | | <226.7> | 223.4 |
| | Se (2) | $4S^1 4p^1$ | 23.283 | 1 | | | 1.98 | 2.495 | | |
| | Te | $5S^1 5p^1$ | 23.882 | 1 | | | 2.11 | 2.264 | 218.5 | 215.6 |
| | Po | $6S^1 6p^1$ | 23.664 | ½ | $r_K$ = 1.50 | 1.516 | | | 146.4 | 146 |
| 7б | F | $2p^1$ | 6.635 | ½ | | | 1.33 | 0.83145 | 80.3 | 79.5 |
| | Cl | $3p^1$ | 8.5461 | 1 | | | 1.81 | 1.2804 | 123.6 | 121.3 |
| | Br | $4p^1$ | 9.3068 | 1 | | | 1.79 | 1.1062 | 106.8 | 111.8 |
| | I | $5p^1$ | 9.9812 | ½ | | | $r^{5+}$ = 0.94 | 1.062 | 102.5 | 106.8 |

## CALCULATIONS AND COMPARISONS

Based on such initial statements and assumptions the value of $P_S$-parameter was calculated (and $\Delta H^0_r$) by the equation (12) for the majority of elements in periodic system – Table 4. The values of covalent, atom and ion radii are basically taken by Belov-Bokiy and partially by Batsanov [20].

The structural coefficient $\gamma$ was taken as equaled to 2 for covalent spatial-energy bonds at the distances of only atom or covalent radii (surface diffusion). Coefficient $\gamma$ was taken as equaled to 1 with ion spatial-energy bonds of elements from subgroups «a» and group 8 in periodic system (internal diffusion). In all other cases this coefficient equals 1 or 1/2 (internal or volume diffusion).

In some cases the values of $\Delta H^0_r$ of the given element was found as average value by its two possible valent states (marked with <...>). The calculations are carried out practically for all elements of six periods irrespectively of their aggregate state that definitely could give higher accuracy than reference data.

Experimental and reference data [2,3] have a relative error within 0.5 – 1.5 (%), but in the given calculations such average error was about 5 %. Probably the search for a more rational technique of registering screening effects for clarifying the effective main quantum number can eventually bring more reliable results for evaluating $\Delta H^0_r$.

## CONCLUSIONS

The dependences of dissociation energy of binary molecules and formation enthalpies of single-atom gases upon initial spatial-energy characteristics of free atoms have been determined.

The corresponding equations, the calculations on which basically agree with the experimental and reference data, have been obtained.

## REFERENCES

[1] Lebedev Yu.A., Miroshnichenko E.A. Thermal chemistry of steam-formation of organic substances. – M.: *"Nauka"*, 1981, 216 p.

[2] Properties of inorganic compounds. *Reference-book/* / Efimov A.I. et al. – L.: "Khimiya", 1983, 392 p.

[3] Rupture energy of chemical bonds. Ionization potentials and affinity to an electron. Reference-book/ / Kondratyev V.I. et al. – M.: *"Nauka"*, 1974, 351p.

[4] Benson S. Thermochemical kinetics. – M.: *Mir,* 1971, 308 p.

[5] Berlin A.A., Wolfson S.A., Enikolopyan N.S. Kinetics of polymerization processes. – M.: *"Khimiya"*, 1978, 319 p.

[6] Korablev G.A. *Spatial-Energy Principles of Complex Structures Formation*, Netherlands, Brill Academic Publishers and VSP, 2005,426p. (Monograph).

[7] Korablev G.A. Spatial-energy criteria of phase-formation processes. – Publishing house *"Udmurt university"*, Izhevsk, 2008, 494 p.

[8] Fischer C.F. Average-Energy of Configuration Hartree-Fock Results for the Atoms Helium to RadonV//*Atomic Data*,-1972, -№ 4, -p. 301-399.

[9] Waber J.T., Cromer D.T. Orbital Radii of Atoms and Ions//*J. Chem. Phys* -1965, -V 42, -№12, -p. 4116-4123.

[10] Clementi E., Raimondi D.L. Atomic Screening constants from S.C.F. Functions, 1.//*J.Chem. Phys*.-1963, -v.38, -№11, -p. 2686-2689.

[11] Clementi E ., Raimondi D.L. Atomik Screening Constants from S.C.F. Functions, 2.//*J. Chem. Phys*.- 1967,- V.47, № 4,- p. 1300 – 1307.

[12] Korablev G.A., Zaikov G.E. Energy of chemical bond and spatial-energy principles of hybridization of atom orbitalls.//*J. of Applied Polymer Science.* USA, 2006., V. 101, № 3, p. 283 – 293.

[13] Thermodynamic properties of individual substances. Reference-book of Academy of Science of the USSR, ed. Glushko V.P., M.: *"Science"*, 1978, v.1, 496 p.

[14] Kofstad P. Deviation from stoichio, diffusion and electrical conductivity in simple metal oxides. M.: *"Mir"*, 1975, 398 p.

[15] Pound G.M. – *S. Phys. and Chem. Ref. Data.* 1972, v.1. p. 135 – 146.

[16] Korablev G.A., Solovyev S.D. Energy of diffusion activation in metal systems//*Bulletin of ISTU*, 2007, №4, p.128-132.

[17] Zaharov M.A. G*rid models of multi-component solid solutions: statistic thermodynamics and kinetics*//Abstract of doctoral thesis. Novgorod State University, 2008, 36 p.

[18] Geguzin Ya.E. Ascending diffusion and diffusion aftereffect//UFN.-1986, v.149, iss.1, p. 149-159.

[19] Batsanov S.S., Zvyagina R.A. Overlap integrals and problem of effective charges. *"Nauka"*, Siberian branch, Novosibirsk, 1966, 386 p.

[20] Batsanov S.S. Structural chemistry. Facts and dependencies. – *M.: MSU* – 2000, 292 p.

In: Chemical and Biochemical Reactions
Editors: G. E. Zaikov, L. Liu et al.

ISBN 978-1-61122-860-1


*Chapter 19*

# TRUTH OF THE TRUTHS AND RESONANCE OF THE RESONANCES

***G. A. Korablev*[1,*]*, N. G. Korableva*[1] *and G. E. Zaikov*[2,†]**
[1]Izhevsk State Agricultural Academy, Izhevsk 426000, Russia
[2]N.M. Emanuel Institute of Biochemical Physics, RAS, 4 Kosygina St., Moscow 119991, Russia

## ABSTRACT

It is ascertained that struggle and opposition between two basic political and economical systems do not have sense since the most progressive system is the combination of the two. One of the possible variants of such formation is pro-posed. It is shown that periods of resonances of solar activity have multipronged importance for our planet.

## TRUTH OF THE TRUTHS

The world is going crazy: people are fighting, poisoning, killing, torturing, jeering… What for?

I thought: A miserable man
What does he want?
Blue is the sky,
Land is enough for all,

But he is fighting still,
What for?
M.U. Lermontov

* E-mail: korablev@udm.net
† E-mail: chembio@chph.ras.ru

Really what for? There is a struggle for power, money, land, spheres of influence, religion, etc. And the results?

But the result is clear and simple enough:
No one loses and no one wins!
Graves and tombs are left after us
Blamed by the Destiny and the Supreme!

It seems the most important in this struggle are large and small, vivid and secret wars between religions, and wars between different political and economical systems.

Only one person – Abraham – was the forefather of three basic religions. His three children unexpectedly chose differ-rent ways and gave birth to Christianity, Islam and Judaism.

But our Universe is one for all and the Earth is also one for all. And their Creator whatever he is called (God, Christ, Allah, Nature…) is also one for all!

In recent hundred years the struggle between political and economical systems has been coming down to the oppositions between capitalism and socialism.

The actual capitalism of the early XX century was wild. Russia has come to a similar system during the last decades. The transition was also «wild» with «shock therapy» and was accompanied by the destruction of the existing state, stable and sound economy, but mainly by multiple human and moral losses. And all these resulted in the Eltsin's default in 90-s that we still cannot recover from.

Before early 30-s the wild capitalism in the USA resulted in Great Depression. Two factors helped to overcome it:

1. Availability of the country with different system in the world – socialistic Russia that purchased industrial goods and even production capacities in the USA.
2. Genial strategist, economist and politician – the President Roosevelt appeared in the right place at the right time. He realized that depression was the result of permissiveness and much free principles of capitalism. He led the country to the capitalism governmentally regulated that was understood by many people as a social capitalism.

Probably it was really so. Only Roosevelt could not implement his idea completely. He passed away and few decades later the USA and the whole world were on the threshold of financial crisis again.

But there is no powerful socialistic empire now, and the presidents of the USA have not been of «Roosevelt» type recently.

Competition between two systems – capitalism and socia-lism shows that socialism though not having defaults and eco-nomical crises, still loses from economical point but nearly always wins from moral and ethical points. Where is the way out? What political and economical system is more rational in all respects?

## Where Is the Truth of the Truths?

Probably everything is not that complicated as it seems. For instance, there are such sciences as physics and chemistry, but on the other hand there are chemical physics and physical chemistry. The most interesting and important lies on the borderline of sciences. The same in economical systems: the way out in arranging economical systems – intermediate.

The first type – socialistic capitalism – Rooselvet's way but with continuation. Some West European countries passed it to some extent.

The second type – capitalistic socialism – was chosen in China. Thus the economy was preserved and billion of popu-lation was fed.

Perhaps Lenin, before he passed away, also came to such conclusion introducing NEP (New Economic Policy) instead of military communism.

The history has already proved that neither capitalism nor socialism are political and economical systems rational enough for the majority of population. As usual, the objective truth is somewhere in the middle between two opposite Truths. Taking into consideration all historic mistakes made during wars, revolutions and counterrevolutions, the humanity gradually and painfully comes to the conclusion on the necessity of building the society with elements of both capitalism and socialism.

If such a strategy is accepted everywhere, global economical links and absence of principal political and econo-mical arguments will minimize the reasons for wars and struggles between countries and peoples. Then the efforts of scientists and politicians will be targeted at the solution of world problems that already exist but are practically not solved:

1. Seeking the ways for treating now incurable diseases.
2. Informal agreement between all the countries and peoples on suspension of diversionary, political and economical counteractions between states and systems.
3. Development of Space shield preventing possible catastrophes of cosmic and universe scope.

And others.

But the implementation of unilateral economic concepts and Truths can again result in global crises. The great poet and prophet Lermontov, not believing in human wisdom, wrote in «Prophecy» in 1830:

The time will come – dark time for Russian land,
Tsar's power be gone and it will be the end…

It is followed by the description of tragic results.

But our people did not listen to the poet's appeals and prophecies. And happened what happened. Besides:

1. In the year of the poet-prophet's $100^{th}$ birthday the $1^{st}$ World War broke out.
2. 100 years after his death the $2^{nd}$ World War broke out.
3. When Lermontov's $150^{th}$ birthday was celebrated (October 1964), the Brezhnev's period was beginning, named developed socialism.

4. 150 years after the poet's death (1991) the transition to «wild» capitalism started in Russia.

Those are only a few coincidences with the poet's jubilees, who wrote 115 years before space flights: «The Earth sleeps in blue glow».

In 2014 we are going to celebrate Lermontov's 200th birthday. This is the year of possible historic catastrophes or shocks, if we all together are not taking any measures.

ut now still:

There's no peace under pines,
No peace under planes and cherries,
There's no peace under woods,
No peace under paradise valleys!
But we must blame ourselves
For breaking peace on Earth!

## On Socialistic Public Capitalism (Individual)

Probably these two definitions – socialistic capitalism and public capitalism have some differences. But in this case we take them as synonyms since socialistic capitalism has to be public. The notion «public capitalism» has been used for a long time and assumes wide and active participation of all personnel in joint-stock companies (JSC).

The main properties of capitalism:

1. Effective competition.
2. Maximum personal interest of everyone.

Eventually they determine the advantages of capitalist system: cost value, quality and range of products, gross national profit, overall progress of economy, etc. But at diffe-rent levels of capitalist economy concentration these original features of capitalism function and reveal themselves diffe-rently. Their greatest efficiency corresponds to small business when the number of JSC members is minimum, the personal interest is clearer and competition is more active. When the company grows bigger, the number of personnel increases – the role of each shareholder gradually diminishes, the competition goes down since new opportunities for regulated activities of different JSC arise. Such process more often takes place in monostructures of big companies of large business.

In his novel «Bread» Mamin-Sibiryak objectively described the competition on the level of small business that always involves the decrease in prices. But today we see neither decrease nor the stabilization of prices but their growth. It means that modern capitalism lost its main positive properties because it became monopolistic, oligarchic. Obviously what intra-industry competition we can talk about if all main pricing issues are solved by a few people. As a result, we have uncontrolled price increase, inflation, exchange and social troubles, and finally – global financial-economic crisis.

In the financial crisis some large banks are being nation-nalized in the USA and Europe, the process is carried out following the market rules. In Russia A. Chubays openly argued that for effective competition it is necessary to break up his monster “RAO EES” into medium companies that will function independently and therefore actively compete between themselves and internationally. The demonopoliza-tion of the corporation was based on civilized market principles.

To continue with such financial and economic methodo-logy and based on the above one of the possible variants of forming socialistic public capitalism (SPC) is proposed for discussion.

1. Medium and small business must be the foundation of SPC. It is necessary to create all the conditions for the organization and functioning of these companies.
2. To demonopolize large economic monsters to small business size but in civilized way, mainly without nationalization, and definitely without revolutions and counterrevolutions.
3. Probably the nationalization can be performed only in some economic and nature intensive industries. For instance, the nationalization of alcohol industry could solve several problems, including the diminishing of counterfeited products.
4. Modern communication and transport techniques allow establishing such contacts in short period in medium business that produce quasi-monopolization in the industry without direct merger of corresponding structures. Thus, the most complicated and important role of governmental and autonomous anti-monopolistic structures.
5. To reconsider the practical side and methodology of shareholding as well as the text part of agreements and contracts to sharply increase everybody’s personal interest.
6. Promptly reacting and effective public organizations for human rights protection are required. Trade unions should act independently without federal and local supervision, and provided with 24-hour channel on radio and TV.
7. To develop multipronged social and pension systems by standard scheme of Western European countries.
8. You should not sneer at the history. Both Red and White struggled for their native country, for better life of Russian people. Let us hope that this dream will come true in a more progressive public formation.

There is only one choice:
either global financial and economic crisis or socialistic public capitalism.

But we don’t want the wild market
And never monopolies press us again.
We vote for such a society
Where everyone feels that he is the Main!

# RESONANCE OF THE RESONANCES

Amazing and strange (at the first glance) coincidences of epochal historic events with the jubilees of the prophet-poet M.U. Lermontov. Probably there are similar examples with other titans of literature, science, art and religion. We want to believe this is not mysticism but actual and substantiated futurology.

In reality the Nature and our Universe definitely celebrate not the jubilees of some prophets and titans in such a way, but their own jubilees that coincide with the birthdays of these selected persons.

Jubilees of the nature are cyclic and are defined for us by the sun cyclic activity (spots on the sun, flashes, torches, etc). These cycles have been studied quite well and are usually well-known. Thus based on different estimations the 11-year cycle is 11,04; 11,1; 11,2; 11,5 years. But actually it has been changing from 9 to 13,5 during the last 250 years (11,25 in the average). The 80-year cycle of maximums of solar spots is deve-loped [1]. Having studies the rings on the wood saw cuts and based on other sources, the scientists assume the existence of several other series of cycles: about 110, 210 and more years. The principle of octave cyclicity is proposed and discussed: 11; 22; 44; 33; 66;... 353,3... [2]. The period between neighboring «ethnogenies» is 32 maximums of cycles, i.e. about 352 years.

If the Universe celebrates Lermontov's jubilees every 100 and 150 years, it means there are cycles of decimal calculus: 100-, 80-, 50- and 1000-year cycles (e.g. Table 1 and Table 2). 10 is found in the equation of the numbers of Wolf's sun spots and 11-year cycle itself is 10 with the correction: 10+1,2=11,2. Similarly: 100+1,2=101,2 and 1000+1,2=1001,2.

In this model the amendment 1,2 can characterize the degree of cycle maximum deviation from the average value. Thus for 11-year cycle the following is usually fulfilled:

$$1{,}2*2>| X-11{,}2 |$$

where X – actual difference between the maximums of neighboring cycles.

The less the cycle decimality, the more the possible deviation from their average values. Based on the rule of «a vertical line» the lower cycle «adjusts» by its duration to the maximum of the upper cycle.

Some examples:

1. Cycle No 2. By the average value of the period: 1769,7+11,2=1780,9. The actual period appeared to be shorter: (8,7) as there is an «adjustment» for 80-year cycle: 1769,7+8,7=1778,4 – Table 2.
2. Cycle No 16. By the average value: 1928,4+11,2=1939,6. Actually a two-apex cycle with maximums 1937,9 and 1938,6 is implemented with a shorter period and in accordance with 80-year cycle (1938,6) – Table 2.

The time of Lermontov's birth (1814,8) corresponds to the period of the solar activity maximum increase that falls at 1816,4. Usually the most powerful solar flashes happen on the activity phase growth or decrease. Thus Lermontov was born in the period of increasing solar activity of cycle No 6. After 9 cycles in 1917,6 a 100-year jubilee of solar activity takes place. This secular period (101,2=1817,6-1816,4) is also an average value.

**Table 1. Maximums of 100-year solar activity**

| | | | | | | | |
|---|---|---|---|---|---|---|---|
| 1 | | 1917,6 | 1816,4 | 1718,2 | 1615,5 | | |
| 2 | **2019,0** | 1917,6 | **1816,4** | 1715,2 | **1614,0** | 1513,0 | 1412,0 |
| 3 | 1310.4 | 1209,2 | 1108,0 | 1006,8 | 905,6 | 804,4 | 703,2 |
| 4 | 602,0 | 500,8 | 399,6 | 298,4 | 197,2 | 96,0 | **-5,2** |
| 5 | -106,4 | -207,6 | -308,8 | -410,0 | -511,2 | -612,4 | -713,6 |

Notes: 1 line – data from [1], 2-5 – calculated numbers.

The above cycle 353,3 occurs from the period of absence of solar spots (1645 year) till the most active solar activity in the late XX century: 1645+353,3=1998,3. Therefore the calculated Lermontov cycles by the step of 101,2 years (1715,2; 1614) slightly differ from actual values of maximums (1718,2; 1615,5) – Table 1.

Here 1816.4 is an initial value [1]. «-» corresponds to the years B.C. Therefore 5,2 year B.C. approximately defines the birth of the legendary personality Jesus Christ, according the literary sources it was within 4th and 8th year B.C.

Using a 1000-year cycle in 1001,2 years from the maximum in 1999,5 [3], we have 1999,5-2*1001,2=-2,9. Besides, 179 complete 11,2-year cycles give the value of 2004,8, by the maximum from 1999,5 this gives the difference about -5,3 years. Thus the general average time of Christ's birth is 4,5 year B.C.

Each cycle is characterized by its physical characteristics, including magnetic field intensity and frequency. The coinci-dences of 11-year and 100-year cycles are the resonance of either cycles or certain frequencies of magnetic fields, the resul-ting characteristic of which is much greater than the initial. If for Lermontov this is the resonance of two cycles, for Christ – resonance of three cycles or resonance of the resonances.

80-year cycle of maximums given in [1] was modified based on modern data [3], the calculation results from the initial value 1778 are shown in Table 2.

The first line contains actual data [1, 3], lines 2–6 – calcu-lated values. Similarly to Table 1, some deviations of these values correspond to the period of absence of sun spots (in 1645–1715). The average value of 80-year cycle is calcu-lated based on the data of the last 300 years and results in 80,3. This value was taken for calculations in lines 2–6. It turned out that the value of the cycle 573.5 roughly corresponds to the time of birth of Prophet Mahomet (Mohammed) (according to literary sources – around 570 year), thus constituting 93 complete 11,2 cycles. This is the reso-nance of two cycles.

Tables 1 and 2 contain rather similar values of cycle maxi-mums: 1615,5; 1614; 1617,4. Practically this is the resonance of the resonances of 11-year, 80-year, 100-year cycles. For Russia it was marked by the end of so-called «troubled period» (1609–1618 years) and beginning of 300-year governing of Romanovs' dynasty (in 1613).

**Table 2. Maximums of 80-year cycles**

| | | | | | | |
|---|---|---|---|---|---|---|
| 1 | 1615,5 | (1693) | 1778 | 1859,8 | 1938,6 | |
| 2 | 1617,4 | 1667,7 | **1778** | 1858,3 | 1938,6 | **2019** |
| 3 | 1537,1 | 1456,8 | 376,5 | 1296,2 | 1215,9 | 1135,6 |
| 4 | 1055,3 | 975,0 | 894,7 | ... | 653,8 | **573,5** |
| 5 | 493,2 | ... | -68,9 | -149,2 | -229,5 | -309,8 |
| 6 | -391,1 | 470,4 | -550,7 | -631 | -711,3 | -791,6 |

There other interesting coincidences for simple resonances with 11-year, 80-year or 100-year cycles according to Tables 1 and 2. For instance, in «Centuries of wise men» [2] B.C. we have: 610 – Shvataketu, 551 – Confucius, 390 – Shan Yan, 556 – Buddha, 550 – Mahavira, 625 – Thales, 550 – Epicharm, 470 – Socrates, 384 – Aristotle, 610 – Anaximander. The calculations are given without 50-year cycles.

From the foregoing examples it is clear that the period of solar activity resonances in some cases coincides with the birth of specifically selected prophets, titans of thought and culture of philosophical and humanitarian field, but is always accompanied by social, economical and political, and ecolo-gical disasters. Global influences of solar activity on flora and fauna of the Earth was proved in the time of A.L. Chizhevsky, N.D. Kondratyev and even earlier. The activity cyclicity and density of the flow of space rays affect not only weather and hydrological phenomena, but also life of plants, animals and people. According to many scientists, all these influence social and political processes in certain ways, this is proved by the corresponding numbers and graphs [4].

The action of proton and X-ray radiation of the sun definitely results in the changes of spatial-energy character-ristics on atomic and molecular level in cells and tissues: P-para-meters [5], based on unique combination of certain inner and outer biophysical-and-chemical factors this sometimes causes the appearance of new and super-new physiological and life functions of an organism. Therefore «the organism intellectual acceleration increases, craving for art and science intensifies, outlook orientation onto Mind and Wisdom takes place» [2]. Geniuses are born not only in the period of cycle active part, but also at other values of solar activity (A.S. Pushkin, A. Ein-stein, etc.). For instance, the great poet A.S. Pushkin was born in the period of solar activity minimum (in 1799). If we calcu-late by the average deviation value 11,2 from the previous maxi-mum of cycle No 4, another intensity maximum would be possible: 1788,1+11,2=1799,3. But it did not happen since, apparently, cycles No 4 and No 5were traditionally adjusting themselves for 100- year cycle No 6 (1816,4) with the periods 17,1 and 11,2, respectfully [1].

Examples and phenomena discussed are only particular cases of global effect of the sun and space onto all living beings on the Earth. The determination of atomic and molecular mechanism of this process is still an unsolved but critical problem. Based on the laws of equilibrium dynamics many physical and chemical interactions are reversible. Here it means the possibility of reverse action when «when combined, the will and mind of millions of people can influence processes in nature» [4].

Just one example:

And when Ukraine has witnessed
The fight of passions never seen,
The Earth everywhere trembled
As previously never been [6].

This happened at 8 o'clock in the morning on December 26, 2004. There are newer examples in the Ukraine. Actually there are thousands of such examples on our planet in different times. This cannot be explained from scientific point based on specific physical and chemical regularities, therefore this research is mainly probabilistic but still some prognostic conclusions can be made:

1. Many prophets and idols of religions specially selected by the Nature (Christ, Mahomet, Buddha, Socrates, Lermontov, etc) were born in the periods of solar activity resonances. The Nature celebrates their jubilees in accordance with its cycles.
2. Such epochs frequently correspond to global ecological, social and political, and other historical changes on our planet.
3. Probably, in general, talent and genius have little depen-dence upon the phase of non-resonance cycle, but are defined by specific features of spatial-energy interactions on atomic-molecular level of each individual.
4. Intensification and optimization of such interactions occur with the increase of cycle resonance degree.
5. Analyzing Tables 1 and 2, we can assume that approxi-mately in 2019 there will be a double resonance of 80-year and 100-year maximums with 11-year cycle of solar activity reduced to them.

## REFERENCES

[1] K.W. Allen. Astrophysical values. *Mir.* – M., 1977. – 448 p.

[2] A.E. Kulinkovich. Prognostication of human history in the 3rd millennium A.D.: Report of the medal winner N.D. Kond-ratyev // *Tendencies and perspectives of sociocultural dynamics.* – M., 1999. – P. 41–77.

[3] Sunspot number, *Royal Observatory of Belgium.* – http//sidc.oma.be/, 1 July 2008.

[4] V.V. Klimenko. Global climate: yesterday, today, tomorrow. *Polit.ru., public lectures, 16*.07.08. – 24 p. – edit@polit.ru.

[5] G.A. Korablev. *Spatial Energy Principles of Complex Structures Formation. Netherlands, Leiden, Brill Academic Publishers and VSP.* 2005. 426p. (Monograph).

[6] G.A. Korablev. Ways of life. – Izhevsk : Publishing house *«Udmurt University»,* 2008. – 115 p.

In: Chemical and Biochemical Reactions
Editors: G. E. Zaikov, L. Liu et al.
ISBN 978-1-61122-860-1


*Chapter 20*

# Morphological Characterization of Cellulose Acetate Based Reverse Osmosis Membranes by Atomic Force Microscopy (AFM)-Effect of Evaporation Time

*N. Gizli and M. Demircioğlu*
Department of Chemical Engineering, Ege University, Bornova, 35100, İzmir, Turkey

## Abstract

The surface structures of asymmetric cellulose acetate membranes prepared by phase inversion method are investigated by atomic force microscopy (AFM). The differences in the surface morphology have been expressed in terms of four types of roughness parameters such as; roughness average ($S_a$), root mean square roughness ($S_q$), surface skewness ($S_{sk}$) and surface kurtosis ($S_{ku}$). The association of the roughness characteristics with the salt rejection performance has been also investigated. The results show that the increase in the evaporation time results in the change of the membrane morphology and rejection efficiency.

**Keywords:** Atomic force microscopy, Reverse osmosis membrane, Morphology, Surface roughness

## 1. Introduction

Atomic Force Microscopy (AFM) gives the topografic images in 3D by scanning a sharp tip over a surface. It is a useful method to characterize the surface structures of membranes due to the possibility of analyzing dry or wet samples without any special pretreatment. AFM imaging of the membrane structure has advantages over both scanning electron microscopy (SEM) and transmission electron microscopy (TEM) since the resolution is higher, the sample preparation is minimum and no electron beam damage can occur [1, 2]. In addition to its imaging capability, AFM can measure the interaction forces which allow characterization of adhesive behavior of polymer film in the nanoscale level. Hence, AFM has become a popular method of investigating the surface structures of polymer and inorganic membranes. AFM characterization of membrane has been focused on the determination the surface morphology and pore structures (pore size, pore size distribution, pore density, surface roughness) [3-10], correlation of membrane structure with membrane properties [1, 11-14], surface adhesion [15] and membrane fouling [16, 17]

It is well known that the morphology and performance of the membranes depend largely on the conditions under which the membranes are prepared [18, 19]. The phase inversion process induced by immersion precipitation is the most common technique to develope asymmetric polymeric membranes [18, 20]. Immersion precipitation method involves mainly four steps, starting with casting phase, then evaporation stage followed by gelation period and finally annealing step. Solvent evaporation is of importance in the case of formation of the asymmetric structure that is the main property of RO membrane [21].

In this study, the effect of evaporation time on resulting pore structure characteristics of both active and support layers have been investigated by AFM analysis. Surface roughness parameters such as; roughness average ($S_a$), root mean square roughness ($S_q$), surface skewness ($S_{sk}$) and surface kurtosis ($S_{ku}$) values are obtained to compare the surface structure differences. Salt rejection performances of the membranes have also been determined to relate them with rougness parameters.

All roughness parameters can be calculated from the AFM image with an AFM software program. However, specific roughness parameters depend on the curvature on the membrane surface and also the size of the cantilever tip, as well as on the treatment of the captured images (plane correction, median fitler, etc.). Thus, the roughness parameters should not be considered as an absolute value that represent surface properties [2].

## 2. Materials and Methods

### 2.1. Materials

Cellulose acetate supplied from FLUKA, with molecular weight of 37000 kg/kgmole and with ≈40% acetyl content was used as polymer. Both extra pure (≈99%) acetone and formamide were supplied from MERCK, used as solvent and swelling agent, respectively.

### 2.2. Preparation of CA Membranes

The immersion precipitation method is applied by following the steps of casting, evaporation, gelation and annealing as described elsewhere [22, 23].

Polymer solution, with the composition of 20 wt % CA, 69 wt % acetone and 11 wt % formamide was cast on a glass plate kept at a temperature of -10°C. After varying period of solvent evaporation (15-130 sec.) in ambient atmosphere (temperature of 25 °C and relative humidity of 55-60%), the membranes were immersed in pure water bath (1°C) for gelation of polymer solution during 90 minutes. Then the membranes were annealed in water bath at 86°C for 6 minutes.

## 2.3. Reverse Osmosis Tests

The laboratory set-up used in permeation experiments has been previously described [23].

The apparatus consists of a 30L feed reservoir, test cell with an effective membrane area of ....$cm^2$, high pressure pump and hydrolic system. Tests were condusted at an operating pressure of 60 bar.

Reverse osmosis performances of the membranes prepared were tested by using a feed with a concentration of 1% NaCl by weight. The concentration of $Na^+$ ion in both feed and permeate were analyzed with Atomic Absorption Spectroscopy (Varian 10 Plus). Salt rejection performance was determined by the equation below:

$$R\% = \left(1 - \frac{C_P}{C_F}\right) \times 100 \qquad (1)$$

where $R$ is solute rejection as per cent, and $C_p$ and $C_F$ are $Na^+$ ion concentrations in permeate and feed streams as ppm, respectively.

## 2.4. AFM Analysis

The surface structure of membranes produced at several evaporation time, was observed by multimode AFM (RT-SHPM) obtained from NanoMagnetic Instruments Co. The membrane surfaces were scanned by aluminum reflex coated silicon probe (Tap 300AI, NanoMagnetics Instruments) having the spring constant of 40 N/m and resonance frequency of 300 KHz in tapping mode. Scan area and speed were chosen as 10x10 micron and 5 micron/s., respectively.

Membrane surfaces were compared in terms of roughness parameters such as; Three roughness parameters such as roughness average ($S_a$), root mean square roughness ($S_q$), surface skewness ($S_{sk}$) and surface kurtosis ($S_{ku}$)

*Average roughness, $S_a$:* the arithmetic average of the absolute values of the measured height deviations from the mean surface taken within the evaluation area. Analytically, is given in Cartesian coordinates by [24]:

$$S_a = (1/Ae) \int_0^{L_y} \int_0^{L_x} |Z(x,y)|\, dxdy \qquad (2)$$

where $Ae$ is the total area over which the values of surface parameters are evaluated and , $Z(x,y)$ is *height function* which is used to represent the point-by-point deviations between the measured topography and the mean surface.

For a rectangular array of $M$ x $N$ digitized profile values $Z_{jk}$, the formula is given by:

$$S_a = \frac{1}{MN}\sum_{k=1}^{M}\sum_{j=1}^{N}\left|Z_{jk}\right| \tag{3}$$

*Root mean square (RMS) roughness, $S_q$:* The root mean square average of the measured height deviations from the mean surface taken within the evaluation area. Analyticaly, $S_q$ is given by:

$$S_q = \left((1/Ae)\int_0^{L_y}\int_0^{L_x} Z^2(x,y)dxdy\right)^{1/2} \tag{4}$$

The digital approximation is:

$$S_q = \left[\frac{1}{\mathrm{MN}}\sum_{k=1}^{\mathrm{M}}\sum_{j=1}^{\mathrm{N}} Z^2{}_{jk}\right]^{1/2} \tag{5}$$

*Surface skewness, Ssk:* a measure of the asymmetry of surface heights about the mean surface.

Analytically, *Ssk* can be expressed as follows:

$$S_{sk} = \frac{1}{(S_q)^3 Ae}\int_0^{L_y}\int_0^{L_x} Z^3(x,y)\,dxdy \tag{6}$$

For digitized profiles it can be calculated from:

$$S_{sk} = \frac{1}{(S_q)^3}\left[\frac{1}{\mathrm{MN}}\sum_{\mathrm{k=1}}^{\mathrm{M}}\sum_{\mathrm{j=1}}^{\mathrm{N}} Z^3_{jk}\right] \tag{7}$$

The surface skewness, related to the third momentum of the height distribution, informs about the asymmetry, that is, its dissimilarity relative to a Gaussian distribution. Surfaces with a positive skewness, such as turned surfaces have fairly high spikes that protrude above a flatter average. Surfaces with negative skewness, such as porous surfaces have fairly deep valleys in a smoother plateau. More random (for example, ground) surfaces have a skewness of about zero. A value of $S_{sk}$ greater than about 1.5 in magnitude (positive or negative) indicates that the surface does not have a simple shape and more advanced parameters are needed to fully describe the surface structure [25].

*Surfa kurtosis, $S_{ku}$:* a measure of the peakedness of the surface heights about the mean surface. Analytically, $S_{ku}$ can be expressed as:

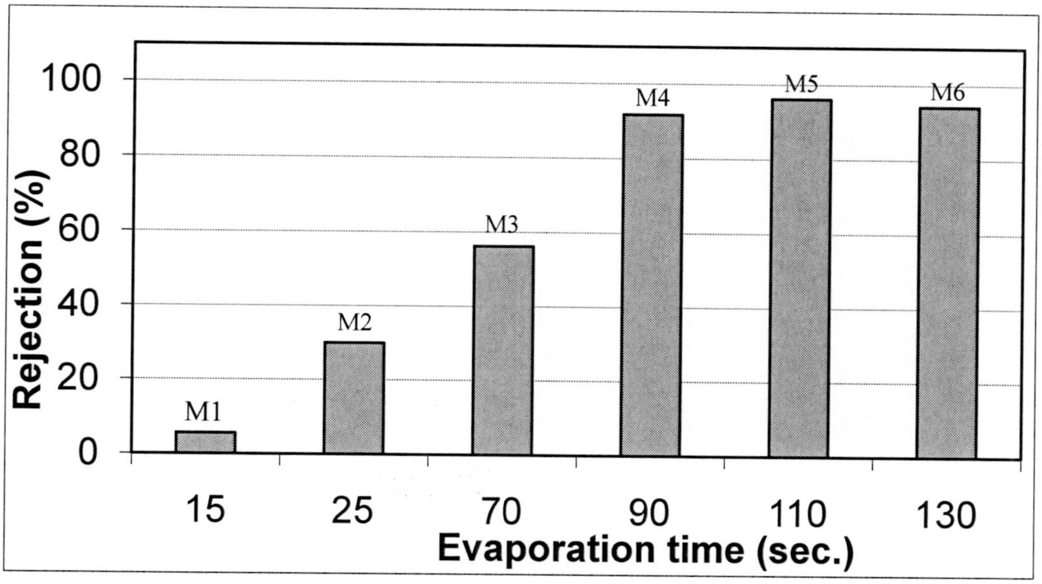


Figure 1. Comparison of performance of the membranes prepared at various evaporation time.

$$S_{ku} = \frac{1}{(S_q)^4 Ae} \int_0^{L_y} \int_0^{L_x} Z^4(x, y) dx dy \tag{8}$$

For a digitized profile, it can be calculated from:

$$S_{ku} = \frac{1}{(S_q)^4} \frac{1}{MN} \sum_{k=1}^{M} \sum_{j=1}^{N} Z_{jk}^4 \tag{9}$$

For the Gaussian distribution, the skewness is equal to zero and the kurtosis is equal to three.

In this study, surface roughness parameters were calculated from the AFM images using the software SPM 1.16.13.1

## 3. Results

Identification (ID) of membranes having different evaporation times was given in the Table 1.

**Table 1. IDs of membranes with their evaporation periods**

| Membrane ID | M1 | M2 | M3 | M4 | M5 | M6 |
|---|---|---|---|---|---|---|
| Evaporation Time (sec) | 15 | 25 | 70 | 90 | 110 | 130 |

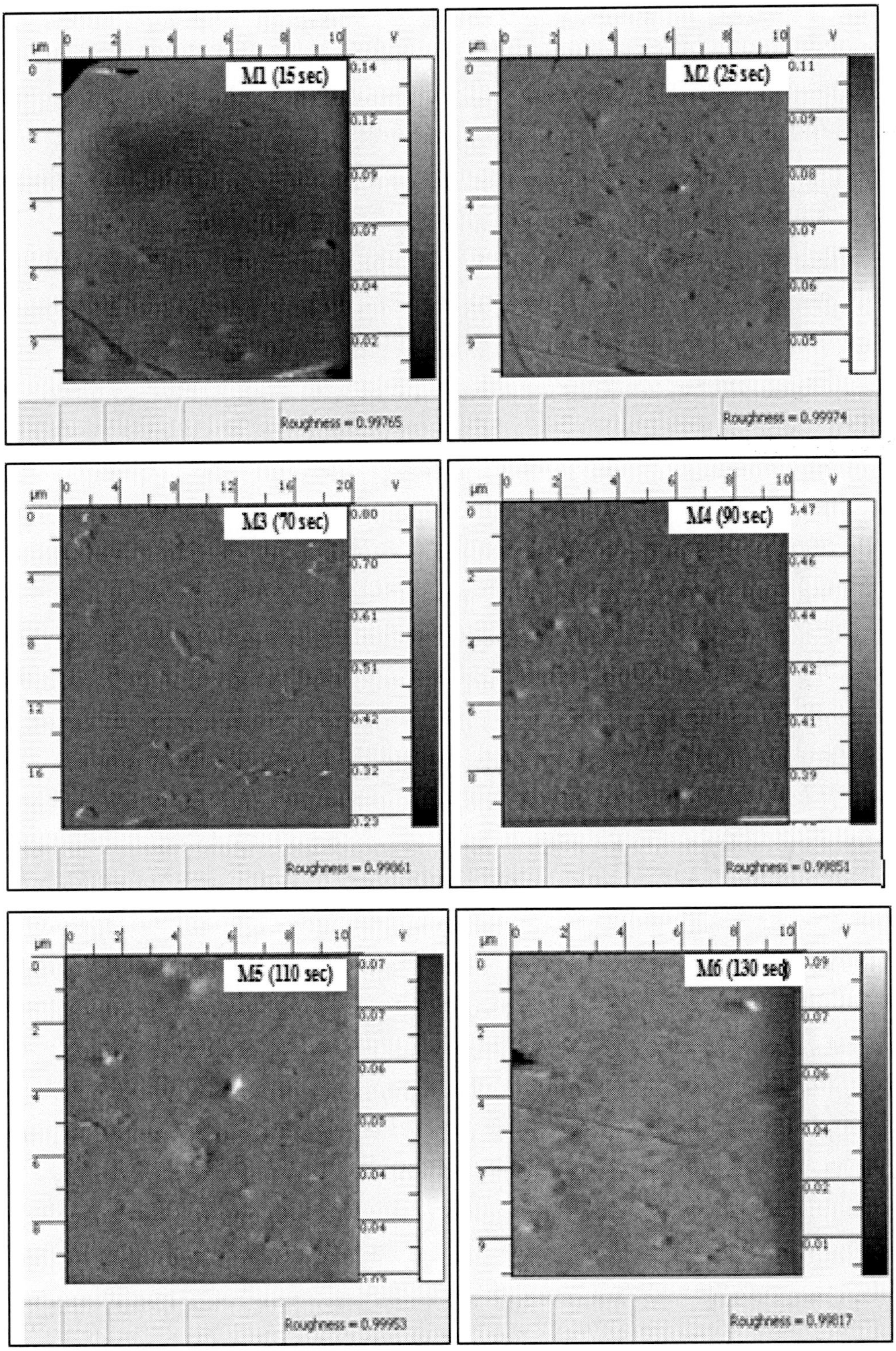


Figure 2. AFM images of active layers of the membranes having the evaporation times of 15, 25, 70, 90, 110 and 130 sec.

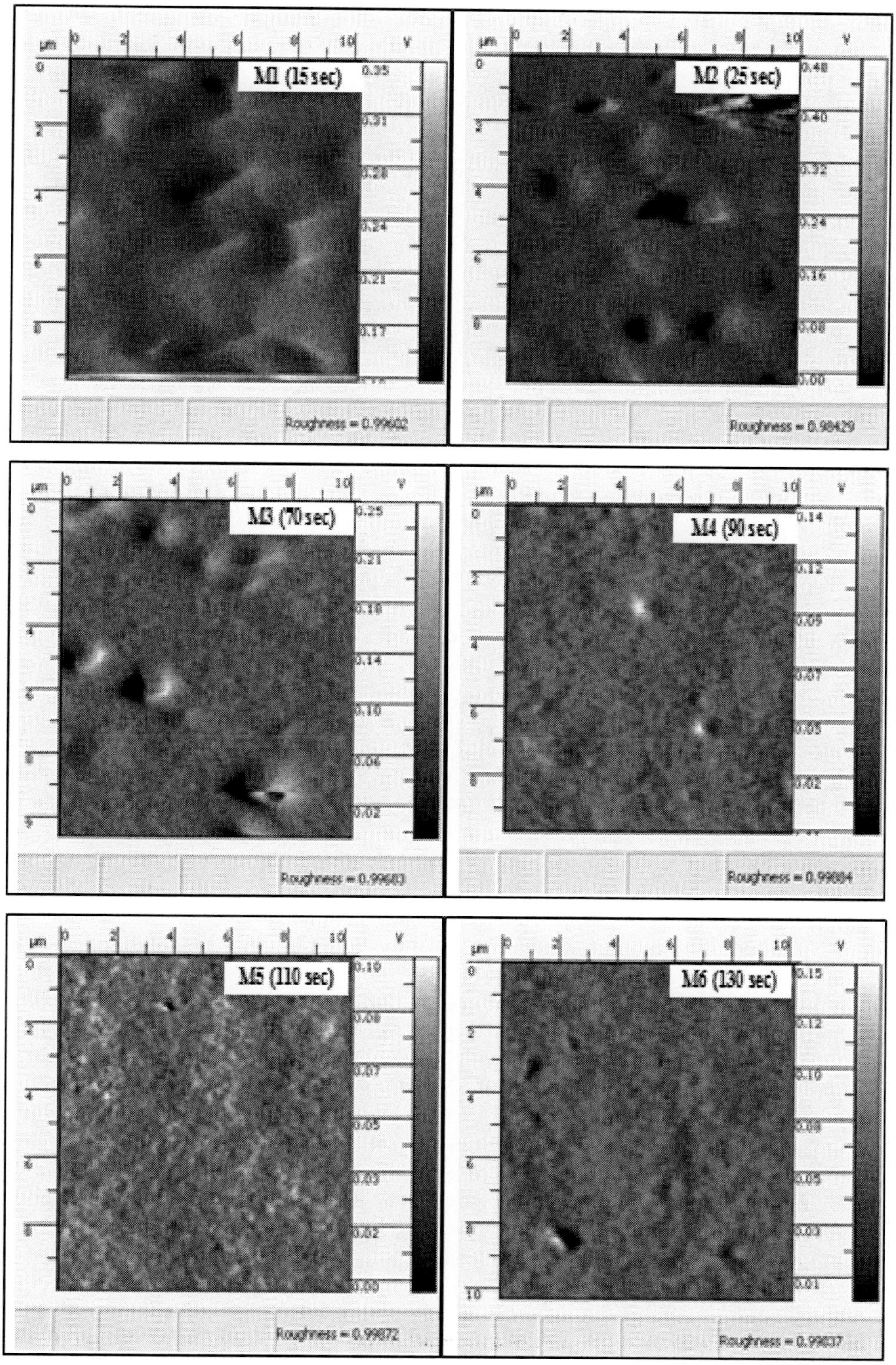


Figure 3. AFM images of support layers of the membranes having the evaporation times of 15, 25, 70, 90, 110 and 130 sec.

**Table 2. Roughness parameters of the membranes**

| | Membrane No | Evaporation time, (sec) | Roughness average, $S_a$ (nm) | Root Mean Square, $S_q$ (nm) | Surface Skewness, $S_{sk}$ (nm) | Surface Kurtosis, $S_{ku}$ (nm) |
|---|---|---|---|---|---|---|
| Active layer | M1 | 15 | 217.75 | 272.10 | -1.41 | 2.83 |
| | M2 | 25 | 203.54 | 239.13 | -0.18 | 2.17 |
| | M3 | 70 | 335.31 | 417.72 | -0.84 | 2.10 |
| | M4 | 90 | 434.48 | 497.85 | -0.24 | 2.12 |
| | M5 | 110 | 571.23 | 655.56 | -0.66 | 2.76 |
| | M6 | 130 | 753.68 | 849.47 | -1.14 | 2.58 |
| Support layer | M1 | 15 | 202.30 | 245.45 | 0.18 | 2.53 |
| | M2 | 25 | 256.80 | 306.50 | 0.31 | 2.10 |
| | M3 | 70 | 261.52 | 295.81 | 0.19 | 2.71 |
| | M4 | 90 | 238.95 | 276.67 | 0.20 | 2.85 |
| | M5 | 110 | 334.48 | 397.85 | 0.24 | 2.12 |
| | M6 | 130 | 349.14 | 416.13 | 0.45 | 2.19 |

## 3.1. Salt Rejection Performances

Salt rejection performance of membranes was given in Figure 1 which helps to show the effect of evaporation time on membrane structure.

As the salt rejection of membrane having 15 seconds of evaporation time, was so low (5.66%), the active layer formation had not started yet. Salt rejection value increased to 30% of salt rejection was obtained by the membrane with 25 seconds of evaporation time. Although salt rejection value of membrane M3 increased to level of 56.23%, it was not able to withstand higher pressures above 60 bar due to the lack of mechanical strength. A well developed structure of active layer has reached at 90 seconds of evaporation for membrane M4. Both the development of active layer formation and the salt rejection performance at 110 seconds of evaporation time (membrane M5) has reached their optimum levels. The decrease in salt rejection of membrane M6 can be explained by the distortion of the active layer.

## 3.2. AFM Analysis

AFM images taken from the active layer (top surface or air side) and support layer (bottom surface or glass side) of the asymmetric membranes produced at various evaporation time are presented in Figure 2 and Figure 3, respectively. As shown in the figures, the images have light and dark regions. The color intensity shows the vertical profile of the membrane surface, with light regions being highest points and the dark points represent the depressions and pores.

Membranes are in asymmetrical structure having a top thin skin layer (active) and a porous support layer. According to kinetics of solvent evaporation time, structural changes develop in both layers.

When the images are compared, it can be seen that active sides are much smoother than the support layer of the membrane. These differences are due to the different nature of the air-solution and of the glass plate-solution interfaces. In the support layer of the membranes (Figure 3) prepared using shorter evaporation times, the average size of the darker depression areas is higher than for the membranes cast at longer evaporation time. The surface morphologies is also illustrated by the values of the statistical parameters (Table2).

## 4. CONCLUSIONS

The morphologies of both surfaces (active and support layer) of asymmetric membranes were investigated by Atomic Force Microscopy.

Increase in the evaporation time results in the change of the membrane morphology. As evaporation time increases, pore diameters of both active and support layers of membrane decrease, which is favorable in RO applications. Salt rejection values increases up to 96% by membranes having evaporation time 110 sec. Further extention of evaporation time results in rejection loss.

Atomic force microscopy images indicated that an increasing in evaporation times leds to a systematic decrease in pore size and an increase in surface roughness. Although roughness average ($S_a$) and root mean square roughness ($S_q$) values of active layer increase with increasing evaporation time, there is no drastic change; in roughness parameters ($S_a$,$S_q$) of support layers.

The negative values of skewness for active layer indicate the porous structure. On the other hand, support layers are more random surfaces since they have skewness of about zero. Both active and support layers have kurtosis (Sku) value scattering between 2.10 - 2.85.

## ACKNOWLEDGMENTS

This work was partially supported by TUBITAK under Grant No. MAG-105M294.

## REFERENCES

[1] D. M. Stamatialis, C. R. Dias, M. N. Pinho, "Atomic force microscopy of dense and asymmetric cellulose-based membranes", *J. Membr. Sci.* 160 (1999) 235-242.

[2] K. C. Khulbe, C. Y. Feng, T. Matsuura, *"Syhntetic Polymeric Membranes-Characterization by Atomic Force Microscopy"*, Springer, 2008.

[3] W. Yoshida, Y.Cohen, "Topological AFM characterization of graft polymerized silica membranes", *J. Membr. Sci.* 215 (2003) 249–264.

[4] W.R. Bowen, N. Hilal, R. W. Lovitt, P. Williams, "Atomic force microscope studies of membranes: surface pore structures of Diaflo ultrafiltration membranes" *J. Colloid Interface Sci.* 180 (1996) 350-359.

[5] W.R. Bowen, N. Hilal, R. W. Lovitt, P. Williams, "Visualisation of an ultrafiltration membrane by non-contact atomic force microscopy at single pore resolution" *J. Membr. Sci.* 110 (1996) 229-232.

[6] W.R. Bowen, T. Doneva, "Atomic force microscopy studies of nanofiltration membranes: surface morphology" *Desalination* 129 (2000) 163-172.

[7] A. Bessières, M. Meireles, R. Coratger, J. Beauvillain, V. Sanchez, "Investigations of surface properties of polymeric membranes by near field microscopy" *J. Membr. Sci.* 109 (1996) 271-284.

[8] S. Singh, K.C. Khulbe, T. Matsuura, P. Ramamurthy, "Membrane characterization by solute transport and atomic force microscopy" *J. Membr. Sci.* 142 (1998) 111-127.

[9] N.A. Ochoa, P. Pradános, L. Palacio, C. Pagliero, J. Marchese, A. Hernández, "Pore size distributions based on AFM imaging and retention of multidisperse polymer solutes: characterization of polyethersulfone UF membranes with dopes containing different PVP" *J. Membr. Sci.* 187 (2001) 227.

[10] N. Hilal, H. AI-Zoubi, N.A. Darwish, A.W. Mohammad, "Characterisation of nanofiltration membranes using atomic force microscopy" *Desalination* 177 (2005) 187-199

[11] J.M. Tan, T. Matsuura, "Effect of nonsolvent additive on the surface morphology and the gas separation performance of poly(2,6-dimethyl-1,4-phenylene)oxide membranes" *J. Membr. Sci.* 160 (1999) 7-16.

[12] S. Al-Jeshi, A. Neville, "An investigation into the relationship between flux and roughness on RO membranes using scanning probe microscopy" *Desalination* 189 (2006) 221–228.

[13] M. Hirose, H. Ito and Y. Kamiyama, "Effect of skin layer surface structures on the flux behaviour of RO membranes" *J. Membr. Sci.,* 121 (1996) 209–215.

[14] S.S. Madaeni, "The effect of surface characteristics on RO membrane performance" *Desalination,* 139 (2001) 371.

[15] W.R. Bowen, N. Hilal, R. Lovitt, C.J. Wright, "A new technique for membrane characterisation: direct measurement of the force of adhesion of a single particle using an atomic force microscope" *J. Membr. Sci.* 139 (1998) 269.

[16] E.M. Vrijenhoek, S. Hong, M. Elimelech, "Influence of membrane surface properties on initial rate of colloidal fouling of reverse osmosis and nanofiltration membranes" *J. Membr. Sci.* 188 (1) (2001) 115-128.

[17] W.R. Bowen, T. Doneva, "Atomic force microscopy studies of membranes: effect of surface roughness on double-layer interactions and particle adhesion" *J. Colloid Interface Sci.* 229 (2000) 544-549.

[18] M. Mulder, "BasicPrinciples of Membrane Technology", Kluwer Academic Publishers, Dordrecht 1997.

[19] T. Mohammadi, E. Saljoughi, "Effect of production conditions on morphology and permeability of asymmetric cellulose acetate membranes" *Desalination* 243 (2009) 1–7.

[20] R. E. Kesting, *Synthetic Polymeric Membranes*, 2nd ed., Wiley&Sons, New York, 1985.

[21] F. G. Paulsen, S. S. Shojaie, W. B. Krantz, " Effect of evaporation step on macrovoid formation in wet-cast polymeric membranes" *J. Membr. Sci.* 91 (3) (1994) 265-282.

[22] R. Kezin, "*The effects of solvent evaporation on asymmetric cellulose acetate membrane structure and performance"*, MSc. Dissertation, Ege University, 2006.

[23] A. E. Başer, "The effect of gelation bath composition on the performance of cellulose acetate (CA) membranes casted with different solvent volume fractions", *MSc. Dissertation,* Ege University, 2006.

[24] An American National Standard ASME B46.1-2002, Surface Texture (surface roughness, waviness, and lay), *American Society of Mechanical Engineers,* U.S.A., 2003.

[25] A. Mendez-Vilas, J.M. Bruque, M.L. Gonzalez-Martın "Sensitivity of surface roughness parameters to changes in the density of scanning points in multi-scale AFM studies. Application to a biomaterial surface" *Ultramicroscopy* 107 (2007) 617–625.

In: Chemical and Biochemical Reactions
Editors: G. E. Zaikov, L. Liu et al.

ISBN 978-1-61122-860-1


*Chapter 21*

# CHARACTERIZATION OF POLY(VINYLCHLORIDE) (PVC) BASED CATION EXCHANGE MEMBRANES PREPARED WITH IONIC LIQUIDS

***Sanem Çınarlı, Nilay Gizli* and Mustafa Demircioğlu***
Ege University Chemical Engineering Department, Bornova-35100, Izmir, Turkey

## ABSTRACT

The aim of this study is to produce poly(vinylchloride) (PVC) based strong acidic cation exchange membranes and to characterize them. Recent studies show hopeful results regarding ionic liquids, which are one of novel alternatives, since they have low volatility, low melting points, thermal stability and nonflammablity. In this research, membranes were prepared by following a heterogeneous method and characterized in physicochemical, morphological and electrochemical ways. Physicochemical characterization consisted of water contents, dry and wet densities, porosities, ion exchange capacities of membranes and isotherm studies. Scanning Electron Microscopy (SEM) analysis was used to perform morphological characterization. Water contents of the membranes were found between 0.3-0.4 while density values were in a range of 0.7-1.2 g/cm$^3$. Porosities were found between 0.2-0.4 and ion exchange capacities changed between 1.2 and 2.8 meq $Na^+$/g-dry. Electrochemical characterization included transport numbers of membranes determined by membrane potential method. They were found around 0.5 for the membranes prepared with dioctyl phthalate (DOP) whereas around 1.0 for the membranes prepared with IL. Diffusivity values of $Na^+$ ions across the membrane were calculated by using Donnan dialysis method and found in order of magnitude -5.

**Keywords:** Cation exchange membrane, ionic liquid, membrane characterization, transport number

* E-mail: nilay.gizli@ege.edu.tr, Tel: +90 232 388 76 00; Fax: +90 232 388 77 76.

# 1. INTRODUCTION

A membrane is defined as a film or layer with a thickness that is small in comparison with the surface. An ion exchange membrane is an ion exchanger of any geometrical shape that can be used as a separation wall between two solutions [1]. Separation membranes have become important parts of the chemical technology because of their growing industrial applications in the areas of biotechnology, nanotechnology and membrane based energy devices besides different membrane based separation and purification processes. These processes are economically practical due to low energy requirements. Advances in membrane technology can make this technology more competitive with traditional, high-energy intensive, environmentally undesirable and costly processes [2].

Ion-exchange membranes are one of the most advanced separation membranes. They are classified into two types by preparation methods as heterogeneous and homogeneous. Homogeneous ones are prepared mostly by polymerization. Heterogeneous ion-exchange membranes may be made by mechanical incorporation of powdered ion-exchange resin into sheets of rubber, polyvinyl chloride and some other extrudable or moldable matrix. Such membranes can be prepared by different methods: (*i*) molding the mixture of ion exchange powder and dry binder polymer under the proper conditions into a sheet, (*ii*) extruding the mixture of ion exchange particles, a binder polymer and a plasticizer as a sheet, (*iii*) dispersing ion exchange particles in a partially polymerized binder polymer and casting into a film, finishing the polymerization, (*iv*) dispersing ion exchange particles in a solution containing a binder and a plasticizer, casting the mixture into a film and evaporating the solvent.

In this study, PVC based strong acidic cation exchange membranes were prepared by following the last method. Suspension polymerized PVC or emulsion polymerized PVC was used as film-forming binder. Dioctyl phthalate or an ionic liquid, which is tetradecyl(trihexyl)phosphoniumdecanoate, was used as plasticizer. Cyclohexanone was used as solvent and commercial strong acidic cation exchange resin was used as ion exchange particles.

A plasticizer can be defined as a low molecular weight material added to polymeric materials to improve their flexibility by lowering glass transition temperature. Plasticizer interacts with the polymer chains on the molecular level as to accelerate its viscoelastic response or increase chain mobility [3]. Dioctyl phthalate is the most commonly used plasticizer for PVC. However, these petroleum-derived products cause endocrine-disruption activity in laboratory rats. Extensive research is going on to find alternative plasticizers for plastic materials. Ionic liquids are one of these few novel alternatives which have shown hopeful results [4]. Ionic liquids, which are salts that are liquid at or near room temperature, exhibit very interesting properties [5, 6]:

- They have wide liquid range.
- Due to their negligible vapor pressure they are not lost into the environment.
- They can solubilize a number of organic molecules and transition metal complexes, which may present reduced solubility in conventional solvents.
- Due to selective solute-ionic liquid interactions it is possible to remove/recover target solutes from reaction media, while keeping other reactants or products.

- Due to their good thermal stability, reactive processes may occur at high temperatures.

Ionic liquids can be based on pyridinum, imidazolium or ammonium. There are also phosphonium based ionic liquids. Phosphonium salts are much more thermally stable than the corresponding ammonium and imidazolium salts. The imidazolium cation also contains protons which are not entirely inert. While they are somewhat acidic, phosphonium salts have no such acidic protons [7].

After preparation step, ion exchange membranes were characterized by physicochemical, morphological and electrochemical ways in this study.

Physicochemical characterization consists of volume fraction of water, density of the membrane, membrane porosity, ion exchange capacity and isotherm studies. Morphological characterization is performed by SEM. Electrochemical characterization of membranes prepared includes the calculation of transport number by measuring membrane potential.

Ion exchange membrane separation processes are used today in several applications from water desalination to electrodialysis to the production of acids and bases by electrodialysis or the elimination of toxic components from industrial effluents. Some processes such as electrodialysis are used on a large scale for the desalination of water or the removal of certain electrolytes from food and chemical products. Other processes such as Donnan dialysis, bipolar membrane water dissociation or the continuous electrodeionization are used on a much smaller scale for the production of ultra pure water or for purification and separation of certain food and pharmaceutical components, or for the treatment of industrial waste streams to recover toxic and valuable materials such as heavy metals or radioactive isotopes [8].

In this research, the membranes were tested by Donnan dialysis. As a result, $Na^+$ ion diffusivities in the membranes were found. The basic applications of the ion exchange membrane process are based on the Donnan membrane equilibrium principle and this process solve two important environmental problems, for the recovery of valuable ions, and the removal of undesirable ions from waste water.

# 2. Theory

## 2.1. Transport Number

In this study, transport number of an ion in cation exchange membranes prepared was calculated by using *Membrane Potential* method.

When an ion exchange membrane separates solutions containing different concentrations of same counter-ions, an electrical potential is generated across the membrane named as membrane concentration potential. Because a cation exchange membrane is permeable to cations, and anions can not pass through the membrane, a difference of electrochemical potential occurs across the membrane, which causes the membrane concentration potential [9].

A liquid junction represents a boundary between two dissimilar solutions across which ions migrate. The differences in mobilities of positive and negative ions result in a liquid junction potential (Figure 1).

Solution I and II contain several ions 1, 2,..., *i*,... at concentrations $C_1$, $C_2$,..., $C_i$,... and they form a liquid junction. The values of $z$ are valences and the values of $t$ are transport numbers. A potential $dE$ will arise from the process occurring in the thin boundary indicated by the dashed line.

If a Faraday current is passed reversibly from Solution I to Solution II, the change in free energy is given by,

$$dG_i = \frac{t_i}{z_i}[(\mu_i + d\mu_i) - \mu_i] = \frac{t_i}{z_i} d\mu_i \quad (1)$$

where $dG_i$ is the change in free energy, $t_i$ is transport number, $z_i$ is valence of $i$ ve $\mu_i$ is the chemical potential of $i$.

For the transfer of all ions, the total free energy change is;

$$\Delta G = \sum_i \frac{t_i}{z_i} d\mu_i = \sum_i \frac{t_i}{z_i} RTd\ln a_i \quad (2)$$

$$\Delta G = -FdE \quad (3)$$

$$dE = -\frac{RT}{F} \sum_i \frac{t_i}{z_i} d\ln a_i \quad (4)$$

where $R$ is gas constant and $T$ is absolute temperature, $a_i$, the activity of $i$, $F$ is Faraday constant and $dE$ is the potential when 1 Faraday current passes through the solutions. Membrane potential is calculated as follows [10];

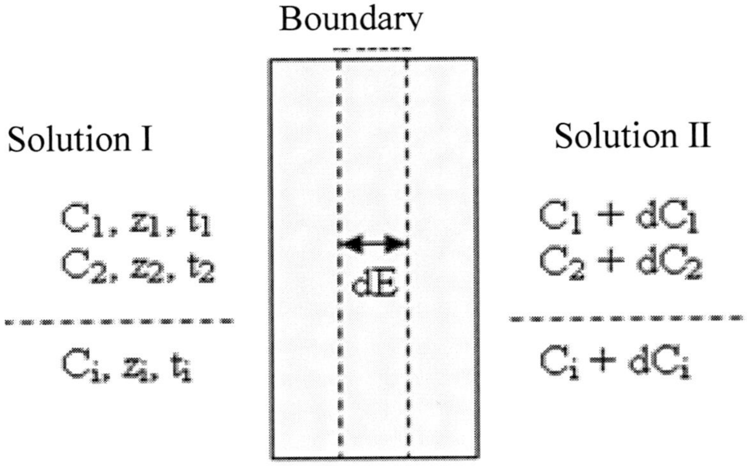


Figure 1. Liquid junction between two different solutions.

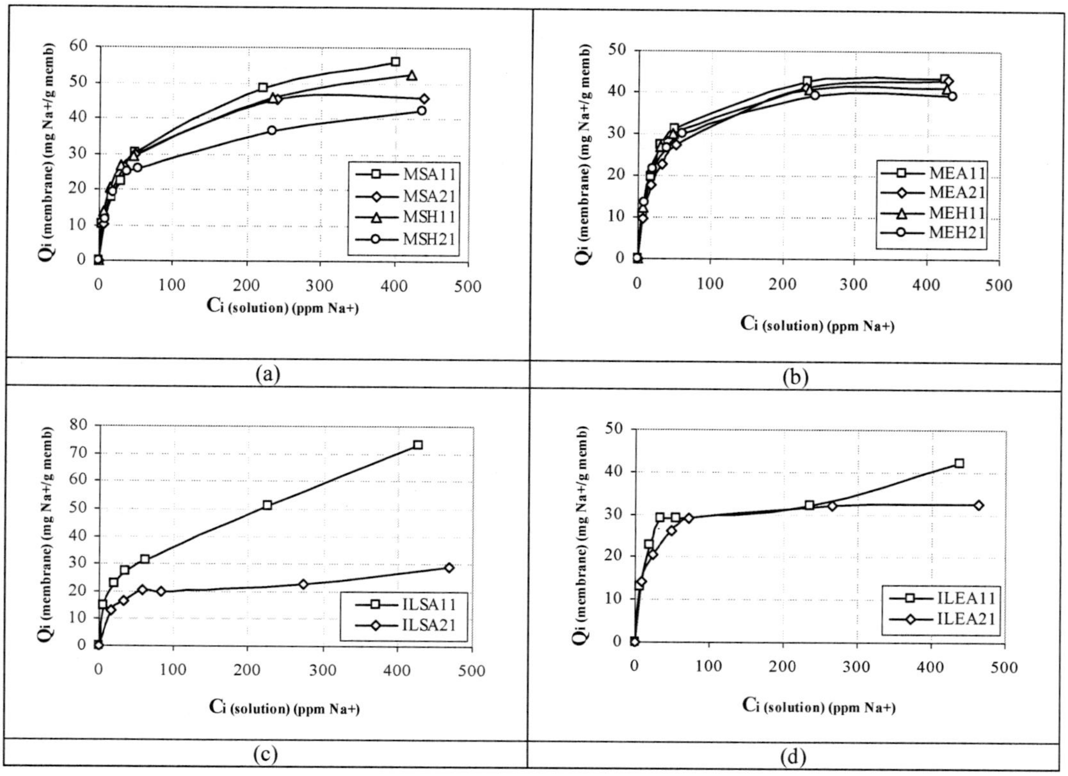


Figure 2. Charts illustrating equilibrium concentration of $Na^+$ in (a) MS-type, (b) ME-type, (c) ILS-type, (d) ILE-type membrane phase and that in solution phase at 30°C.

$$\Delta E = -\frac{RT}{F}\int_{I}^{II}\sum_{i}\frac{t_i}{z_i}\,d\ln a_i \tag{5}$$

When an ion exchange membrane separates solution having an activity of $a_1$ from solution having an activity of $a_2$ of the same electrolyte ($a_1 > a_2$), the membrane potential is given by,

$$\Delta E = -\frac{\bar{t}_+}{z_+}\frac{RT}{F}\ln\frac{(a_+)_2}{(a_+)_1} - \frac{\bar{t}_-}{z_-}\frac{RT}{F}\ln\frac{(a_-)_2}{(a_-)_1} \tag{6}$$

For 1:1 electrolyte;

$$\frac{(a_+)_2}{(a_+)_1} = \frac{(a_-)_2}{(a_-)_1} = \frac{a_2}{a_1} \tag{7}$$

$$\Delta E = -(\bar{t}_+ - \bar{t}_-)\frac{RT}{F}\ln\left(\frac{a_2}{a_1}\right) \tag{8}$$

$$\Delta E = (1 - 2\bar{t}_+)\frac{RT}{F}\ln\left(\frac{a_2}{a_1}\right) \tag{9}$$

where $\Delta E$ is the potential for 1:1 electrolytes, $\bar{t}_+$ and $\bar{t}_-$ are transport numbers of cations and anions in the membrane phase and $a_1$ and $a_2$ are the activities of electrolytes [9]. The activities ($a_i$) are calculated by activity coefficients by using Debye-Hückel Theory of Electrolyte Solutions Eq. (10) [11].

$$\log_{10}\gamma_{\pm} = -0.510 z_+ |z_-| \frac{(I_m)^{1/2}}{1 + (I_m)^{1/2}} \tag{10}$$

Where $\gamma_{\pm}$ is the mean ionic activity coefficient, $z_i$ is valence of ion $i$ and $I_m$ is ionic strength. From the slope of the graph ($E$ versus ln $(a_2/a_1)$) transport numbers are calculated.

## 2.2. Diffusivity

The performance of cation exchange membranes prepared was analyzed by diffusion test with Donnan dialysis. Two-compartment cell of equal volumes was used and cation exchange membrane was placed between these compartments.

When a salt dissociates in solution, it is ions rather than the molecules which diffuse. In the absence of an applied electrical potential, the diffusion of a single salt may be treated as molecular diffusion [12]. As a result of the determination of cation concentration, molecular diffusivities are calculated by two-cell method [13]:

$$D = \frac{K\delta V}{2\varepsilon St}\ln\frac{c_{A2}^{0} - c_{A1}^{0}}{c_{A2}^{t} - c_{A1}^{t}} \tag{11}$$

where $K$ is tortuousity factor, $\delta$ is thickness of the membrane, $V$ is equal volume of each compartment, $\varepsilon$ is the fraction of the area which is open to mass transfer in the membrane, $S$ is cross-sectional area of the membrane, $c_{A1}^{0}, c_{A2}^{0}, c_{A1}^{t}, c_{A2}^{t}$ are molar concentrations of component $A$ in the cells at $t = 0$ and at $t = t$.

The constants of the experimental set-up can be collected under one constant $\alpha = (K\delta V / 2\varepsilon S)$, and this constant can easily be determined by conducting an experiment with a liquid pair whose molecular diffusivity is already known.

# 3. EXPERIMENTAL

## 3.1. Materials

Two kinds of polymers were used as binder for cast solution of membrane. First material used was suspension polymerized polyvinyl chloride (S-PVC) whereas the second one was emulsion polymerized polyvinyl chloride (E-PVC). Dioctyl phthalate or an ionic liquid, which is tetradecyl(trihexyl)phosphonium decanoate (IL), were used as plasticizer. For all types of membranes prepared, polystyrene crosslinked with divinyl benzene, macroporous, and strongly acidic cation exchange resins (CXR) was used as resin and cyclohexanone (CH) as solvent.

S-PVC was supplied from PETKIM and E-PVC from SOLVIN. CH was supplied from Lab-Scan, DOP from Akdeniz Kimya, IL from CYTEC and CXR from PUROLITE (C-150). The properties are given in Table 1.

Doctor blade was used to cast the solution over a glass manually. Two-compartment Plexiglas cell of equal volumes was used for performance tests. Two-compartment glass test chamber of equal volumes and Ag/AgCl electrodes (saturated KCl) were used for electrochemical characterization. The concentrations of cations were determined by Atomic Absorption Spectrophotometer (AAS, *Varian 10 Plus A*). Thicknesses of membranes were measured by micrometer (*Chuan Brand, 0-25 mm - 0.01 mm*).

**Table 1. Properties of materials used for the preparation of membranes**

| Material | Properties | Unit | Range/Value |
|---|---|---|---|
| CXR | Polymer structure | | Polystyrene crosslinked with DVB |
| | Functional groups | | $R\text{-}SO_3^-$ |
| | Particle size | mm | 0.3-1.2 |
| | Total exchange capacity | eq/L | 1.8 |
| | Maximum operating temperature | °C | 130 |
| S-PVC | K-number | | 69-72 |
| E-PVC | K-number | | 73 |
| DOP | Molecular weight | kg/kmol | 390.56 |
| | Boiling point | °C | 385 |
| | Density | kg/L | 0.986 |
| CH | Molecular weight | kg/kmol | 98 |
| | Boiling point | °C | 156 |
| | Density | kg/L | 0.945 |
| IL | Appearance | | Paste |
| | Molecular weight | kg/kmol | 655.11 |
| | Melting point | °C | 20 |
| | Boiling point | °C | >180 |
| | Density | kg/L | 0.887 |

## 3.2. Method

### *3.2.1. Membrane Preparation*

The common step for the preparation of all types of membranes was to grind the CXRs to a particle size around 45 µm. The resins powdered were dried at 60°C in an atmospheric oven for 24 hours. Two kinds of formulations were used in the preparation of membranes. In the first formulation, the ratio of DOP amount to PVC amount was chosen as 1:1 (*RLow*), while in the second as 2:1 (*RHigh*), where *Low* and *High* refer to the relative amount of plasticizer per unit amount of polymer. In the drying of membranes cast, two different methods were applied. In the first method, the solution cast on the glass surface was left to dry by ambient air while in the second by heating at 75°C in an oven. *RLow* for the membranes based on DOP is DOP:PVC:CH:CXR=1:1:6:3, while *RHigh* DOP:PVC:CH:CXR=2:1:6:3. *RLow* for the membranes based on IL is PVC:IL:CH:CXR=1:1:8:3, whereas *RHigh* is IL:PVC:CH:CXR=2:1:8:3. Preparation procedures are given in Table 2.

### *3.2.2. Membrane Characterization*

Characterization of membranes was performed in three ways: Physicochemical, morphological and electrochemical

**Table 2. Materials and preparation procedures of cation exchange membranes**

| Membrane Name | | Ratio | Preparation Procedure |
|---|---|---|---|
| MS-type | MSA11 | *RLow* | The mixture of S-PVC+DOP+CH was prepared. CXR was added and was heated to 45°C during mixing. The final mixture was cast on the glass surface and was left one day to dry by ambient air. |
| | MSA21 | *RHigh* | |
| | MSH11 | *RLow* | The mixture of S-PVC+DOP+CH was prepared. CXR was added and was heated to 45°C during mixing. The final mixture was cast on the glass surface and was dried at 75°C in an oven for 10 minutes. |
| | MSH21 | *RHigh* | |
| ME-type | MEA11 | *RLow* | The mixture of E-PVC+DOP+CH was prepared. CXR was added and was not heated during mixing. The final mixture was cast on the glass surface and was left one day to dry by ambient air. |
| | MEA21 | *RHigh* | |
| | MEH11 | *RLow* | The mixture of E-PVC+DOP+CH was prepared. CXR was added and was not heated during mixing. The final mixture was cast on the glass surface and was dried at 75°C in an oven for 10 minutes. |
| | MEH21 | *RHigh* | |
| ILS-type | ILSA11 | *RLow* | IL and S-PVC were mixed. CH was added and mixed. CXR was added on this mixture. The mixture was heated to 45°C. The final mixture was cast on the glass surface and was left one day to dry by ambient air. |
| | ILSA21 | *RHigh* | |
| ILE-type | ILEA11 | *RLow* | IL and E-PVC were mixed. CH was added and mixed. CXR was added on this mixture. The final mixture was cast on the glass surface and was left one day to dry by ambient air. |
| | ILEA21 | *RHigh* | |

**Table 3. Physicochemical characteristics of membranes cast**

| **Name** | ***IEC*** (meq $Na^+$/g-dry) | $\phi_w$ | $\rho_d$ (g/cm$^3$) | $\rho_w$ (g/cm$^3$) | $\tau$ |
|---|---|---|---|---|---|
| MSA11 | 2.085 ± 0.055 | 0.3859 ± 0.0002 | 0.94 ± 0.03 | 1.05 ± 0.03 | 0.37 ± 0.01 |
| MSA21 | 1.662 ± 0.055 | 0.3810 ± 0.0003 | 0.85 ± 0.03 | 0.94 ± 0.03 | 0.34 ± 0.01 |
| MSH11 | 2.108 ± 0.055 | 0.3991 ± 0.0003 | 0.82 ± 0.03 | 0.94 ± 0.03 | 0.35 ± 0.01 |
| MSH21 | 1.520 ± 0.043 | 0.3435 ± 0.0003 | 0.97 ± 0.03 | 1.08 ± 0.03 | 0.34 ± 0.01 |
| MEA11 | 1.589 ± 0.067 | 0.4094 ± 0.0003 | 1.06 ± 0.04 | 1.20 ± 0.04 | 0.42 ± 0.01 |
| MEA21 | 1.656 ± 0.060 | 0.3912 ± 0.0003 | 1.02 ± 0.04 | 1.18 ± 0.04 | 0.40 ± 0.01 |
| MEH11 | 1.401 ± 0.053 | 0.3986 ± 0.0003 | 1.13 ± 0.04 | 1.24 ± 0.04 | 0.43 ± 0.01 |
| MEH21 | 1.692 ± 0.065 | 0.3290 ± 0.0003 | 1.01 ± 0.04 | 1.09 ± 0.04 | 0.33 ± 0.01 |
| ILSA11 | 2.802 ± 0.053 | 0.3470 ± 0.0003 | 0.85 ± 0.03 | 1.05 ± 0.04 | 0.31 ± 0.01 |
| ILSA21 | 1.674 ± 0.051 | 0.3116 ± 0.0004 | 0.93 ± 0.09 | 1.16 ± 0.05 | 0.30 ± 0.02 |
| ILEA11 | 2.019 ± 0.072 | 0.3048 ± 0.0004 | 0.86 ± 0.04 | 1.02 ± 0.04 | 0.27 ± 0.01 |
| ILEA21 | 1.269 ± 0.132 | 0.3018 ± 0.0005 | 0.71 ± 0.03 | 0.86 ± 0.03 | 0.24 ± 0.01 |
| Resin (commercial) | 1.44 | 0.5000 | 1.25 | - | - |
| Resin (after grinding) | 2.324 ± 0.046 | - | - | - | - |

*Physicochemical Characterization*

Physicochemical characterization was performed including volume fraction of water, density of the membrane, membrane porosity, ion exchange capacity and isotherms [14].

For the determination of volume fraction of water, membrane was immersed in 50 mL of distilled water for 24 hours and then weighed. Wet membrane was dried at 60°C for 30 min, weighed, and the volume fraction of water was determined by [15];

$$\varphi_w = \frac{W_w - W_d}{W_w} \times 100 \qquad (12)$$

Dry and wet membranes of known areas and thicknesses were weighed for dry and wet densities, respectively.

$$\rho_d = \frac{W_d}{V_d}, \ \rho_w = \frac{W_w}{V_w} \qquad (13)$$

where $\rho_d$ and $\rho_w$ are dry and wet densities in g/cm$^3$, respectively and $V_d$ and $V_w$ are volumes of dry and wet membrane in cm$^3$, respectively.

Membrane porosity was calculated in the way below [2]:

$$\Delta V = \frac{(W_w - W_d)\rho_d}{\rho_{water} W_d}$$
$$\tau = \frac{\Delta V}{1 + \Delta V} \qquad (14)$$

where $\Delta V$ means the volume increase of the membrane upon absorption of the water per unit volume of dry membrane in cm$^3$-water/cm$^3$-dry membrane, $\rho_{water}$ is the density of water in g/cm$^3$ and $\tau$ is membrane porosity.

For the determination of ion exchange capacity (*IEC*) [16], the membrane was cut into three pieces of size 3 cm x 3 cm. They were immersed in 100 mL of 1 M HCl for 24 hours. After they were washed with distilled water for 4 hours, they were equilibrated with 100 mL of 0.01 M NaCl for 24 hours. *IEC* was calculated by the decrease of $Na^+$ concentration in NaCl solution.

$$IEC = \frac{M_{i,Na^+} - M_{e,Na^+}}{W_d} \qquad (15)$$

where $M_{i,Na^+}$ and $M_{e,Na^+}$ are the initial and equilibrium amount of $Na^+$ in meq $Na^+$, respectively. $W_d$ is the weight of dry membrane in g and *IEC* is weight based ion exchange capacity in meq $Na^+$/g.

In these studies, equilibrium relation between the concentration of ions in membrane and in solution at constant temperature (30°C) was found by using six different concentrations (25, 50, 75, 100, 300 and 500 ppm $Na^+$). The membrane was cut of size 3 cm x 3 cm. They were immersed in 100 mL of 1 M HCl for 24 hours. Its surface was washed with distilled water. 100 mL of the solutions were prepared and blank samples ($C_{0i}$) were taken for AAS. The membranes were held in the solutions at 30°C for 24 hours.

The solutions were taken after 24 hours and the metal ion concentration was measured by AAS. The concentration of $Na^+$ in the membrane ($Q_i$) versus that in the bulk solution ($C_i$) was plotted (Fig. 2).

*Electrochemical Characterization*

In this study, the transport number of $Na^+$ in cation exchange membrane prepared was found by measuring membrane potential in two-compartment test chamber of equal volumes [17]. The membrane was cut with diameter of 4 cm and was immersed in 100 mL of 1 M HCl for 1 hour. Then, it was placed tightly between the compartments. One half of the test chamber was filled with 150 mL of 0.01 M NaCl ($C_1$) solution. Magnetic stirrer bars were put. Other half was filled with 145 mL of 0.001 M NaCl ($C_2$) solution. The electrodes were connected to the multimeter. The concentration in the chamber half which was filled with 0.001 M NaCl solution was increased by adding 1 M NaCl solution and temperature was measured. The membrane potential values were read after each addition of 1 M NaCl solution. Membrane potential ($E$) versus ln ($a_2/a_1$) was plotted.

*Morphological Characterization*

Morphological characterization was performed by Scanning Electron Microscopy (*JEOL JSM-6060*). Membrane samples were coated with gold layer to make them conductive. Surface and cross-sectional views were examined.

### *3.2.3. Performance Study*

The performances of the membranes were analyzed by Donnan dialysis by using two-compartment cell of equal volumes. Membrane was equilibrated with 1 M HCl solution. It was placed between two compartments. Concentrate part was filled with $Na^+$ solution and permeate part with pure water. While both of the parts were continuously being mixed, samples were taken at specified time intervals. Cation concentrations were determined by AAS. The concentration of both of the parts versus time was plotted and diffusivities were determined from these data.

## 4. Results and Discussion

For the preparation of membranes, firstly plasticizer, then, solvent and finally, CXRs were added and mixed until the appearance of the cast mixture became homogeneous. The commercial CXR which has the particle size of 0.3-1.2 mm was ground to 45 μm since resin becomes finer, it is distributed more uniformly through the mixture resulting in a more homogenous membrane product [18].

### 4.1. Physicochemical Characterization

The results of physicochemical characterization were listed in Table 3.

Mean volume fraction of water for MS-type membranes was found as $0.3774 \pm 0.0001$, while that for ME-type membranes as $0.3821 \pm 0.0002$. Negligible difference in volume fraction of water can be explained by the resin ratio of membranes which was kept constant at 3:1 (CXR:PVC) for all types of membranes studied. Average water contents of IL based membranes were found as $0.3293 \pm 0.0003$ and $0.3033 \pm 0.0004$ for ILS-type and ILE-type membranes, respectively, showing that there was no remarkable difference in water content of the membranes. Same arguments could be proposed for both DOP and IL plasticized membranes by taking into consideration corresponding average values of water contents, $0.3797 \pm 0.0001$ and $0.3163 \pm 0.0002$.

Average dry density value for MS-type membranes was found as $0.90 \pm 0.02$ g/cm$^3$, while those for ME-type membranes as $1.05 \pm 0.02$ g/cm$^3$. ME-type membranes had greater density values with respect to MS-type membranes because ME-type ones had more regular and stiff structure than MS-type membranes.

Mean dry density values prepared with IL were found as $0.89 \pm 0.05$ g/cm$^3$ and $0.79 \pm 0.03$ g/cm$^3$ for ILS-type and ILE-type membranes, respectively.

MS-type membranes had a mean wet density value about 1.00 ± 0.02 g/cm$^3$, while ME-type ones had around 1.18 ± 0.02 g/cm$^3$. ME-type membranes had greater density values with respect to MS-type membranes because ME-type membranes had more rigid structure resulting in less thickness than MS-type ones.

Mean wet density values prepared with IL were found as 1.10 ± 0.03 g/cm$^3$ and 0.94 ± 0.03 g/cm$^3$ for ILS-type and ILE-type membranes, respectively. All types of membranes having same plasticizer to PVC ratio had similar decrease or increase trend in dry and wet density values.

The membrane porosity value of MS-type membranes was around 0.35 ± 0.01, while that of ME-type ones was about 0.39 ± 0.01.

Average porosity of ILS-type membranes was 0.30 ± 0.01 and that of ILE-type ones was 0.26 ± 0.01. Porosities of the membranes contained IL were lower than that of the membranes contained DOP regarding as the average values, 0.28 ± 0.01 and 0.37 ± 0.00, respectively. Molecular size of IL is greater than that of DOP. Free volume of IL and thus porosity of the membranes prepared with IL is expected to be larger than DOP regarding molecular sizes of them. However, density values of the membranes prepared with IL was not so different from that with DOP and porosity values of this type of membranes were lower than the membranes prepared with DOP. This situation can be explained by strong interactive forces in IL, which has an ionic structure, and resin particles in the membrane attract each other causing a decrease in porosity.

Ion exchange capacity (IEC) depends basically on the content of ion exchange resin in heterogeneous ion exchange membranes, the amount of which was kept constant in all formulations in this study.

An average value of IEC was found as 1.844 ± 0.026 meq $Na^+$/g-dry for MS-type membranes while that was found as 1.585 ± 0.031 meq $Na^+$/g-dry for ME-type ones.

Average IECs of ILS- and ILE-type membranes were found as 1.920 ± 0.037 meq $Na^+$/g-dry and 1.644 ± 0.075 meq $Na^+$/g-dry membranes, respectively. IEC decreased with increasing IL amount because the resins in the membrane having the formulation with *RHigh* ratio were covered by plasticizer more than that with *RLow* ratio.

When the mean IECs of all types of membranes were considered, it was seen that the average value of the membranes prepared with IL, 1.782 ± 0.042 meq $Na^+$/g-dry, was higher than that with DOP, 1.714 ± 0.020 meq $Na^+$/g-dry.

The resin ground had an IEC of 2.324 ± 0.046 meq $Na^+$/g-dry, which is greater than that of the resin in commercial size. The rise in IEC of CXR ground can be explained by the decrease in particle size and porosity of resin after grinding.

In isotherm studies, equilibrium relation between the concentration of ions in membrane and in solution at constant temperature (30°C) was investigated by using six different $Na^+$ concentrations.

Equilibrium relation between the concentration of $Na^+$ in MS-type membrane phase and in solution phase was shown in Figure 2(a). A steep change in all the isotherms was clearly seen up to the solution concentration of 50 ppm $Na^+$. Although there was a separation among isotherms in richer concentrations than 50 ppm $Na^+$, membranes having low PVC:DOP ratio (MSA11, MSH11) had similar behavior.

**Table 4. Transport numbers of membranes**

| *Name* | *MSA11* | *MSA21* | *MSH11* | *MSH21* | *MEA11* | *MEA21* | *MEH11* | *MEH21* | *ILSA11* | *ILSA21* | *ILEA11* | *ILEA21* |
|---|---|---|---|---|---|---|---|---|---|---|---|---|
| $t_{Na+}$ | 0.54 | 0.56 | 0.55 | 0.61 | 0.53 | 0.54 | 0.54 | 0.55 | 0.97 | 0.99 | 0.98 | 0.95 |

Sorption behavior of ME-type membrane phase was shown in Figure 2(b). Initial profile had a sharp increase up to solution concentration of 50 ppm $Na^+$ for all types of membranes. Around solution concentration of 250 ppm $Na^+$, all the isotherms reached a plateau level between 40 and 45 ppm $Na^+$.

Equilibrium relation of ILS-type membrane phase was shown in Figure 2(c). Sorption equilibrium curves for membranes prepared with ionic liquids remarkably differed from each other. Initial steep changes in sorption capacities of membranes were around 6 and 16 ppm $Na^+$ for ILSA11 and ILSA21, respectively.

The sorption behavior of the membrane ILSA11 was highly interesting since there was continuous rise in equilibrium curve, whereas the increase in sorption capacity for ILSA21 was around 10 ppm $Na^+$ within the same range of solution concentration.

Equilibrium behavior of ILE-type membrane phase was shown in Figure 2(d). ILEA11 and ILEA21 had a sharp change in sorption capacity reaching 15 ppm $Na^+$ with the solution concentration of 9 ppm $Na^+$. The plateau value reached by ILEA21 is 32 ppm $Na^+$ but ILEA11 showed an increase like ILSA11 at a solution concentration of 236 ppm $Na^+$.

Larger sorption capacities for the membranes having low PVC : IL ratio (ILSA11, ILEA11) was in accordance with IECs for the same membranes.

## 4.2. Electrochemical Characterization

While in an electrolyte solution current is carried by both cations and anions with different portions, in ion exchange membranes current is carried preferentially by counter-ions. The fraction of current that is carried by a certain ion is expressed by transport number of that ion [8].

**Table 5. Diffusivity values of membranes prepared**

| *Name* | *D (cm²/s) (x10⁵)* |
|---|---|
| MSA11 | 4.36 ± 0.79 |
| MSA21 | 1.97 ± 0.35 |
| MSH11 | 3.49 ± 1.18 |
| MSH21 | 2.62 ± 0.31 |
| MEA11 | 5.68 ± 1.23 |
| MEA21 | 4.11 ± 1.81 |
| MEH11 | 1.59 ± 0.37 |
| MEH21 | 3.17 ± 1.44 |
| ILSA11 | 4.19 ± 0.95 |
| ILSA21 | 1.31 ± 0.90 |
| ILEA11 | 13.0 ± 4.40 |
| ILEA21 | 5.70 ± 2.80 |

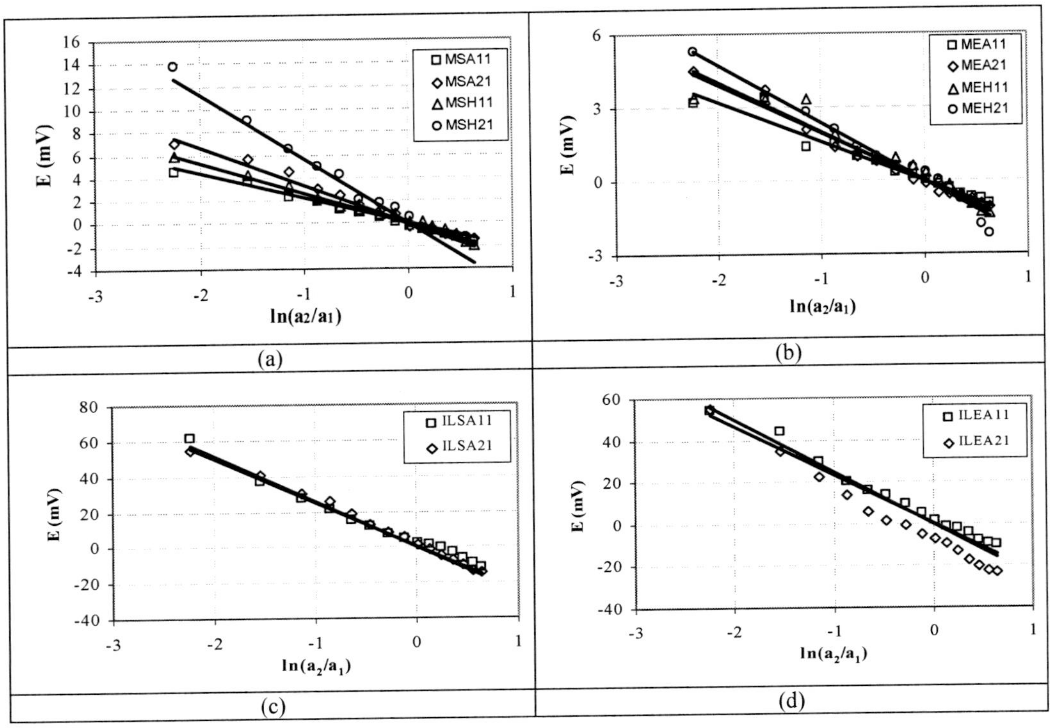


Figure 3. Charts illustrating the membrane potential versus $\ln(a_2/a_1)$ for (a) MS-type (b) ME-type (c) ILS- and ILE-type membranes.

In this study, the transport number of $Na^+$ in the membrane was found by the measurement of membrane potential.

Curves illustrating membrane potential versus $\ln(a_2/a_1)$ values for each membrane were shown in Figure 3. Transport numbers of cation $Na^+$ in membranes, as the slope of trend line decreased, with an increase were given in Table 4.

It was obviously seen from Table 4 that transport numbers of IL based membranes were larger than those of DOP based ones since IL, which is a conductive material, was combined with CXR that is a selective material, resulting in an increased mobility of ions passing through the membrane.

## 4.3. Morphologic Characterization

Morphological characterization was performed by SEM analysis to investigate surface and cross-sectional views of the cation exchange membranes prepared.

Surface views of MS-type membranes dried by ambient air were shown in Figure 4. In these images, the surfaces facing the glass, on which the membranes were cast, and the surfaces open to air were illustrated. On the glass side of MSA11 and MSA21, they had organized structure with very small pores. However, they had large cavities on the air sides. It may be because of solvent evaporation during drying. When MSA11 and MSA21 were compared, the pore size on the air side of MSA21 was smaller than that of MSA11. The reason could be doubled plasticizer ratio used in MSA21.

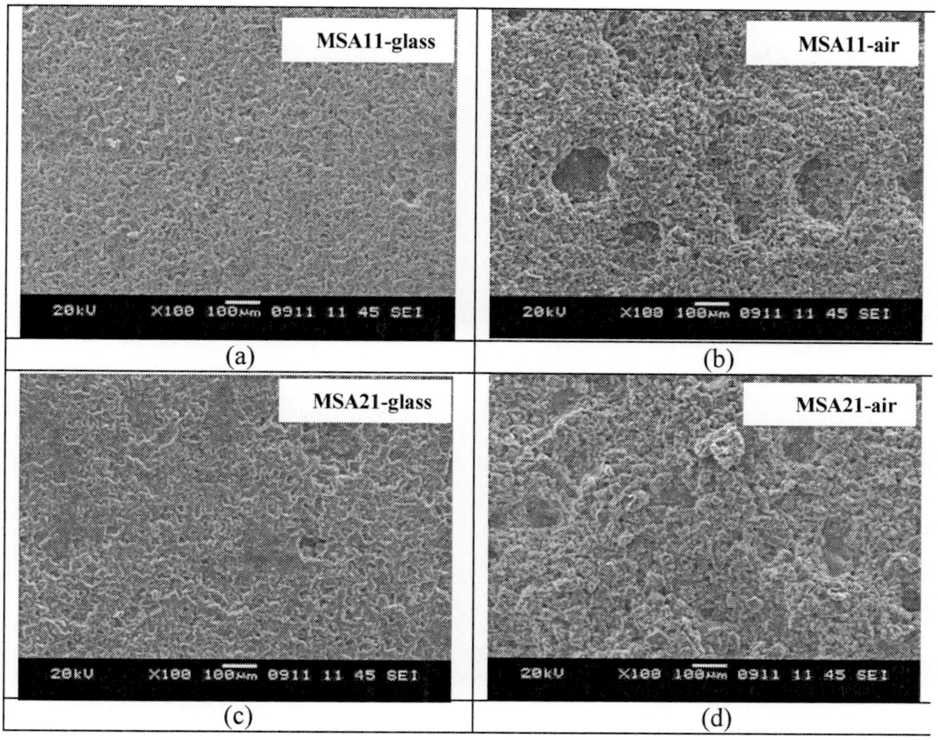


Figure 4. SEM images of MS-type cation exchange membranes dried by ambient air.

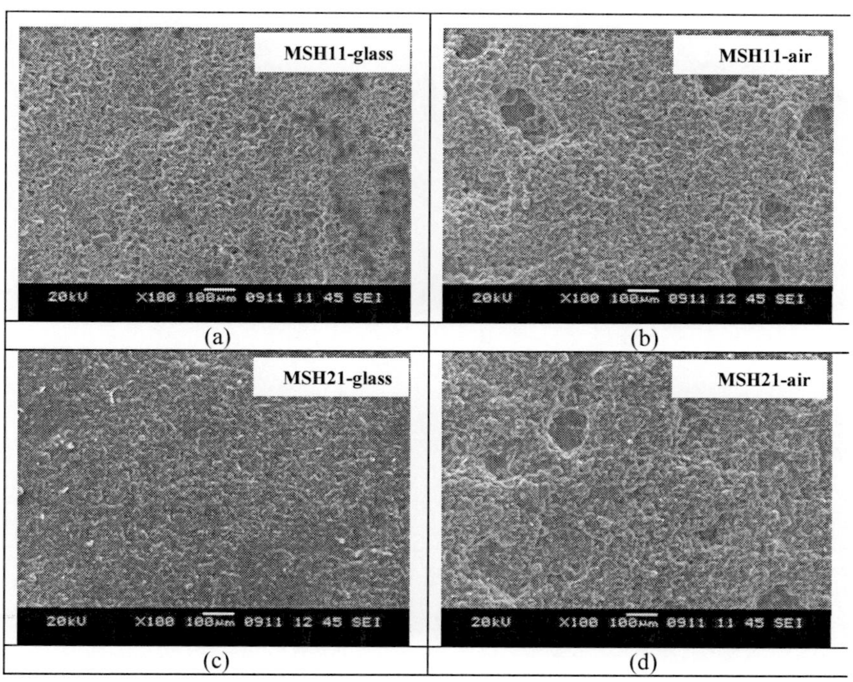


Figure 5. SEM images of MS-type cation exchange membranes dried by heating.

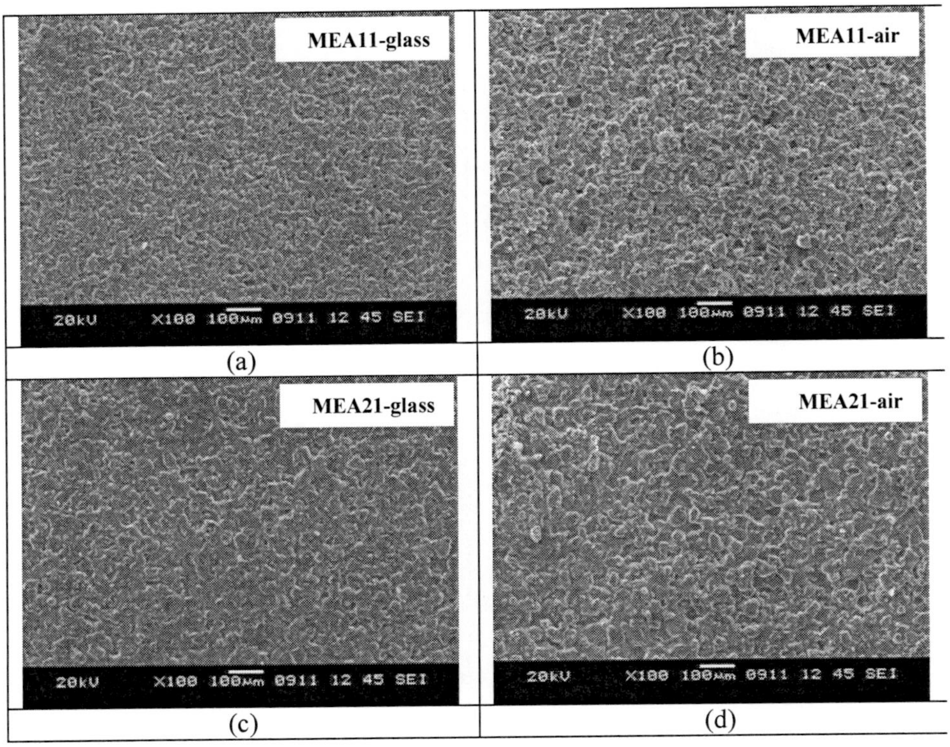


Figure 6. SEM images of ME-type cation exchange membranes dried by ambient air.

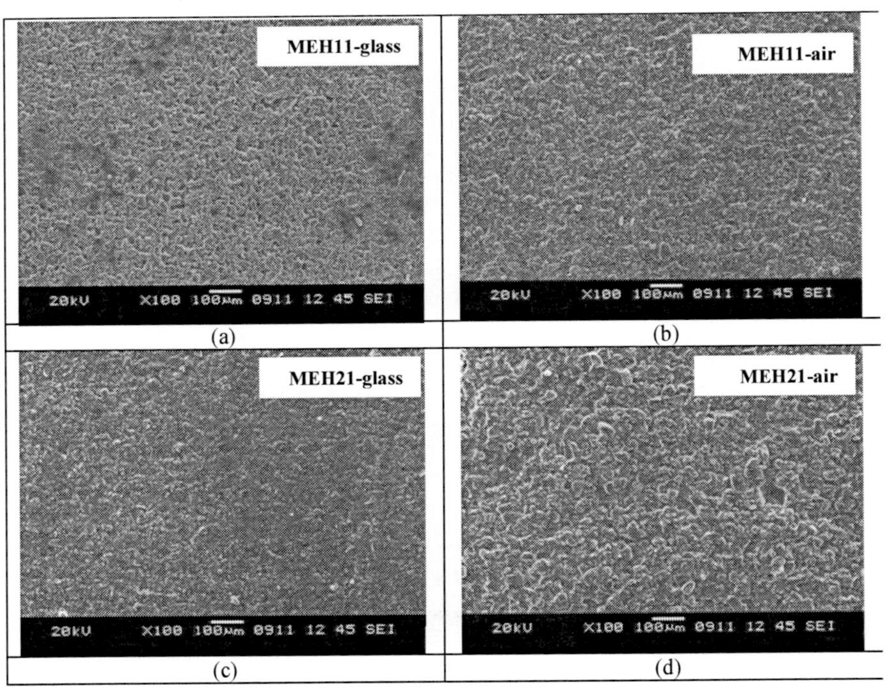


Figure 7. SEM images of ME-type cation exchange membranes dried by heating.

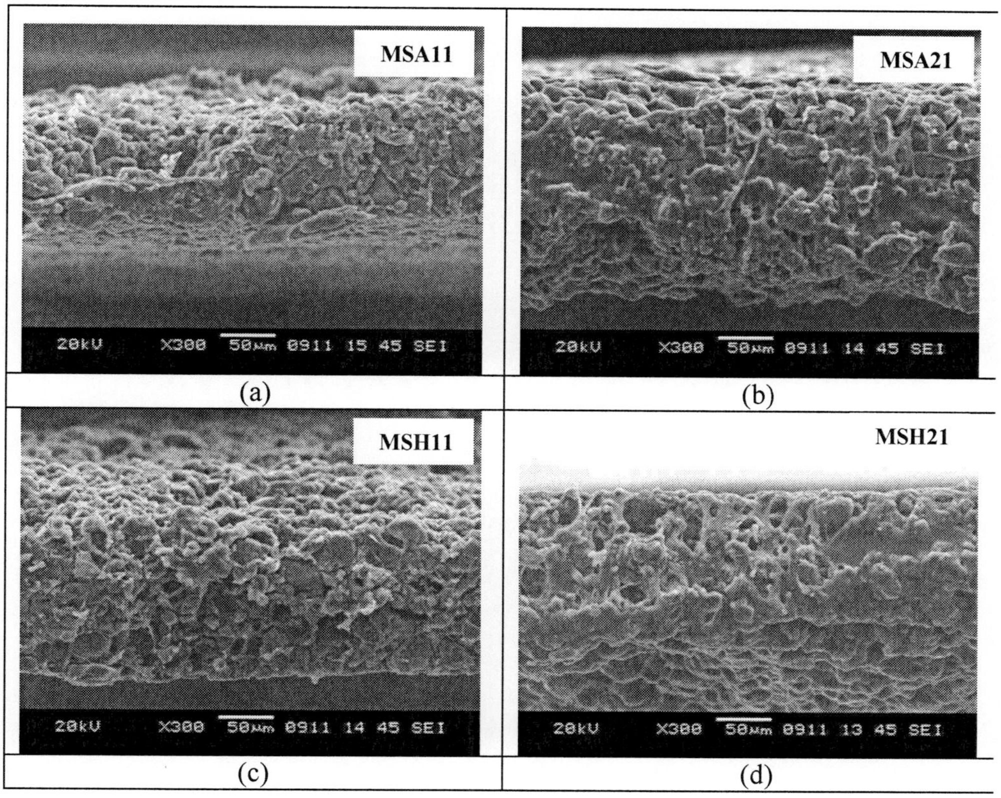


Figure 8. Cross-sectional views of MS-type cation exchange membranes.

Surface views of MS-type membranes dried by heating in an oven were shown in Figure 5. On the glass sides, membranes had an ordered structure, but they had large cavities on their air sides. When the plasticizer ratio was doubled as in MSH21, the pores on the glass side were closed.

In the images of ME-type cation exchange membranes given in Figure 6, on the glass sides, membranes had more regular structure than on the air sides of them. The air side of MEA11 had some pores, while that of MEA21 did not have any pores.

From Figure 7, there was not so much difference between glass and air side for ME-type membranes dried by heating.

Surfaces facing glass in ME-type membranes were similar to that in MS-type ones. However, surfaces open to air in ME-type membranes had no cavities and they were smoother than those in MS-type ones.

Cross-sectional views of membranes prepared with DOP were shown in Figure 8 and Figure 9. CXR particles were attached to each other and to PVC particles by DOP. As a result, the membranes prepared with lower amount of DOP were more brittle than those with higher one. At the first glance, thickness of the membrane MSA11 is less than that of membrane MSA21. The distribution of the resins was more balanced in the membranes having a ratio of *RHigh* than those of *RLow* due to the larger amount of plasticizer. In the comparison of the drying condition, it can be stated that more regular structure is obtained in drying by heating.

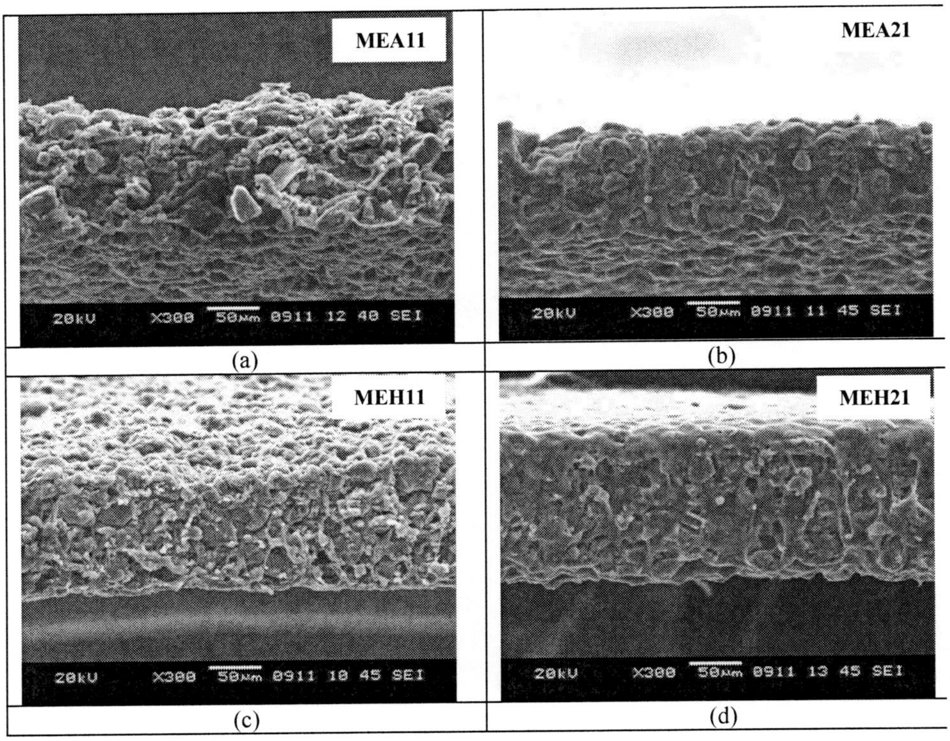


Figure 9. Cross-sectional views of ME-type cation exchange membranes.

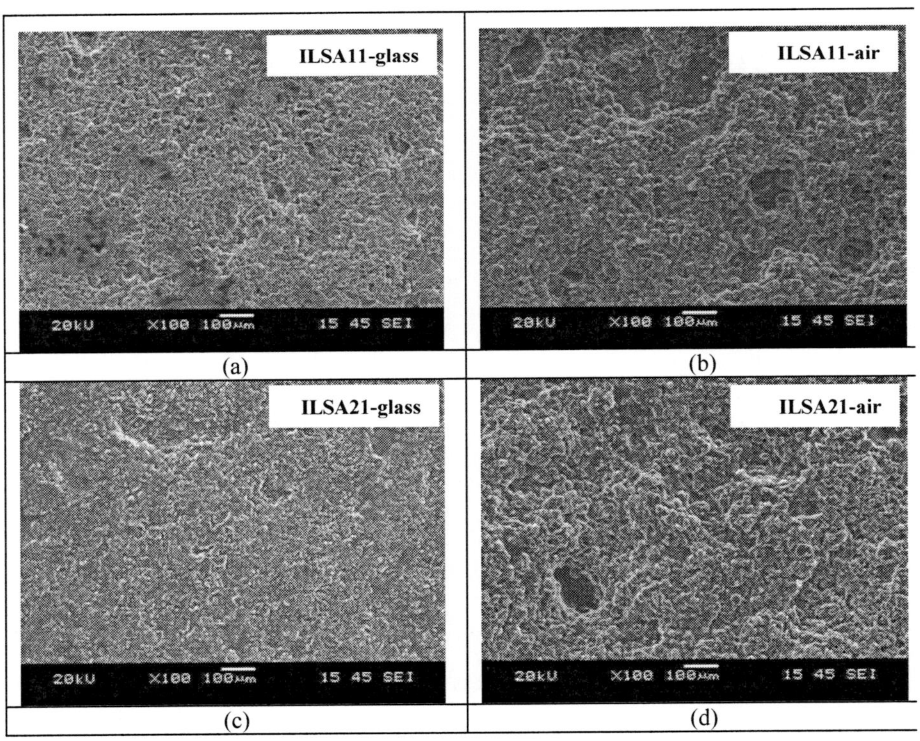


Figure 10. SEM images of ILS-type cation exchange membranes.

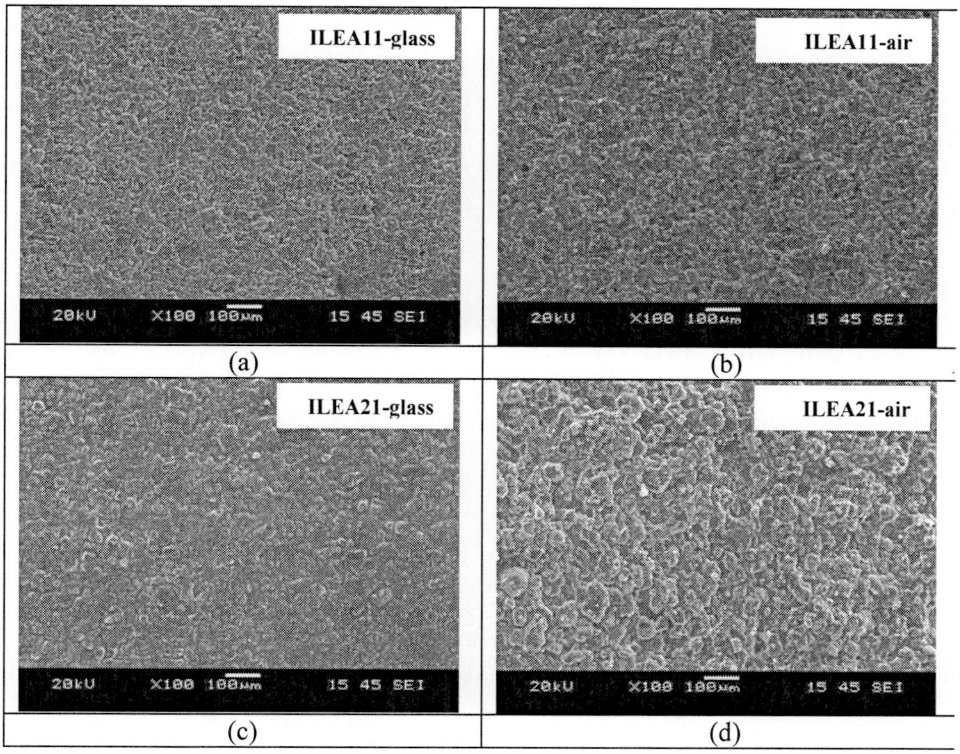


Figure 11. SEM images of ILE-type cation exchange membranes.

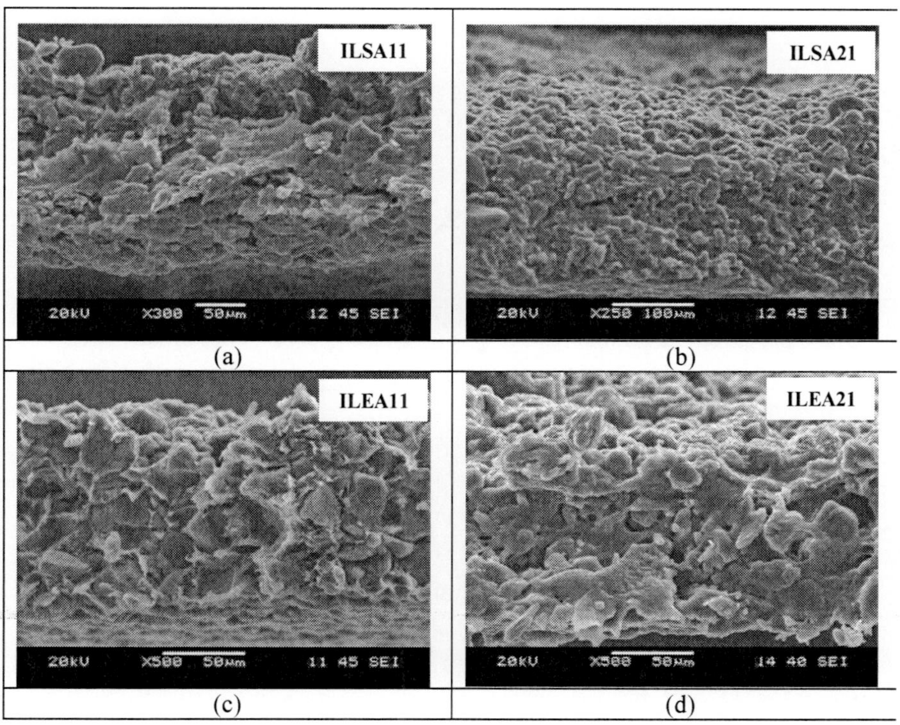


Figure 12. Cross-sectional views of IL-type membranes.

Surface views of membranes prepared with IL were shown in Figure 10 and Figure 11. On the glass side of them, membranes had organized structure with small pores. Nevertheless, ILS-type membranes had larger pores on the air sides. When ILSA11 and ILSA21 were compared, there were fewer pores on the air side of ILSA21 than that of ILSA11. The cause can be higher plasticizer amount of ILSA21. At the air side of ILEA11 had very small pores, while ILEA21 did not have pores.

The surfaces facing the glass of ILS-type membranes were rougher than those of ILE-type ones. The surfaces open to air of ILS-type membranes had cavities while ILE-type membranes were more uniform (Figure 10 and Figure 11).

Cross-sectional views of IL based membranes were shown in Figure 12. They were more disordered than the membranes prepared with DOP. Moreover, ILE-type membranes had more uniform view with respect to ILS-type ones.

### 4.4. Performance Study

After getting the graphs of concentration versus time, the diffusivity values were calculated by using Donnan dialytic method. In calculations, a constant ($\alpha$) was determined by conducting experiments with a liquid pair whose molecular diffusivity was already known. The molecular diffusivity taken was $D$=1.17x10$^{-5}$ cm$^2$/s at the concentration of NaCl, $C$ = 0.4 M [12]. The resulting diffusivity values were shown in Table 5.

All diffusivity values were around (3.47 ± 1.10) x 10$^{-5}$ except for ILEA11, which had a value of (13.0 ± 4.40) x 10$^{-5}$. IEC of this membrane was also high (Table 3). The values were about the level of molecular diffusivities in liquids in order of magnitude -5.

## 5. Conclusion

In this research, PVC based strong acidic cation exchange membranes were prepared by heterogeneous method and then characterized by physicochemical, morphological and electrochemical ways.

First parameter was the type of polymeric binder. Two kinds of PVC, S-PVC and E-PVC, were used to prepare cation exchange membranes. Second parameter was the kind of plasticizer from DOP and IL. Third one was drying procedure after casting the solution as air drying and drying by heating in an oven at 75°C.

For the preparation of membranes, commercial CXR was ground to 45 μm. As resin becomes finer, is distributed more uniformly through the mixture resulting in a more homogenous membrane.

Thicknesses of membranes were varied with respect to polymers used for preparation. ME-type membranes had a thickness lower than MS-type ones.

There was no notable difference in volume fraction of water among all types of membranes.

ME-type membranes had greater density values with respect to MS-type ones because ME-type membranes had more stiff structure than MS-type membranes.

Porosities of the IL based membranes were lower than that of the DOP based ones.

In this study, the amounts of ion exchange resins were kept constant in all formulations. For IL based membranes, IEC decreased with increasing IL amount. The average value of IL based membranes was higher than DOP based ones.

A sharp change in all the isotherms for the membranes were seen at dilute concentrations. There was a separation among isotherms for MS-type membranes in richer concentrations. At richer concentrations, all the isotherms for ME-type membranes reached plateau levels. Sorption equilibrium curves for IL based membranes differed remarkably from each other.

Membranes were characterized in morphological ways by SEM. The surfaces (facing the glass) of ME-type membranes were similar to those of MS-type. However, the surfaces open to air of ME-type membranes had no cavities and they were smoother than MS-type ones. Glass sides of the ILS-type membranes were rougher than those of ILE-type membranes. Air sides of ILS-type membranes had cavities while ILE-type ones were more uniform. In cross-sectional views of the IL based membranes, they were more disordered than DOP based ones.

The transport numbers of IL based membranes was larger than that DOP based ones.

The performances of membranes were tested by Donnan dialysis. All diffusivity values were similar except ILEA11, which had a higher value. The values were in the range of molecular diffusivities in liquids.

By looking at the experimental results, a strongly acidic heterogeneous cation exchange membranes can be prepared with tetradecyl(trihexyl)phosphonium decanoate, which is a phosphonium based ionic liquid. IL based membranes prepared had similar physicochemical and morphological properties and they had also better electrochemical properties than the membranes prepared with DOP, which is a traditional plasticizer. Hence, ion exchange membranes with higher conductivity and higher electrochemical properties can be produced by using ionic liquids.

## REFERENCES


[1] A.A. Zagorodni, *Ion Exchange Materials Properties and Applications*, Elsevier, 2007.

[2] R.K. Nagarale, G.S. Gohil, V.K. Shahi, Recent developments on ion-exchange membranes and electro-membrane processes, *Adv. Colloid Interface Sci.* 119 (2006) 97-130.

[3] G. Wypych, Handbook of Plasticizers, *ChemTec Publishing,* Toronto-New York, 2004.

[4] M. Rahman, C.S. Brazel, The plasticizer market: An assessment of traditional plasticizers and research trends to meet new challenges, *Prog. Polym. Sci.* 29 (2004) 1223-1248.

[5] R.E.D. Sesto, C. Corley, A. Robertson, J.S. Wilkes, Tetraalkylphosphonium-based ionic liquids, J. *Organometallic Chemistry* 690 (2005) 2536–2542.

[6] P. Izak, S. Hovorka, T. Bartovsky, L. Bartovska, J.G. Crespo, Swelling of polymeric membranes in room temperature ionic liquids, *J. Membr. Sci.* 296 (2007) 131–138.

[7] CYTEC *Industries, CYPHOS IL 103 Phosphonium ionic liquid,* http://www.cytec.com/specialty-chemicals/PDFs/PhosphoniumSalts/CYPHOSIL103.pdf

[8] H. Strathmann, *Ion-Exchange Membrane Separation Processes,* Elsevier, 2004.

[9] T. Sata, Ion Exchange Membranes Preparation, Characterization, Modification and Application, *The Royal Society of Chemistry,* UK, 2004.

[10] N. Lakshminarayanaiah, *Transport Phenomena in Membranes,* Academic Press, New York, 1969.
[11] I.N. Levine, *Physical Chemistry,* McGraw-Hill International Editions, 1995.
[12] T.K. Sherwood, R.L. Pigford, C.R. Wilke, *Mass Transfer, McGraw-Hill,* International Student Edition, 1975.
[13] C.J. Geankoplis, *Transport Processes and Unit Operations,* Prentice-Hall Inc., 1993.
[14] G.S. Gohil, V.K: Shahi, R. Rangarajan, Comparative studies on electrochemical characterization of homogeneous and heterogeneous type of ion-exchange membranes, *J. Membr. Sci.* 240 (2004) 211-219.
[15] J.K. Fang, H.C. Chiu, J.Y. Wu, S.Y. Suen, Preparation of polysulfone-based cation exchange membranes and their application in protein separation with a plate-and-frame module, *React. Funct. Polym.,* 59 (2004) 171-183.
[16] P.V. Vyas, P. Ray, G.S Trivedi, S.K. Adhikary, R. Rangarajan, Studies on exchange equilibria of cations between cation-exchange membranes and electrolytic solutions, *J. Colloid Interface Sci.* 246 (2002) 366-371.
[17] A. Canas, J. Benavente, Electrochemical characterization of an asymmetric nanofiltration membrane with a NaCl and KCl solutions: influence of membrane asymmetry on transport phenomena, *J. Colloid Interface Sci.* 246 (2002) 328-334.
[18] P.V. Vyas, B.G. Shah, G.S. Trivedi, P. Ray, S.K. Adhikary, R. Rangarajan, Studies on heterogeneous cation exchange membranes, *React. Funct. Polym.* 44 2000 101-110.

In: Chemical and Biochemical Reactions
Editors: G. E. Zaikov, L. Liu et al.

ISBN 978-1-61122-860-1


*Chapter 22*

# STRUCTURE AND PROPERTIES OF POLYETHYLENE COATINGS MODIFIED BY BISMALEIMEDES

***T. R. Deberdeev*, O. V. Stoyanov and R. Ya. Deberdeev***
Kazan State Technological University, Kazan 420015, Russia

## INTRODUCTION

In the work [1], the regularities of chemical crosslinking of HDPE by some maleimide derivatives (MI) have been investigated. It was shown that the crosslinking occurs in the process of polymer oxidation by the air oxygen. In this case, the network junctions are formed directly by the MI molecules.

The goal of this work is the investigation of the structure-mechanical characteristics of HDPE chemically crosslinked by MI.

## EXPERIMENTAL PART

*Materials.* The powdered non-stabilized high density polyethylene (HDPE) type 16803-070 (GOST 16337-77) prepared by AO "Kazanorgsynthez" was used. The particle size of the powder has diameter of 315 jam. The HDPE characteristics are shown in Table 1.

As the crosslinking agents, the re-crystallized N.N'-μ-phenylenedimaleimide (PDMI) produced by Tambov NPO "Pigment" and Kharkov Khimreactives plant; dithiodi-phenylenedimaleimide (TPDMI) produced by Tambov VNIIKhimpolymer; and hexamethylenedimaleimide (HMDM1). The last one was synthesized according to procedure described in [2]. The maleimides characteristics are shown in Table 2.

The HDPE and MI compositions were prepared by mechanical mixing of components. The coating samples were formed by spreading compositions on substrate with consecutive rolling and kept in a heated oven at different temperatures (T) and times (t). The temperature in the oven was held constant with accuracy of ±1 °C. The coating thickness was different and the samples were prepared by using special frames. The sample cooling was performed in air at room temperature. As a substrate, the fluoroplast 40П was used.

## Test Methods

The X-ray investigations were performed by using DPOH-2.0 diffractometer, the nickel filter was applied to the C-K radiation. The state of crystallinity was determined according to Germans and Vaidinger described in [3]. The effective sizes of crystallites were calculated using the diffraction curves according to Sherrer [3].

The heat-physical investigations of the samples were conducted according to procedure of differential scanning calorimetry by using the micro-calorimeter DCM-2M at heating rate 12.5 K/min. The preciseness of the heat effects determinations was ±4 %. The results were treated according the procedure published in [4].

**Table 1. Characteristics of used HDPE**

| | |
|---|---|
| Density, g/ $cm^3$ | 0.920 |
| Crystallir\ity state, % | 42 |
| Melting point, °C | 108 |
| Double bonds content >C=C< per 1000 carbon atoms | 0.3 |
| -$CH_3$ groups content per 100 carbon atoms | 3.4 |
| Melt flow index, g/10 min | 7.0 |
| Tensile strength, MPa | 11.0 |
| Elongation at break, % | 450 |
| Yield stress, MPa | 11.0 |

The sample elastic deformation heat effects evaluation was conducted by using the deformation micro-calorimeter [5,6]. The measurements and results treatment were conducted according to procedure described in [7,8]. The sample density was determined by the floating method at 20±0.3°C (GOST 15139-69). The error of measurements was less than 0.1 %. The stress-strain properties were evaluated according to the GOST 11262-80 procedure by using the "Instron" tester. The adhesion strength of coatings prepared by the powder technology was evaluated by film separation from the counterpart under 90° in the Instron tester. The rate of separation was 10 mm/min. The samples for adhesion tests of coatings were prepared by blowing a polymer powder on a pipe size 100x100x90 mm carefully cleaned by pellet stream and preheated to 220°C. The time to prepare the polymer layer thickness 0.7 - 1.0mm was 40 sec. The adhesion strength under normal pulling force was determined according to procedure published in [9].

The integral coefficient of hexane diffusion was determined according to procedure published in [10].

The microhardness of coatings was evaluated by using microhardness tester ПМТ-3 according to the Vickef's method (a four-edge pyramid with the top angle 136°).

**Table 2. Characteristics of crystalline crosslinking agents**

| MI type | Structural formula | Empirical formula | $T_{melt}$, °C |
|---|---|---|---|
| PDMI | 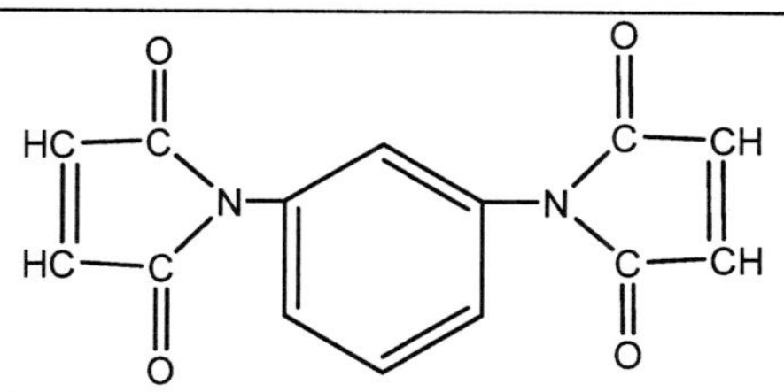 | $C_{14}H_8N_2O_4$ | 204 |
| TPDMI | 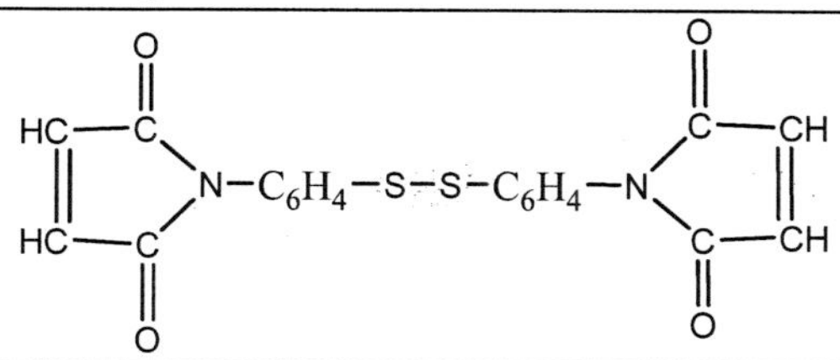 | $C_{20}H_{12}N_2S_2O_4$ | 167 |
| HMDMI | 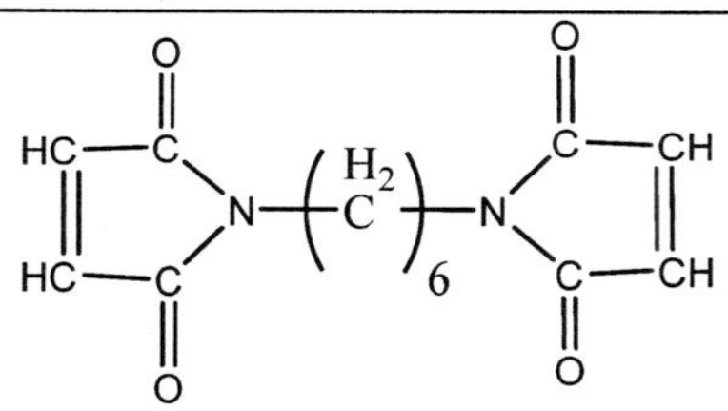 | $C_{14}H_{16}N_2O_4$ | 162 |

The load, P, was 10 g, the loading time - 40 sec. The microhardness (H) was calculated according to the formula:

$$H = P/(ad)^2,$$

where d - is the diagonal of the indentation impression in μm; a - is equal to 0.306 (transformation coefficient).

# Results and Discussion

## Structural Features of Chemically Crosslinked Polyethylene Coatings

It is known that formation of crosslinked network leads to structure changes in polyethylene. To evaluate the changes, we studied some parameters which directly or indirectly characterize the structure organization of polymer in the coating films.

It was shown before [1] that MI forms directly the network chemical junctions in PE. Because of this, the state of crystallinity, a, of the films crosslinked with PDMI and HMDMI determined by the X-ray, IRS and DSC methods has a tendency to some smal decrease with increase of the crosslinking agent concentration (Table 3). However, even with incorporation of large enough amounts of the agent, the amorphicity of the polymer is relatively small. It should be noted that the effect of PDMI and HMDMI on the α is practically the same-the differences in α values are in the region of data scatter. Thus, the found in [1] significant differences in the network junctions distribution along the film thickness do not affect the character of *integral* change of α.

With incorporation of Ml into PE, some narrowing of the melting temperature range and the maximal melting temperature, $T_{max}$ decrease take place. This witnesses the reduction of crystallite size. The obtained result is confirmed by the data of X-ray analysis and evaluation of the crystallites for PE crosslinked by PDMI. The crystallite size, $L_{110}$ decreases with increase of the PDMI concentration, but the $L_{200}$ does not show any regular changes (Table 3).

In case of the TPDMI, there is a reversed picture - the a and melting temperature slightly decrease with decrease of the MI concentration what is justified [1] taking into account the decrease of the crosslink density with increase of the TPDMI content (Table 4).

**Table 3. Structural characteristics of HDPE crossliriked with PDMI and HMDMI**

| MI type and % | Melting range, °C | | | ΔH, kJ/kg | Crystallinity state, % | | | Effective size of crystallites, A | |
|---|---|---|---|---|---|---|---|---|---|
| | $T_{in}$ | $T_{end}$ | $T_{max}$ | | DSC | IRS | X-ray | $L_{110}$ | $L_{200}$ |
| 0 | 72 | 124 | 113 | 81 | 33 | 37 | 42 | 151 | 81 |
| HMDMI* | | | | | | | | | |
| 0.5 | 72 | 115 | 18 | 73 | 32 | 34 | - | - | - |
| 1.0 | 69 | 116 | 109 | 78 | 31 | 35 | - | - | - |
| 2.0 | 70 | 115 | 109 | 75 | 31 | 35 | - | - | - |
| 4.0 | 72 | 1115 | 108 | 74 | 30 | 35 | - | - | - |
| 6.0 | 70 | 118 | 109 | 61 | 27 | 27 | - | - | - |
| PDMI* | | | | | | | | | |
| 0.5 | 72 | 115 | 109 | 78 | 32 | 34 | 40 | 143 | 81 |
| 1.0 | 70 | 117 | 109 | 78 | 32 | 35 | 36 | 136 | 92 |
| 2.0 | 70 | 115 | 109 | 76 | 32 | 34 | 37 | 135 | 79 |
| 4.0 | 71 | 116 | 109 | 74 | 3 | 33 | 39 | 135 | 90 |
| 6.0 | 70 | 117 | 110 | 64 | 27 | 27 | 39 | 135 | 79 |

* T=200°C, t=30min.

**Table 4. Structural characteristics of HDPE modified by TPDMI (T=200°C)**

| TPDMI, % | t=60min | | | t=180min | | |
|---|---|---|---|---|---|---|
| | Crystallinity state,% | | $T_{max}$, °C | Crystallinity state,% | | $T_{max}$, °C |
| | IRC | DSC | | IRS | DSC | |
| 0 | 33 | 32 | 112 | 28 | 31 | 108 |
| 0.5 | 31 | 29 | 109 | 24 | 26 | 106 |
| 1.0 | 33 | 30 | 109 | 28 | 27 | 107 |
| 2.0 | 33 | 31 | 111 | 29 | 27 | 107 |
| 4.0 | 34 | 32 | 111 | 32 | 28 | 109 |
| 6.0 | 34 | 32 | 111 | 32 | 28 | 109 |

Simultaneously with the modifier concentration increase, it was found the film density, *d,* increase (Figure l). This can be explained by condensation of the amorphous regions due to incorporation of "heavy" MI molecules into the polymer macromolecule chains. The straight-line type of the dependency indicates that the density of the partially crosslinked PE is additively summed from the polymer density and the density of the modifier, and that the changes in the phase structure are insignificant. The reduction of the crystallinity state is compensated by incorporation into polymer chains of the "heavy" fragments.

The *d* increase with T and t confirms the role of oxygen combining in the network formation [1] and is related, probably, to both the general amount of chemical junction increase and enhancing of their functionality.

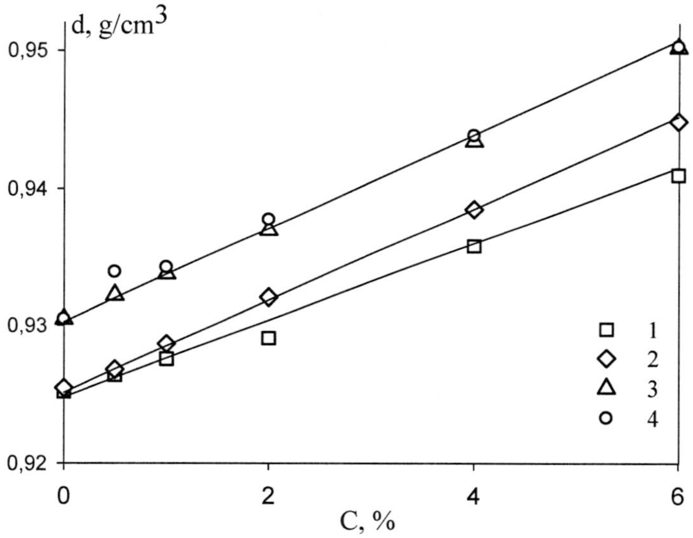


Figure 1.Dependence of film density on PDMI content; t=30 min ; T=180°C (I); T=200°C (2); T=240°C (3).

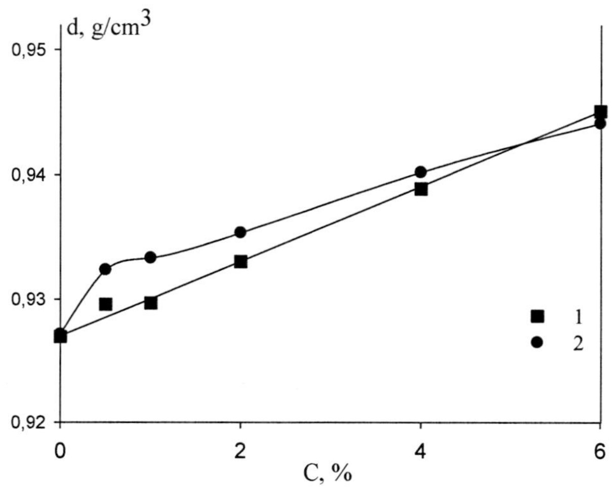


Figure 2. Dependence of film density, d, on PDMI content; T=200°C; t=20 min (1); t=40 min (2).

It should be noted some *d* values deviation from the additive dependency in the area of small additions of PDMI corresponding to the low density of PE network (Figure l). However, the deviations are in the experiment error. At the same time, increase of t to 40 min. at 200°C causes already a significant deviation of *d* from the linear dependence (Figure 2)as a "shoulder" indicating the anomalous sharp increase of packing density in the amorphous polymer regions. Note, that such sensitive to structure parameter as the hexane diffusion coefficient, D, also indicate on the anomalous changes in the amorphous PE regions. The changes are, probably, also related with the packing density increase. The curves of the concentration dependence of D have extremum (Figure 3). The minimum of D is in the region of relatively small P D M I concentrations analogous results were obtained for the systems with HMDMl and TPDMI (Figure 4).

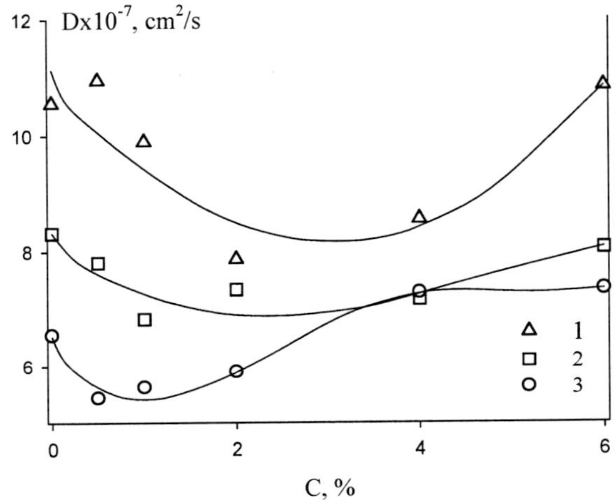


Figure 3. Dependence of the integral diffusion coefficient, D, of hexane on the PDMI concentration, C. T=200°C, t=30 min ; film cooling in water at 20°C (1); T=200°C, t=30 min ; film cooling in air at 20°C (2); T=200°C, t=40 min ; film cooling in water at 20°C (3).

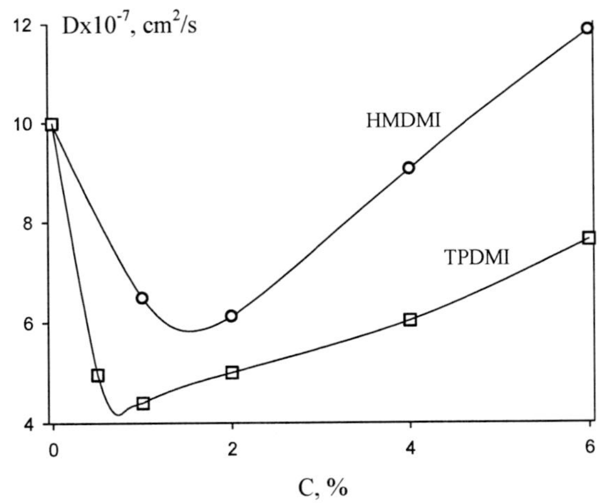


Figure 4. Dependence of the integral diffusion coefficient, D, of hexane on concentration of HMDMI (T=220°C, t=30 min) and TPDMI (T=220°C, t=30 min).

The additional proof of the fact that the assumed structure changes are real.zed in the amorphous regions is the absence of the linear correlation between the specific melting heat, ΔH and the PE density (Figure 5).

The character of diffusion coefficient changeson the TPDMI concentration shows that at the large enough time of coating film formation, the TPDMI behavior is analogous tj behavior of PDMI as it expected [1].

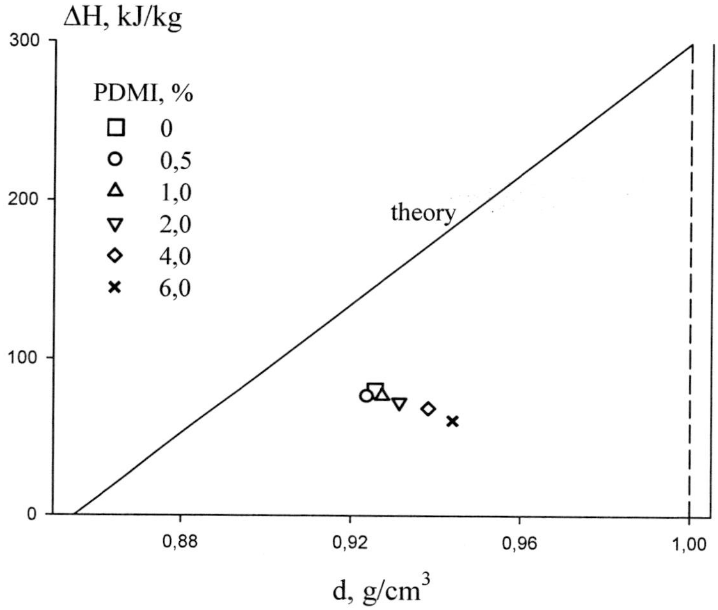


Figure 5. Relationship between specific melting heat, ΔH, abd density of PE coating films modified be different amounts of PDMI (T=200°C, t=30 min).

The original PE diffusion coefficient decrease at significant time of coating film formation occurs due to realization in it more dense macromolecule packing in the polymer amorphous regions at cooling. This aspect has been discussed in details in [8].

At chemical junction formation by the MI molecules and small changes in state of crystallinity, it can be reasonably expected the increase of the polymer elastic modulus, E, with crosslinking agent concentration increase. This assumption is confirmed by the experiment (Figure 6).In this case we again see that for the PE modified with PDMI there is a significant local E maximum in the area of small crosslinking agents concentrations. It should be noted that while the thermal expansion coefficient, β, of films crosslinked with small amount of PDMI (2-6, %) changes antibatically compared with E, the β increases in the area of concentrations of 0 to 2 % (Figure 6). The simultaneous increase of E and β in combination with other results allow to assume that some looseness of the structure as the result of insignificant amorphization at formation of weakly crosslinked polymer is accompanied by release of the load-carrying polymer chains, from one side, and formation of more packed amorphous regions taking intermittent place between fully crystallized and fully amorphous phases.

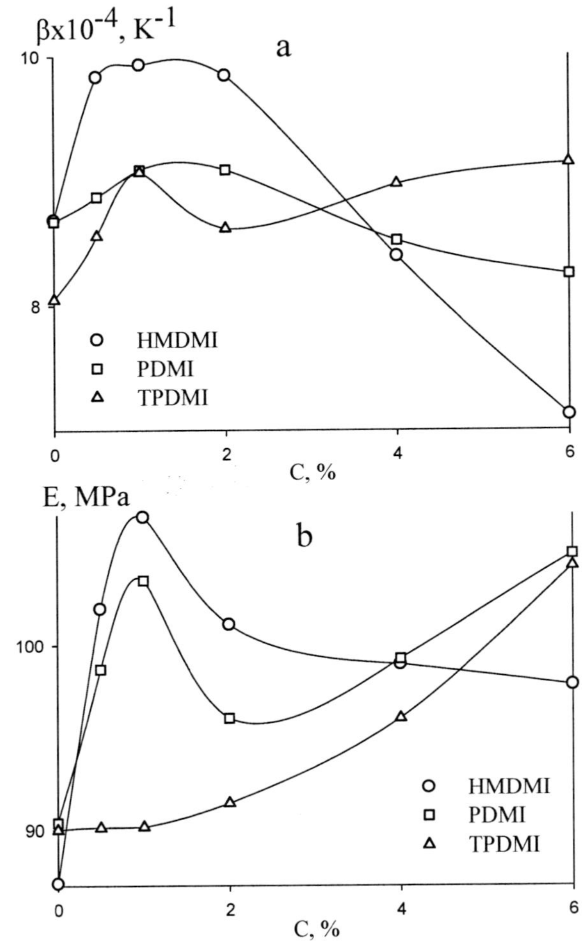


Figure 6. Dependence of the coefficient of thermal For PE modified with TPDMI at long formation times, expansion, β, (a) and elastic modulus. E, (b) on MI.

The distinctive feature of combination of the analogous concentration dependencies for PE modified by HMDMI is absence of the local maximum E in the range of small crosslinking agent concentrations (Figure 6). This result witnesses the weak crosslinking ability of HMDMI compared with PDMI. The traditional antibatic change of E and β takes place at high HMDMI concentrations what is related with increase of stiff network junctions concentration on the phone of insignificant amorphization of the system. For PE modified with TPDMI at long formation times there are local maxima of E and β analogously content compositions with PDMI. The further decrease of E and increase of β can be explained by reduction of network density with TPDMI concentration increase [1] (Figure 6).

Thus, the main polymer structure changes occur in the case of low concentration of maleimides when the crosslinking density is relatively low. It should be noted that the anomalous structure changes are characteristic for low-crosslinked PE independently on the type of crosslinking. The changes are determined, mainly, by the nature of the crosslinks, and to some extent by the process conditions [9-12].

## Physical Properties of Chemically Crosslinked Polyethylene Coatings

Formation of PE chemical network in the course of coating film preparation leads to physical properties changes along with the structure changes. Study of the physical properties changes is very important to establish the relationship between the crosslinking agent concentration, technological process features and the properties. The non-uniformity of the crosslink density along the film thickness when the crosslinking occurs predominantly in the surface layers [1] leads to microhardness of the layers increase (Table 5) caused by the layers' crosslink density increase.

It should be noted that the outer surface layer has higher hardness compared with the inner one. This is in agreement with described before [1] distribution of crosslinks along the film thickness. The tensile strength, σ, of coating films modified with PDMI does not show any tendency to significant changes compared with the original values in whole range of crosslinking agent concentrations. However, the elongation at break, ε, significantly depends on the modifier content and the time-temperature conditions of film preparation (Figure 7). So, in the investigated temperature range, the concentration dependencies at relatively low **t** are extreme, the maximums are realized in the area of small additions of PDMI. The ε increases in 1.5 - 3.0 times. The time, ***t***, increase leading to crosslink density increase reduces the elongation at break what it should be expected. The found ε increase is related to the inhibition of the heat oxidation [1] and increasing of the load-carrying chains concentration at crosslinking. In this case, the network density is relatively small what does not limit significantly the chain mobility.

The concentration dependencies of σ and ε for films modified by HMDMI have different character: the σ maximum is observed in the area of small crosslinking agent concentrations analogously to ε (Figure 7). Such type of PE dependencies on the crosslink density for PE is well known [13]. Their extreme character is caused by occurring in polymer competing processes-increase of the load-carrying polymer chains and limit of macromolecule segments mobility along with partial amorphization and chain destruction at crosslink density increase. The existence of σ maximum for systems with HMDMI and absence of any for compositions with PDMI can be explained by the difference in the chemical structure of the modifiers directly forming the network junctions in PE. It is possible that the more stiff structure of PDMI retards to the higher extent the relaxation processes in PE, while the methylene chain of HMDMI allows more mobility of the network chains.

The results of the PE modified with TPDMI stress- strain properties evaluation are in quality agreement with the made assumptions: σ of MI modified PE also does not show significant changes compared with the original material (Figure 7). Some decrease of σ and ε for the unmodified PE at formation time 180 min. is probably the result of heat- oxidation destruction. At the same time, ε increases with TPDMI content increase (Figure 7). This result is easily explained by taking into account occurring in this case the crosslink density decrease and the antioxidative action [ 1 ].

It should be noted that the films modified with HMDMI show higher mechanical properties compared with the films modified with PDMI and TPDMI.

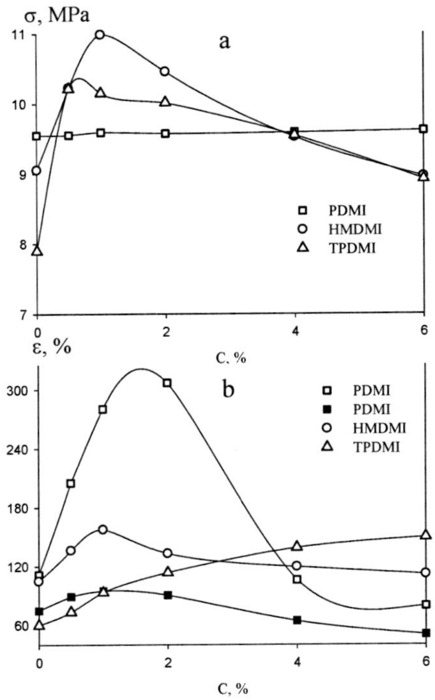


Figure 7. Dependence of tensile strength, σ (a), and elongation at break, ε (b), on content MI. PDMI: T=200°C, t= 20min (□);T=200°C, t= 30min (■). HMDMI: T=200°C, t= 30min (○). TPDMI: T=200°C, t= 180min (Δ).

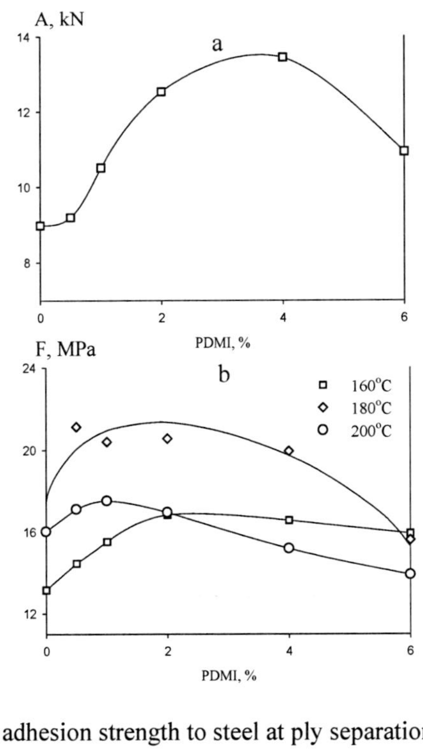


Figure 8. Dependence of PE adhesion strength to steel at ply separation, A, (a) and at pulling normally from steel plate, F, (b) on PDMI content.

**Table 5. Microhardness of outer and inner IIDPE film surfaces crosslinked by different type and amount of modifiers***

| MI type | Film side | Maleimide content, % | | | | | | t, min |
|---|---|---|---|---|---|---|---|---|
| | | 0 | 0.5 | 1.0 | 2.0 | 4.0 | 6.0 | |
| HMDMI | Outer | 1.13 | _ | 2.32 | 2.50 | 2.55 | 2.60 | 30 |
| | Inner | 1.09 | | 2.07 | 2.20 | 2.27 | 2.30 | |
| PDMI | Outer | _ | _ | 3.02 | 3.30 | 3.38 | 3.40 | 30 |
| | Inner | - | - | 2.70 | 2.95 | 3.07 | 3.09 | |
| TPDMI | Outer | _ | 3.02 | 1.66 | 1.60 | 1.64 | 1.62 | 30 |
| | Inner | - | 2.46 | 1.30 | 1.30 | 1.31 | 1.30 | |
| TPDMI | Outer | _ | 3.84 | 3.40 | 3.35 | 3.37 | 3.30 | 180 |
| | Inner | - | 3.42 | 2.90 | 2.85 | 2.81 | 2.85 | |

* T=200°C, film thickness 1 mm.

## Adhesion Properties of Chemically Crosslinked Polyethylene Coatings

Adhesion of coatings to counterpart is a veiy important property. The data in Figure 8 show the adhesion strength dependence of a PE film on PDMI content. It can be seen that the dependence is typical to the one of crosslinked PE [7]. The curves have extremes, the maximal values of adhesion strength, A, is in the range of 2-4% of the MI content. The maximal value of the adhesion strength at pilling F, corresponds to different MI contents depending on the coating film formation time. The differences in the maximum locations obtained by using the two test methods can be explained by differences in the type of sample deformation.

The adhesion strength increase is based on the strength enhancement in the bordering layers as the result of the chemically crosslinked network formation and the stabilizing effect. The adhesion strength decrease at the PDMI content increase can be explained in this case by the rheological factor: polymer crosslinking begins and proceeds with high rate in the sample upper layers what retards the monolithization of the melt and can cause defects formation on the boundary PE-metal in the border polymer layers. The support of this assumption is in the character of F change at the PDMI content of *6%* at different temperatures: F(T=160 °C) >F( 180 °C) >F(200 °C).

Thus, the chemical modification of PE by maleimides leads to significant improvements of coating films adhesion properties.

## References

[1] Stoyanov O. V., R. Ya. Deberdeev, and G.E. Zaikov // *Russian Polymer News,* v.5, No.l, 2000, pp.6-15.

[2] Kutyrev A.A., Mansurov D.E., Cherkasova O.A., Stoyanov O.V., Deberdeev R.Ya., and N.A. Mukmeneva. Depos. In *ONIITEChim,* Cherkasy, 1993, Preprint No. 85-KhP93.

[3] Martynov M.A., and K.A. Bylegshanina, "*X-ray spectroskopy of polymers* "(Russ.), L. Chimiya, 1972, 216 p.

[4] Godovskii Yu.K.,*Thermophysical methods in polymer investigations* "(Russ.), M., Chimiya, 1976, 216 p.

[5] Mirontsov L.I., Privalko V.P., and A.I. Antonov, "Composite polymeric materials ", Kiev, *Naukova Dumka,* No. 18,1983,pp.3-10.

[6] Godovskii Yu.K., "Heat-physics of polymers", M. *Chimiya,* 1982, 280 p.

[7] Deberdeev R Ya., Shmakova O.P., Stoyanov O. V., and V.P. Privalko, "Composite polymeric materials, Kiev, *Naukova Dumka,* No. 38, 1988, pp. 16-21.

[8] Shmakova O.P., Deberdeev R Ya., Stoyanov O. V., Kulevtsov G.N.. Pleshchinskaya N.I., and A.A. Polvanskii. // *Coatings and lacquers and their applications.* No.2, 1987, pp.33-35.

[9] Stoyanov O. V., and RYa. Deberdeev // *Vysokomol. Soed.,* Ser. B.,v.29, No.I, pp.22-24.

[10] Stoyanov O.V., Deberdeev R.Ya., Shmakova O.P., and N.V. Svetlakov, // *Vysokomol. Soed.,* A, v.28, No.l, 1986, pp.375-380.

[11] Aslanyan V.A.,Vardanyan V.I., andFelenyan // *Plast Massy* (Russ.), No. 4. 1989, pp. 34-35.

[12] Aslanyan V.A.,Vardanyan V.I., and Felenyan // *Plast Massy* (Russ.), No. 10, 1989, pp. 30-32.

[13] Sirota A.G. "*Structure modification and properties ofpolyolefins*" (Russ.), L. Chimiya, 1984, 152 p.

In: Chemical and Biochemical Reactions
Editors: G. E. Zaikov, L. Liu et al.

ISBN 978-1-61122-860-1


*Chapter 23*

# THE CALCULATION OF THE CRITICAL CONVERSION OF CROSSLINKED EPOXYAMINE POLYMERS

***T. R. Deberdeev*[1,*], *V. I. Irzhak*[2], *R. Ya. Deberdeev*[1] and *O. V. Stoynov*[1]**

[1]Kazan State Technological University, Kazan, 420015, Russia
[2]Institute of Problems of Chemical Physics, Russian Academy of Sciences, Chernogolovka, 142432 Russia

## INTRODUCTION

The simplest and commonest method for the description of the correlation between the structure of a network polymer and the kinetics of a chemical reaction is based on the statistical approach. According to this approach- the structure is defined as a single-valued function of the depth of conversion and the initial composition of the system. This method of calculating the structural parameters was described in most detail in monographs [1- 2] and reviews [3- 4]. However as is well known [2-4] when functional groups change their reactivity (substitution effect) during the process (for example during polycondensation with unequal reactivities of primary and secondary amino groups) the polymer structure becomes dependent on the reaction route- that is- the above unambiguity of the structure- conversion relationship is not satisfied. In this case- a rigorous kinetic calculation for all structural elements is required. One should mention that this fact is not yet universally recognized and probabilistic approaches are rather popular (see- for example- [5]).

Evidently the known formulas for the critical conversion value derived on the basis of probabilistic approaches cannot characterize systems with a substitution effect. An exact solution of the kinetic problem seems to be rather complicated even when effective computational means are used. However this solution may be considerably simplified by using the concept of bond blocks [3]. Note that the pioneering communication on this subject was recommended for publication by N.S. Enikolopov [6].

* E-mail: deberdeev@mail.ru

The concept of bond blocks involves a combination of the kinetic and statistical approaches. This combination allows one to propose a generalized method for the estimation of the gel point- so that the known formulas appear as special cases under certain conditions. This work concerns the derivation of this equation.

## Concept of Bond Blocks

For the sake of convenience- let us consider a particular system composed of a bifunctional epoxy oligomer and a pentafunctional amine. The presence of primary and secondary amino groups in the latter compound is a prerequisite for the substitution effect.

Just as the statistical calculation- the concept of bond blocks uses the functional fractions of reactive groups rather than the molar concentrations of the reagents: $n_i$ stand for the functional fractions of the corresponding groups capable of reacting with the amine whose functional fraction is denoted as $n_a$:

$$n_i = \frac{N_i}{\sum_j N_j}$$

where $N_i$ is the molar concentration of the corresponding groups at a given moment (at a given conversion); at the same time $N_i = i \cdot m_i$ - where $m_i$ is the molar concentration of the reagent containing i active functional groups. $\alpha_i$ are the probabilities of the reactions for the corresponding groups (depths of conversion). The balance condition is the following:

$$\sum_i N_i \alpha_i = A\alpha_A$$

where the subscript i refers to the epoxy reagents and the subscript A denotes the amine.

All the values of $\alpha_i$ and $n_i$ are found by solving the corresponding sets of kinetic equations.

The network structure is determined by the concentration of junctions of various functionalities, and the junctions are represented by fragments of the curing amine agent. Such structural elements comprise bond blocks, whose concentration should in this case be calculated by the kinetic method, that is, by solving the corresponding kinetic equations. The number of oligomeric fragments attached to a junction characterizes its functionality.

In the case of the bifunctional epoxy oligomer-pen- tafunctional amine system (e.g., the amine may be represented by diethylenetriamine, or DETA), one should distinguish between primary and secondary amino groups from the very onset of the reaction. Using digits in parentheses to denote the numbers of primary (first digit) and secondary (second digit) hydrogen atoms, we can list the following network-comprising groups: A(41) is the initial amine, junction of zero functionality; A(40) is the amine with a reacted secondary amino group, functionality 1, tail fragment in the network; A(22) is the product of the reaction of one primary amino group, functionality 1, tail fragment in the network; A(21) is a junction of functionality 2, central fragment; A(03) is a junction of functionality 2, central fragment; A(20) is a trifunctional junction; A(02) is a trifunctional junction; A(01) is a tetrafunctional junction; and A(00) is a pentafunctional junction.

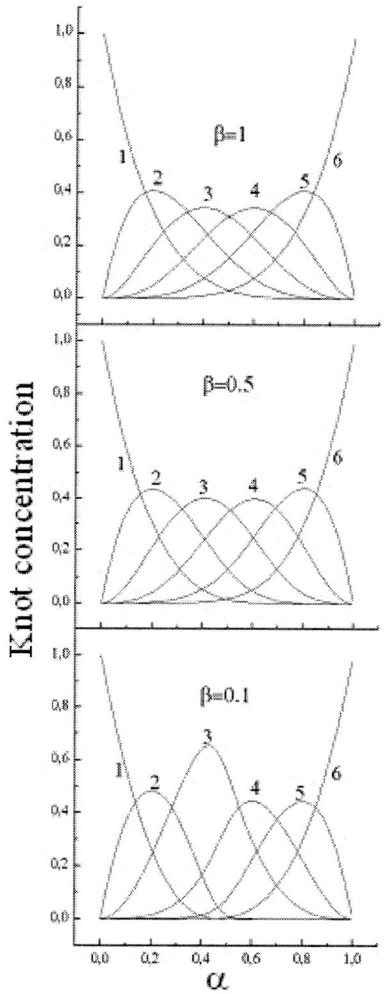


Figure 1. Concentration of (1) nonfunctional, (2) mono- functional, (3) bifunctional, (4) trifunctional, (5) tetrafunctional, and (6) pentafunctional network junctions vs. conversion at β = (a) 1, (b) 0.5, and (c) 0.1.

If the ratio of the kinetic constants of the secondary and primary amino groups is denoted as $\beta = k_2/k_1$, the set of equations for all the amino groups is written as:

$$dB/dt = -B\{4A(41)+4A(40)+2A(22)+2A(21)+2A(20)+$$
$$+\beta\,[A(41)+2A(22)+A(21)+3A(03)+2A(02)+A(01)]\};$$
$$dA(41)/dt = -(4+\beta)BA(41);$$
$$dA(40)/dt = \{\beta A(41) - 4A(40)\}B;$$
$$dA(22)/dt = \{4A(41) - 2(1+\beta)A(22)\}B;$$
$$dA(21)/dt = \{4A(40)+2\beta A(22) - (2+\beta)A(21)\}B;$$

$$dA(03)/dt = \{2A(22) - 3\beta A(03)\}B;$$
$$dA(02)/dt = \{3\beta A(03)+2A(21) - 2\beta A(02)\}B;$$
$$dA(20)/dt = \{\beta A(21) - 2A(20)\}B;$$
$$dA(01)/dt = \{2\beta A(02)+2A(20) - \beta A(01)\}B;$$
$$dA(00)/dt = \beta A(01)B; \tag{1}$$

Figure 1 shows how the concentration of junctions with various functionalities changes in the course of the reaction. It is characteristic that the concentration-conversion curves corresponding to different p values appear similar: the maximum concentration of junctions with a given functionality corresponds to the same depth of conversion, irrespective of the relative reaction rate constant of the secondary amino group. At the same time, this maximum value itself changes depending on the kinetic parameters of the polycondensation reaction.

Since the reactivity of epoxy groups is assumed to remain unchanged in the course of the reaction, the calculation of the network structure may be performed within the statistical approach [3] using the calculated concentrations of the junctions.

According to the theory of branching processes [1], the calculation algorithm is the following.

Let function u(x) characterize the probability of the generation of an oligomer chain in the next generation by the same chain in the preceding generation. According to the theory,

$$u(x) = x\left(1 - \alpha + \alpha \sum n_{ai} u^{i-1}(x)\right) \tag{2}$$

where $x$ is an arbitrary variable, a is the depth of conversion with respect to epoxy groups, and $n_{ai}$ is the proportion of i-functional junctions.

Differentiation of Eq. (2) leads to the expression for the gelation condition:

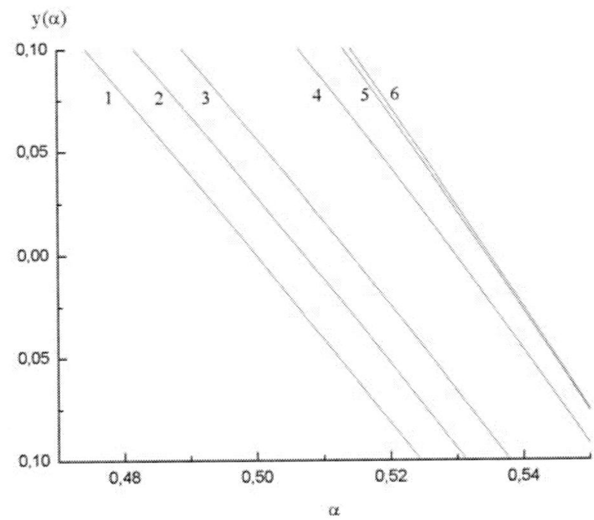


Figure 2. Graphical solution of equation for the critical conversion at β = (1) 1, (2) 0.7, (3) 0.5, (4) 0.2, (5) 0.1, and (6) 0.01.

$$\frac{\partial u(x)}{\partial x} = \left(1 - \alpha + \alpha \sum_i n_{ai} u^{i-1}(x)\right) + x\alpha \sum_i (i-1) n_{ai} u^{i-2}(x) \quad (3)$$

At $x = 1$, we have u(1) = 1 and

$$u'(1) = \frac{1}{1 - \alpha \sum (i-1) n_{ai}} \quad (4)$$

Hence, the critical conversion with respect to epoxy groups is determined as the solution to the equation

$$\alpha \sum_i (i-1) n_{ai} = 1 \quad (5)$$

Formally Eq. (5) coincides with the critical condition $\alpha_c(\bar{f}_w - 1) = 1$ for f-functional polycondensation, with the only difference that the weight-average functionality $\bar{f}_w$ changes during the process in our case; i.e., this parameter is a function of the depth of conversion a rather than a characteristic of the initial system.

**Table 1. Dependence of size of critical conversion (gel-point) from reactionary ability of a secondary amine (system: diepoxyde – DETA, equifunctional ratio)**

| β | 1 | 0.7 | 0.5 | 0.3 | 0.1 | 0.01 |
|---|---|---|---|---|---|---|
| $\alpha_c$ | 0.5 | 0.507 | 0.514 | 0.524 | 0.535 | 0.535 |

Figure 2 shows the graphical solutions of equation $y(\alpha) = 1 - \alpha \sum_i (i-1) n_{ai}$ for various β. The $\alpha_c$ value is determined by the condition y(a) = 0.

Below the values of the critical conversion $\alpha_c$ are presented as functions of the kinetic constant β.

As is seen, the solution depends on β, a conclusion that does not in any way follow from the statistical theory [1]. However, it is not difficult to show that we arrive at the known formula when the requirements of the statistical approach are met for the dependence of $n_{ai}$ on the conversion.

Indeed, according to the statistical approach,

$$A_i = A \frac{f!}{i!(f-i)!} \alpha_A^i (1 - \alpha_A)^{f-i} \quad (6)$$

where $A_i$ is the concentration of the i-functional junction and f is its maximum possible functionality (for DETA, $f = 5$).

Since $n_{ai} = \frac{iA_i}{\sum_i iA_i}$, substitution of formula (6) into Eq. (5) yields the following result:

$\alpha(f-1)\alpha_A = 1$ This expression coincides with the known formula for copolycondensation of bifunctional and f-functional reagents.

Note that the critical conversion at low (but not zero!) values of the p constant approaches a constant value of $\alpha_c$ = 0.535. This is likely to be related to the fact that the polycondensation process becomes stagewise. At the first stage, the lower the p value, the higher the conversion of primary amino groups. The depth of conversion reaches 40%. The second stage involves the accumulation of polyfunctional junctions in amounts providing the network formation. Evidently, the value of the additional conversion must be quite definite in this case.

As follows from Figure 3, the ratio β of the constants exerts a marked effect on the gel point if the reagents are present in nonequivalent amounts.

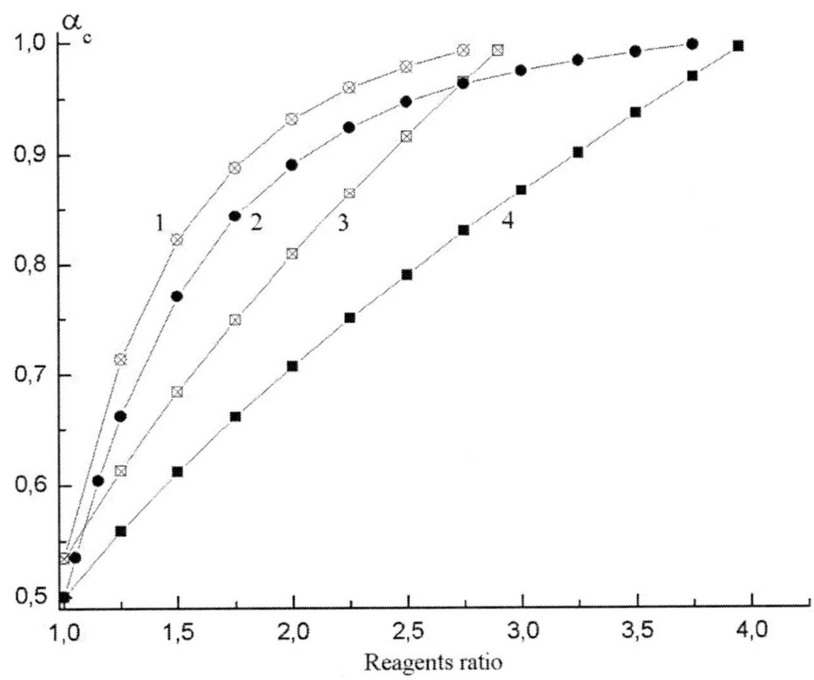


Figure 3. Critical conversion at various ratios between the (A) amine and (B) epoxy oligomer reagents and β = (1, 3) 0.1 and (2, 4) 1.0. (1, 2) A/B ratio and (3, 4) B/A ratio.

## Calculation of Gel Point for Ternary Systems

### Monofunctional Modifying Agent

Let us initially consider a monofunctional epoxy compound as the modifying agent for the system under study [7, 8]. In this case, the kinetic scheme is similar to the above set of equations (1). First, however, one should take into account the second epoxy component (B1) and, accordingly, additional kinetic constants; second, junctions should be distinguished not only by the types of their amino groups but also by the types of the added epoxy oligomers. Indeed, the functionality of the junction in this case will depend not on the total number of reacted hydrogen atoms in the amine but only on the number of these atoms that have reacted with a bifunc tional reagent.

Let the first and second subscripts denote the numbers of the added bi- and monofunctional (tail) oligomers, respectively. Then the $A_{01}(40)$ junction contains one tail and thus has zero functionality. In addition to $A_{00}(41)$, junctions $A_{01}(22)$, $A_{02}(21)$, $A_{02}(03)$, $A_{03}(20)$, $A_{03}(02)$, $A_{04}(01)$, and $A_{05}(00)$ also have zero functionality. Monofunctional junctions are $A_{10}(40)$, $A_{10}(22)$, $A_{11}(21)$, $A_{11}(03)$, $A_{12}(20)$, $A_{12}(02)$, $A_{13}(01)$, and $A_{14}(00)$; bifunctional junctions are $A_{20}(21)$, $A_{20}(03)$, A21(20), A21(02), A22(01), and A23(00); and trifunc- tional junctions are $A_{30}(20)$, $A_{30}(02)$, $A_{31}(01)$, and $A_{32}(00)$. Junctions $A_{40}(01)$ and $A_{41}(00)$ are tetrafunctional, and junction $A_{50}(00)$ is pentafunctional.

In this case, the condition of gelation assumes the following form:

$$\alpha \sum_i (i-1) n_{ai} = 1 \qquad (5a)$$

where, just as above, the subscript i stands for the functionality of the junction but with allowance for the presence of a monofunctional "chain terminator"; that is, the functionality is equal only to the number of the added bifunctional oligomers

The strong dependence of the critical conversion on the concentration of the monofunctional modifying agent (see table) is due to a decrease in the functionality ofjunctions, as is shown in Figure 4. In contrast to the data presented in Figure 1, the curves depicting the changes in the concentrations of junctions of different functionalities are shifted to the right along the abscissa axis; the higher the content of the modifying agent, the more pronounced is the shift. The final state of the system was independent of the kinetic constants in the preceding case; however, the presence of a monofunctional modifying agent, as follows from Figure 3, exerts a strong effect on the structure of the formed network, so that the gel point cannot be achieved at all at a certain concentration of the modifier.

Similar results are also obtained when cyclocarbon- ate is used as the modifying agent [7-9].

The specific kinetic feature of cyclocarbonate is that its functional group reacts with a primary amino group at a ratio of one group per two hydrogen atoms. The molecular structure of DETA initially contains two primary and one secondary amino groups. Therefore, depending on the ratio between the kinetic constants, various reactions of amine hydrogen atoms are possible from the very beginning: reaction of primary group with the carbonate, reaction of primary group with the epoxy group generating a secondary hydrogen atom, and finally reaction of secondary atom with the epoxy group. The kinetic calculation makes it possible to estimate the concentrations of amino groups of various structures:

A(41), A(40), A(22), A(21), A(20), A(03), A(02), A(01), A(00), $A_1(21)$, $A_1(20)$, $A_1(02)$, $A_1(01)$, $A_1(00)$, $A_2(01)$, $A_2(00)$.

As above, the digits in parentheses denote the numbers of primary and secondary hydrogen atoms in the amine fragment. The subscript at A stands for the number of carbonate groups that have reacted with the amine.

Evidently, the first of the above groups have reacted neither with the epoxy nor with the carbonate group, A(40) and A(22) have reacted with one epoxy group, A(21) and A(03) have reacted with two epoxy groups, A(20) and A(02) have reacted with three groups, A(0i) has reacted with four groups, and, finally, A(00) has reacted with five epoxy groups. $A_1(21)$ and $A_2(0l)$ contain no fragments of the epoxy oligomer; $A_1(20)$, $A_1(02)$, and $A_2(00)$ have reacted with one epoxy group; $A_1(01)$ has reacted with two epoxy groups; and $A_1(00)$ has reacted with three epoxy groups. The functionality of a junction is equal to the number of the added bifunc- tional epoxy chains.

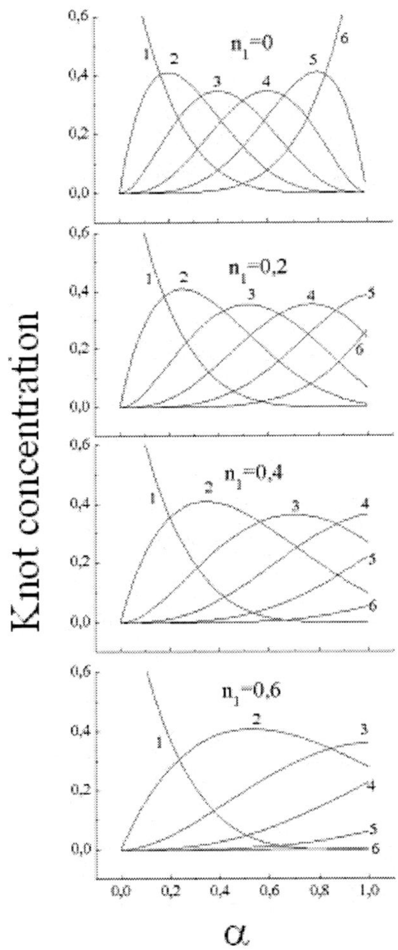


Figure 4. Concentration of (1) nonfunctional, (2) mono- functional, (3) bifunctional, (4) Afunctional, (5) tet- rafunctional, and (6) pentafunctional network junctions vs. conversion at $\beta = 1$ and different $n_1$.

**Table 2. Dependence of critical conversion on the content $n_1$ of the monofunctional component for various relative rate constants of the reactions of the modifying agent with primary ($k_1$) and secondary ($\beta_2$) amino groups**

| $n_1$ | $k_1 = 1$ | $k_1 = 5$ | $k_1 = 1$ | $k_1 = 0.5$ | $k_1 = 0.5$ |
|---|---|---|---|---|---|
| | $\beta_1 = \beta_2 = 1$ | $\beta_1 = \beta_2 = 1$ | $\beta_1 = \beta_2 = 0.5$ | $\beta_1 = \beta_2 = 1$ | $\beta_1 = \beta_2 = 0.5$ |
| 0 | 0.500 | 0.500 | 0.514 | 0.500 | 0.514 |
| 0.1 | 0.527 | 0.527 | - | 0.527 | 0.542 |
| 0.2 | 0.559 | - | 0.572 | - | 0.574 |
| 0.3 | 0.597 | - | - | 0.597 | 0.613 |
| 0.4 | 0.645 | 0.645 | 0.655 | - | 0.661 |
| 0.5 | 0.707 | - | - | 0.707 | 0.722 |
| 0.6 | 0.790 | - | 0.794 | - | 0.804 |
| 0.65 | 0.845 | 0.845 | - | - | 0.857 |
| 0.7 | 0.913 | - | 0.912 | 0.913 | 0.923 |
| 0.74 | 0.981 | 0.981 | 0.980 | - | 0.987 |

Note: $\beta_1$ is the relative rate constant of the reaction between the secondary amino group and the epoxy group of the basic oligomer.

The critical conversion values calculated through Eq. (5) are presented in Figure 5.

Even if the reactivity of the cyclocarbonate is fairly high, its ultimate conversion is far from unity. Actually, even having a relatively low reactivity, the epoxy component may react with a hydrogen atom even at early stages of the process and thus disturb the initial equifunctional ratio between the amine and the cyclocarbonate. Evidently- the lower the reactivity of the cyclocarbonate- the more pronounced is this factor. Since- under the conditions of incomplete cyclocarbonate conversion- the reaction of the epoxy fragment is complete- the critical conversion but slightly depends on the ratio between the rate constants of the competing reactions.

The relative reactivity of the secondary amino group has a minor effect on the critical conversion value (Figures 6- 7) but a notable effect on the ultimate depth of conversion with respect to the cyclocarbonate. As the rate constant of the second addition decreases- theprobability of the reaction between the epoxy component and the primary amino group increases and the initial equifunctional cyclocarbonate: amine ratio is disturbed to an increasing extent and at earlier stages.

## Bifunctional Modifying Agent

In contrast to a monofunctional modifying agent- a bifunctional modifier has no effect on the critical conversion- because neither component terminates network chains and the functionality of junctions no longer depends on which of the components is attached to an amine fragment. In this case- the critical conversion is determined according to the following equation:

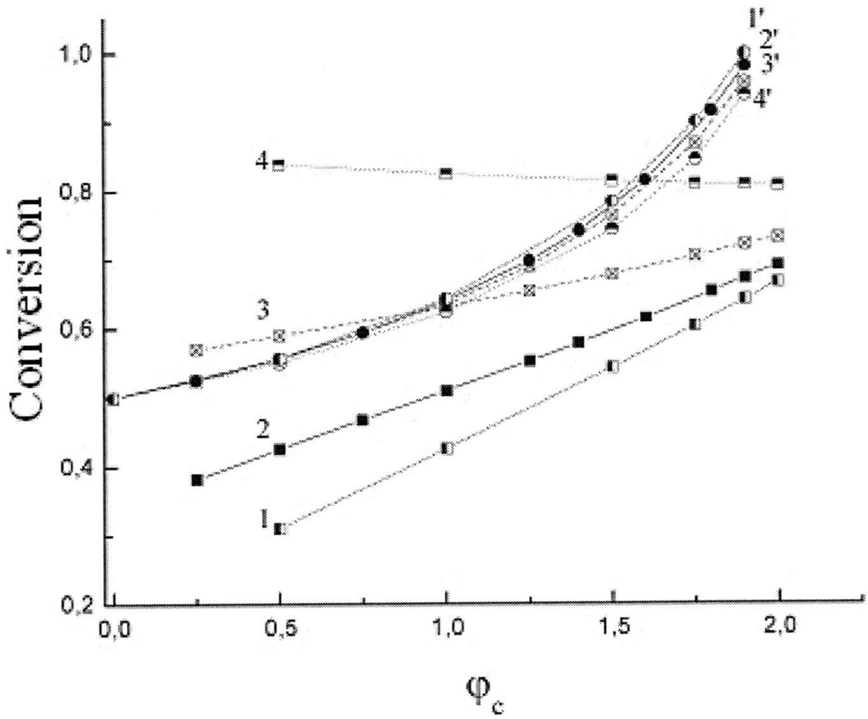


Figure 5. (1-4) Ultimate conversion of cyclocarbonate groups and (1'-4') critical conversion with respect to epoxy groups vs. relative concentration $\varphi_c$ of cyclocarbonate groups at the equifunctional total ratio. Ratio of rate constants of the cyclocarbonate and epoxy groups $k_c/k_b$ = 0.5 (1- 1');1 (2- 2); 2 (3- 3'); 5 (4- 4').

$$(\alpha_b n_b + \alpha_2 n_2)\sum_i (i-1)n_{ai} = 1 \tag{5b}$$

where the subscript b refers to the basic oligomer- the subscript 2 denotes the modifier- and $(\alpha_b n_b + \alpha_2 n_2) \equiv \alpha_{eff}$. The difference in the reaction rate constants of epoxy groups pertaining to different oligomers- as well as their relative concentrations- has an effect on $a_{eff}$ rather than on the gel point.

Obviously- this conclusion is valid only when we deal with a modifier similar to a bifunctional epoxy oligomer- i.e.- when one hydrogen atom of the amine is consumed in the reaction of the functional group. When a reagent similar to bifunctional cyclocarbonate is used as the modifier- the situation is changed. The reason is not termination of network chains- as in the case of monofunctional agents- but the change in the limiting functionality of a junction. Actually- since each carbonate group reacts with two hydrogen atoms- the presence of one carbonate fragment attached to the amine reduces the limiting functionality of the junction by unity- whereas the presence of two fragments reduces it by two units. Therefore- as is shown in Figure 7- the critical conversion increases with increasing concentration of a bifunctional modifying agent but this increase is less pronounced than in the case of a monofunctional agent.

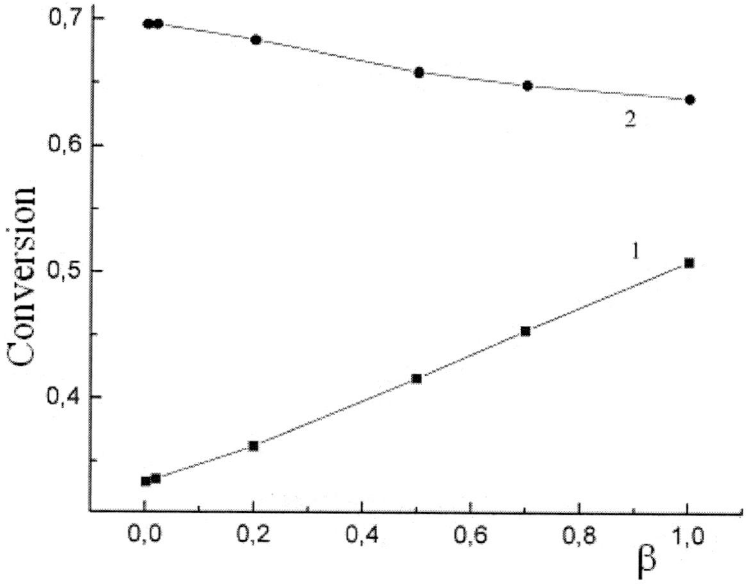


Figure 6. (1) Ultimate conversion of cyclocarbonate groups and (2) critical conversion with respect to epoxy groups vs. β value.

The above examples demonstrate that Eq. (5) offers a virtually universal approach for the estimation of the critical conversion value. The only exception concerns systems in which all reagents change their reactivity during the process. For such systems- a rigorous solution of the kinetic problem is required similar to the one obtained for three-dimensional polymerization (see for example [10-12]).

The proposed approach to the problem of estimating the critical conversion value also makes it possible to characterize the structure of a network polymer: the content of the gel fraction and the concentration of elas- tically active chains.

According to the theory of branching processes, the network structure is determined by the chain termination parameter *v*. This parameter is found by solving Eq. (2) with allowance for u(1) = v. Thus,

$$v = 1 - \alpha + \alpha \sum n_{ai} v^{i-1} \tag{7}$$

or, after some simplification,

$$n_{a5} v^3 + (n_{a5} + n_{a4}) v^2 + (n_{a5} + n_{a4} + n_{a3}) v + 1 - n_{a1} - \alpha^{-1} = 0 \tag{8}$$

The solution of this cubic equation (for a pentafunc- tional curing agent, such as DETA) makes it possible not only to express *v* as a function of the depth of conversion a but also to derive the condition for finding the gel point. Indeed, since *v* = 1 at this point [1], Eq. (8) yields formula (5).

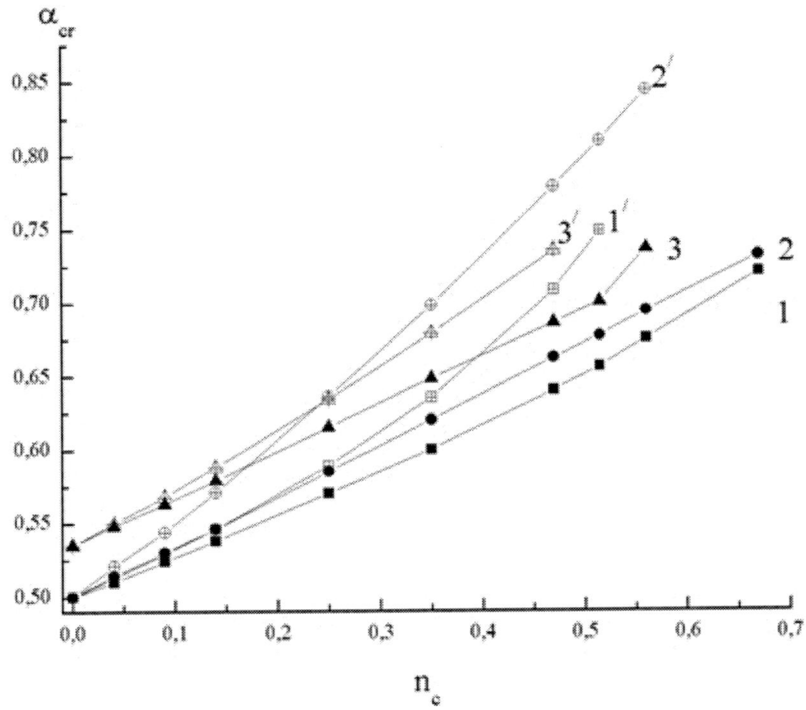


Figure 7. Critical conversion vs. concentration of (1-3) bifunctional and (1-3') monofunctional cyclocarbonate at various constants: (1, 1') k = 1, β= 1, (2, 2') k = 5, β = 1, and (3, 3') k =1, β = 0.1.

## REFERENCES

[1] V. I. Irzhak, B. A. Rozenberg, and N. S. Enikolopyan, *Polymer Networks: Synthesis, Structure, Properties* (Nauka, Moscow, 1979) [in Russian].

[2] S. I. Kuchanov, *Methods of Kinetic Calculations in Polymer Chemistry* (Khimiya, Moscow, 1978) [in Russian].

[3] V. I. Irzhak, *Usp. Khim.* Vol .66, №6,598 (1997).

[4] S. I. Kuchanov, *Adv. Polym. Sci.* Vol .152, 157 (1999).

[5] H. Tobita, *J. Polym. Sci., Part B: Polym. Phys.* Vol.36, 2423 (1998).

[6] M. L. Tai and V. I. Irzhak, *Dokl. Akad. Nauk SSSR* Vol.259, 856 (1981).

[7] R. M. Garipov, S. Yu. Sof'ina, E. N. Mochalova, et al., *Plast. Massy,* No. 7, 21 (2003).

[8] R. M. Garipov, T. R. Deberdeev, I. A. Chernov, et al., *Plast. Massy* No. 8, 6 (2003).

[9] R. M. Garipov, V A. Sysoev, V. V. Mikheev, et al., *Dokl. Akad. Nauk* Vol.392, №4, 649 (2003).

[10] G. V. Korolev, T. F. Irzhak, and V. I. Irzhak, *Vysokomol. Soedin., Ser.* A Vol.43, №6, 970 (2001).

[11] G. V. Korolev, T. F. Irzhak, and V. I. Irzhak, *Khim. Fiz.* Vol.21, №1, 58 (2002).

[12] G. V. Korolev, T. F. Irzhak, and V. I. Irzhak, *Vysokomol. Soedin., Ser.* A Vol.44, №1, 5 (2002).

In: Chemical and Biochemical Reactions
Editors: G. E. Zaikov, L. Liu et al.
ISBN 978-1-61122-860-1


*Chapter 24*

# 16TH INTERNATIONAL CONFERENCE FOR RENEWABLE RESOURCES AND PLANT BIOTECHNOLOGY (NAROSSA-2010)

***G.E. Zaikov, L.L. Madyuskina and M.I Artsis***
N.M. Emanuel Institute of Biochemical Physics Russian Academy of Sciences,
4 Kosygin str., Moscow 119334, Russia
chembio@sky.chph.ras.ru

NAROSSA® 2010 conference was held in Maritim Hotel (Magdeburg, Germany) in the period June 7 – 8, 2010. The organizers of conference were the association Pilot Pflanzenöltechnologie Magdeburg e. V. (PPM) and the sponsor of this conference was OHMI Consulting GmbH.

Prof. Dr. Melvyn F. Askew (Census, Wolverhampton, United Kingdom), Prof. Dr. Grete Bach (ifn Forschungs- und Technologiezentrum GmbH, Elsteraue/OT Tröglitz, Germany), Dr. Ute Bauermeister (Gesellschaft für Nachhaltige Stoffnutzung mbH, Halle/Saale, Germany), Dr. Peter Bloß (Kunststoff-Zentrum in Leipzig gGmbH, Leipzig, Germany), Dieter Bockey (UFOP e.V., Berlin, Germany), Dr. Susanne Brandt (Landesbauernverband Sachsen-Anhalt e.V., Magdeburg, Germany), Thorsten Breitschuh (NAROSSA e.V., Magdeburg, Germany), Dr. Waldemar Buchwald (Institute for Natural Fibres and Medicinal Plants, Poznan, Poland), Dr. Matthias Gohla (Fraunhofer IFF - Institut für Fabrikbetrieb und -automatisierung, Magdeburg, Germany), Torsten Graf (Thüringer Landesanstalt für Landwirtschaft, Dornburg, Germany), PhD Paul P. Kolodziejczyk (Biolink Consultancy, New Denver, Canada), Prof. Gotthard Kunze (Leibniz-Institut für Pflanzengenetik und Kulturpflanzenforschung IPK, Gatersleben, Germany), Prof. Dr. Lothar Mörl (Otto-von-Guericke-Universität, Magdeburg, Germany), Prof. Dr. Ingo Schellenberg (Hochschule Anhalt (FH), Kompetenzzentrum Life Sciences, Bernburg, Germany), Dr. Heike Schimpf (Landesanstalt für Landwirtschaft, Forsten und Gartenbau / KoNaRo, Bernburg, Germany), Prof. Grzegorz Spychalski (Institute for Natural Fibres and Medicinal Plants, Poznan, Poland), Dr. Ralph Thomann (Institut für Getreideverarbeitung GmbH, Bergholz-Rehbrücke, Germany), Dr. Maria Wladyka-Przybylak (Institute for Natural Fibres and Medicinal Plants,

Poznan, Poland), Prof. Dr. Gennady E. Zaikov (N. M. Emanuel Institute of Biochemical Physics, Moscow, Russia) were members of Scientific Committee of this conference.

The program of the conference included follow main topics:

Plant Breeding for non-food Application
Biotechnology
In vitro culture
Smart breeding
Renewable Resources for Material Applications
Biobased composites (Natural fibres, Nano particles etc.)
Biobased polymers
Biobased additives
Processing, Characterisation and Application of Secondary Plant Metabolites
Biomass conversion technologies
Biorefineries
Cosmetics and pharmaceuticals
4. Biomass and Efficient Energy Generation
Energy crops and organic residues
Thermal, thermochemical and biochemical conversion technologies
Gas cleaning and product upgrading
Sustainable Development in Environment, Economy and Sociology
Acceptance
Sustainability in production and logistics
Regional material, nutrient and energetic cycles

About 100 participants from 27 research centers from Germany, France, Russia, Poland, Czech Republic, Guinea, Indonesia took part in this conference.

Scientific program of the conference included 3 plenary lectures and 33 oral presentations which were divided on two sessions.

The topic of first plenary lecture was "Greeting of the patron" (H. O. Aeikens, Minister for Agriculture and the Environment of Saxony-Anhalt, Magdeburg, Germany). The second plenary lecture was about feedstocks and (bio) technologies for biorefineries (J. Venus, Leibniz-Institut für Agrartechnik Potsdam-Bornim e.V., Germany). Socially sustainable development of the rural areas was discussed in the third lecture (G. Spychalski, Director of the Institute of Natural Fibres and Medicinal Plants, Poland).

The first session of conference had 18 oral presentations. P. Gerth, T. Bagusch, M. Poschke, J. Zimprich (Centre of Competence Engineering Sciences/Renewable Materials, University of Applied Sciences, Magdeburg, Germany) spoke about biomaterials for consumer products. A. Breier, K. Gliesche, P. Govindarajulu, C. Rentsch, B. Rentsch (Leibniz-Institut für Polymerforschung Dresden e.V., Germany; Universitätsklinikum Carl-Gustav-Carus, TU Dresden, Germany; Fa. Catgut GmbH, Markneukirchen, Germany) gave presentation about tissue engineering executed on textile scaffolds embroidered with degradable surgical thread materials. The lecture of M.I. Artsis, A.P.Bonartsev', A.L.Iordanskii, G.A.Bonartseva, G. E.Zaikov (General Institute of Chemical Physics, Russian Academy of Sciences, Moscow, Russia; A.N. Bachs Institute of Biochemistry, Russian Academy of Sciences, Moscow, Russia; Faculty of Biology, Moscow State

University, Moscow, Russia) was devoted to biodegradation and medical application of microbial poly(3-hydroxybutyrate). The results about flammability of polymeres reinforced with lignocellulosis raw materials was included in the talk of M. Wladyka-Przybylak, K. Bujnowicz (Institute of Natural Fibres and Medicinal Plants, Poznan, Poland). J.-P. Krause (Institute of Nutrition Science, University of Potsdam, Germany) spoke about correlations between interfacial rheology and film forming behaviour of plant proteins.

The lecture of P. L. Mikolajczak, A. Bogacz, M. Karasiewicz, P. M. Mrozikiewicz (Dep. of Pharmacology and Biotechnology, Institute of Natural Fibres and Medicinal Plants, Poznan, Poland; Dep. of Pharmacology; Laboratory of Experimental Pharmacogenetics, Dep. of Clinical Pharmacy and Biopharmacy Poznan University of Medical Science, Poznan, Poland) was devoted to role of puerarin and daidzin as secondary plant metabolites - study on alcoholism. The information about PCR-based detection endophytic bacteria appearing in in vitro cultures of different clones of Urtica dioica L. using specific PCR-primers, derived from bacterial 16S rDNA sequences was included in the lecture of C. Schneider, B.Eng. Sven Wartenberg, B.Eng. Jasmin Dufrenne, B. Biskupek-Korell (Institut für Pflanzenkultur e.K., Schnega, Germany; FH Hannover, Fakultät 2, Abt. Bioverfahrenstechnik, Germany). M. Chudy, G. Mankowska (Dep. of Hemp Breeding, Agronomy and Seed Production; Dep. of Biotechnology *both:* Institute of Natural Fibres and Medicinal Plants, Poznan, Poland) spoke about genetic resources of Cannabis sativa L. in the collection of the gene bank at INF&MP in Poznan.

The lecture of B. Biskupek-Korell, S. Becker, J. Dufrenne, P. Rauscher, C. Schneider (FH Hannover, Fakultät 2, Abt. Bioverfahrenstechnik, Hannover, Germany; Institut für Pflanzenkultur e.K., Schnega, Germany) was devoted DNA fingerprinting and charakterisation of genetic variation of different clones of Urtica dioica L. via RAPD and RAPD-derived SCAR markers. The problems of detection of phytopathogenic RNA-viruses were discussed by K. Florschütz, A. Schröter, M. Körner, G. Kunze (Leibniz-Institute of Plant Genetics and Crop Plant Research (IPK), Gatersleben, Germany). M. Strybe, N. Kryszak (Dep. of Hemp Breeding, Agronomy and Seed Production; Harvesting Technology and Fibre Plants Evaluation Dep.; Institute of Natural Fibres and Medicinal Plants, Poznan, Poland) spoke about the effect of harvesting time on cellulose content in fibrous hemp.

The application of biotechnology in flax and hemp breeding was discussed by K. Wielgus, A. Luwanska, G. Mankowska, J. Przewozna (Department of Biotechnology, both: Institute of Natural Fibres and Medicinal Plants, Poznan, Poland).

Five last oral lectures were devoted to the next problems: public acceptance of biomass power plants – results of recent environmental psychological surveys (J. Zoellner, P. Schweizer-Ries, I. Rau, Institut für Psychologie, FG Umweltpsychologie, Otto-von-Guericke-Universität Magdeburg, Germany); regional value added of bioenergy activities: a material flow approach (R. Offermann, T. Seidenberger, D. Thrän, DBFZ German Biomass Research Centre, Leipzig, Germany); sustainable intra-logistics with renewable energies and fuel cells (H. Zadek, R. Schulz, Institute of Logistics and Material Handling Systems, University "Otto-von-Guericke", Magdeburg, Germany); prevention of disposal of greenhouse gases from digested residues by optimal use of the nitrogen fertilizer potential (U. Bauermeister, GNS Gesellschaft für Nachhaltige Stoffnutzung mbH und BMVerfahrenstechnologie, Halle/Saale, Germany); criteria system of sustainable agriculture (CSSA-KSNL) (G. Breitschuh, H. Eckert, I. Matthes, J. Strümpfel, G. Bachmann, T. Breitschuh, Thuringian Ministry for agriculture and environment, Erfurt, Germany; Thuringian state institute for agriculture, Jena, Germany; VAVB e.V., Jena, Germany).

The second session of conference had 15 oral presentations. R. Thomann, C. Luft, S. Icking, H. Hiob, R. Schneeweiß (Institut für Lebensmittel und Umweltforschung e.V., Nuthetal, Germany; Görlitzer Kornbrennerei und Spirituosenfabrik, Seyda, Germany; Institut für Getreideverarbeitung GmbH (IGV), Nuthetal, Germany) gave presentation about economical and ecological aspects in the production of ethanol from grain with the use of roller mills. M. Seiffert, F. Langer, F. Scholwin (German Biomass Research Centre, Leipzig, Germany) spoke about provision pathway for biomethane. The title of presentation of R. Pätz (Hochschule Anhalt, Germany) was “Simple kinetics of methane formation using model substrates”. The lecture of D. Backofen, M. Adam, H. Tschöke, M. Könnig, J. Schmidt (Institute of Mobile Systems; Institute of Fluid Dynamics and Thermodynamics; Otto-von-Guericke-University, Magdeburg, Germany) was devoted to spray formation of alternative Diesel fuels under engine-like conditions. The information about energetic efficiency of four arable crop species was presented by J. Kolodziej, J. Pudelko, J. Mankowski (Institute of Natural Fibres and Medicinal Plants, Poznan, Poland) and the information about the potential of Sorghum varieties as bioenergy crops was included in the presentation of A. Hartmann, K. Zeise, M. Fritz (Technologie- und Förderzenturm Straubing / Sachgebiet Rohstoffpflanzen und Stoffflüsse, Straubing, Germany).

Multiscale metabolic modeling: an integrative systems biology approach for biomass research was discussed in the lecture of M.-R. Hajirezaei, M. R. Ghaffari, R. Lemke, B. H. Junker, J. Müller, B. Usadel, R. Wünschiers, F. Schreiber (Leibniz Institute of Plant Genetics and Crop Plant Research (IPK), Gatersleben, Germany; SunGene GmbH, a BASF Plant Science Company, Gatersleben, Germany; Martin-Luther-Universität Halle-Wittenberg, Yield and Ecophysiology, Halle/Saale, Germany; Max-Planck-Institute for Molecular Plant Physiology, Integrative Carbon Biology, Golm, Germany). M. Ernst, M. Ouzonova, T. Presterl, P. Westhoff, R. Meyer, A. Melchinger, J. M. Montes, C. Riedelsheimer, C. Grieder, J. Selbig, A. Larhlimi, M. Stitt, R. Sulpice, A. Czedik-Eysenberg, L. Willmitzer, T. Altmann (Leibniz Institute of Plant Genetics and Crop Plant Research (IPK), Gatersleben, Germany; KWS Saat AG, Einbeck, Germany; Heinrich-Heine-Universität Düsseldorf, Germany; Universität Hohenheim, Germany; Universität Potsdam, Germany; Max-Planck-Institut für molekulare Pflanzenphysiologie, Golm, Germany) spoke about structural genomic research for energy maize. The problems of agricultural residues as a renewable source of bio-energy with special focus on cereal straws were discussed by S. Narra, C. Glaser, P. Ay, H.-J. Gusovius, C. Stollberg (Chair of Mineral Processing, Brandenburg University of Technology, Cottbus; Leibniz Institute for Agricultural Engineering Potsdam-Bornim, Germany; Process Technology of Biogenous Resources, Faculty of Engineering, University of Applied Sciences Wismar, Poel, Germany). M. Lochynska, G. Oleszak (Dep. of Silkworms Breeding and Mulberry Cultivation; Experimental Farm in Petkow; Institute of Natural Fibres and Medicinal Plants, Poznan, Poland) discussed the problems of multi-use of the white mulberry (Morus alba L.).

G. Fleck, N. Stadermann (Pilot Pflanzenöltechnologie Magdeburg e.V., Germany; Maschinenfabrik Reinartz, Neuss, Germany) gave presentation under the title “Technology development of decentralized processing of Jatropha-seeds”. The lecture of S. Traoré, A. Diarra, M. Traoré, S. Magassouba (Polytechnic Institute, University of Conakry, Guinea; FSE, Dep. of Economics, University of Sonfonia, Conakry, Guinea; Institute of medical plants, Dubreka, Guinea) was devoted to survey of Jatropha curcas energy potential for an efficient production of biodiesel as renewable resource. The information about non-food crops as a feedstock for modern bio-based industry was included in the lecture of K. Heller, M. Mackiewicz-Talarczyk, P. Baraniecki (Institute of Natural Fibres and Medicinal Plants, Poznan, Poland).

The last oral presentations were devoted to the next problems: oilseed Bbio-refinery – a new concept of value-added processing (F. Pudel, Pilot Pflanzenöltechnologie Magdeburg e.V., Germany); evaluation of different fractionation methods for the extraction of pure fatty acids (M. Rohrbeck, Pilot Pflanzenöltechnologie Magdeburg e.V., Germany); IFT Performance of MES surfactant from Palm Olein for EOR application (E. Hambali Surfactant and Bioenergy Research Center, Bogor Agricultural University, Indonesia); the development of Palm Methyl Ester Sulfonic Acid (MESA) to increase oil well recovery (S. Mujdalipah, E. Hambali, E. Zulchaidir (Surfactant and Bioenergy Research Center, Bogor Agricultural University, Indonesia; PT. Findeco Jaya, Pulogadung, Jakarta).

Poster session included 17 presentations about some aspects of plant biotechnology and renewable resources. Particularly there was the poster of Russian scientists: "Mechanism of stable radical generation in lignin under the action of nitrogen dioxide" (E. Ya. Davydov, I. S. Gaponova, S. M. Lomakin, G. B. Pariiskii, T. V. Pokholok, Gennadi E. Zaikov, N.M. Emanuel Institute of Biochemical Physics, Russian Academy of Sciences, Moscow, Russia).

The problems which were discussed during this conference are very important for pure as well as for practical application.

In: Chemical and Biochemical Reactions
Editors: G. E. Zaikov, L. Liu et al.

ISBN 978-1-61122-860-1


***Chapter 25***

# DEBERDEEV RUSTAM JAKUBOVICH: 50 YEARS IN SCIENCE

## *G. E. Zaikov*[*]

N. M. Emanuel Institute of Biochemical Physics,
Russian Academy of Sciences, 4 Kosygin str., Moscow 119991, Russia

Professor Rustam Yakubovich Deberdeev was born in 1940 in Arsk city, Tatarstan. He graduated from a secondary school in 1958 in the city Naberezhnye Chelny and used to work as a metalworker and a driver. He entered military service and entered the Kazan State Technological University after its completion in 1962. (formerly titled as the Chemical and Technological Institute).

R.Ya. Deverdeev graduated from University in 1967 and succeeded in positions of a laboratory staff member, engineer, junior researcher, assistant professor, associate professor and finally, a full professor. He has been heading the Chamber of processing technology of polymers and composite materials since 1990.

R.Ya. Deberdeev defended his master degree thesis (Ph.D.) in 1972. Later on, in 1988, he defended his doctoral degree thesis. He obtained the academic rank of a professor in 1990.

Professor R.Ya. Deberdeev is the recognized specialist in synthesis of monomers, oligomers, polymers and composite materials, development of processing technologies so as energy-and-resource saving and eco-friendly technologies.

He supervised both fundamental and applied research works concerning influence of physical and chemical structure of components on performance of anticorrosive coatings and constructional items based on thermoplastic and thermosetting polymers. These works have resulted in working out of new corrosion prevention-targeted materials to protect pipelines, water treatment equipment of thermal power plants, chemical equipment, production accessories etc. His materials and technologies were successfully introduced into facilities of the Energy Ministry in former USSR, JSC "Pigment" (Saint - Petersburg), Almetievsk pipe plant, JSC "Kamaz" (Naberezhnye Chelny), production association "Elaz" (Elabuga-city),

[*] E-mail: Chembio@sky.chph.ras.ru

JSC «Nizhnekamskneftechim» (Nizhnekamsk), JSC "Orgsintez" (Kazan), JSC "Novotec-polymer" (Novokuybyshevsk) etc. between 1980 and 2008.

The research works of Professor R.Ya.Deberdeev in the recent years are mainly focused on unconventional approaches to synthesis of monomers, oligomers and polymers and to working out of energy- and resource saving conditions. He has worked out technical solutions for industrial level production of ethylene-propylene copolymers based on the new approach to behaviour of rapid chemical reactions and movement of interacting media in a restricted volume. The production technology was deployed in the synthetic caoutchouc facility of JSC «Nizhnekamskneftexim». R.Ya. Deberdeev has also contributed to working out of the technology allowing to produce the brand-new divinyl synthetic rubber. This technology has been successfully deployed at JSC «Nizhnekamskneftexim» Production capacity is 100 000 tones per year. The production process of halogenated elastomers, which is a brand-new for world science and industry, has been worked out. It is based on mixing intensification of emulsions or solutions in a restricted volume, promoting greatly the rate at which a target product is produced. The specific features of this technology are resource and energy efficiency, ecological safety and compact technological equipment.

Another research is carried out in the field of synthesis of N-oxides and aromatic C-nitrosocompounds and their application in composite systems as radical process initiators, various modifiers with promising application prospective including low-temperature systems for vulcanization of composites based on unsaturated rubbers. The research resulted in estimations of dependency between structure and vulcanizing activity and interpretations of thermo decay of vulcanizing agents, vulcanization stage mechanisms and the development of corresponding composite materials.

The evolution of a society sets new challenges for the chemical industry even in well-known and robust technology areas such as polyvinylchloride composites. Research works carried out in the course of recent years provide ways to create metal-containing greases, plasticizers and multifunctional stabilizers for PVC compositions. The behavior of such materials in the new eco-friendly compositions and related materials has been analyzed. The technologies developed to obtain multifunctional additives are resource-saving, green and are widely applied in the production industry.

R.Ya. Deberdeev created and patented the new unique and simple process for polymer processing technology in the recent years. It deals with production of marked polymer piping through control of the rough tube bar in the viscous flow state and the marking ribbons in the rubber state.

The example of interesting research of the group under his supervision is the creation of a brand-new energy- and resource-efficient green technology of synthesis of polymeric sulphur with some principal differences from existing ones. The molten sulphur in dispersed to the nanoscale state and then polymerized into a macromolecular chain with new properties. The implementation of this technology will allow obtain a worldwide priority and unlimited polymeric sulphur production capabilities.

The field of study of R.Ya. Deberdeev which also attracts interest is devoted to a synthesis of polymer composite electrets from large-scale industrial polymers such as polyethylene, polypropylene, polystyrene and ethylene-propylene copolymers etc. He has worked out theoretical models of electret composites and was the first one to separately estimate parameters of homo - and heterocharges, to find high-temperature relaxation agents and other examples and law patterns of their behaviour, which stipulate performance and durability of various electret composites. Composite electrets have been synthesized and tried. Their large-scale application will allow, in some cases, to exclude addition of preserving agents into food products thus improving the ecology of a human living environment.

Lateral thinking of R.Ya.Deberdeev and his connections with Russian science schools allow him to provide target–oriented training of the chemical technology engineers. He has made collaboration agreements in the field of education and staff training with the Institute of Chemical Physics by N.N. Semenov, Institute of Problems of Chemical Physics (Chernogolovka city), Institute of metal-polymer systems (Gomel city, Belorus), JSC «Polymiz» (Kazan), JSC "Orgsintez" (Kazan) et. al. The department under his supervision is at the top of the Russian Federation rating of the final training university departments providing graduates specialized in «Processing technology of plastics and elastomers» and «Technology and design of packing materials».

The studies of R.Ya. Deberdeev inspired transition to the new research area in the Kazan State Technological University – the modelling and designing of technological processes for synthesis of oligomers and polymers. He supervised more than 25 Master Degree postgraduates (Ph. D.'s) and 3 Doctors of sciences.

Professor R.Ya. Deverdeev is the author of more than 350 monographs, textbooks and articles published in such journals as "RAS reports", "Macromolecular compounds", "Plastic masses", «Journal Polymer Science», «International Journal Polymeric Materials», «Russian Polymeric News» et. al. His developments are protected by more than 100 patents and author's sertificates.

R.Ya. Deberdeev has been the head of a group with similar scientific interests for many years. This group provides solutions for urgent scientific and technical challenges of the national industry.

R.Ya. Deberdeev is socially recognized both in science and education and has such honours as the «Honoured scientist of the Republic of Tatarstan and the Russian Federation», the laureate of the Republic of Tatarstan State Award, and the Award of the Russian Federation Government. He is honoured by certificates of the American Bibliographic Institute for the outstanding success in research activities.

## REFERENCES

Berlin Al.Al., Deberdeev R.Ya., Peruxin Yu.V., Garipov R.M. «Polioksimetileny» (Polyoxymethylenes), Moscow: *Nauka Publisher,* 2008, 286 p.

Mezhikovskij S.M., Arinshtejn A.A., Deberdeev R.Ya. «Oligomernoe sostoyanie veshestva» (Oligomeric State of Substance), Moscow: *Nauka Publisher,* 2006, 235 p.

Mezhikovskii S.M., Arinstein A.E., Deberdeev R.Ya. Oligomeric State of Substances *Nova Science Publishers*,Inc.: New York, 2009, 256 p.

Zakharov V.P., Berlin Al.Al., Monakov Yu.B., Deberdeev R.Ya. «Fiziko-himicheskie osnovy protekanija bystryh zhidkofaznyh processov» (Physico-chemical foundations of rapid liquid-phase processes behavior) Moscow: *Nauka Publisher,2008*, 348 p.

# INDEX

## A

## B



## C

## D

## Q

## T